Lindsay Childs

A Concrete Introduction
to Higher Algebra

Springer-Verlag

New York   Heidelberg   Berlin

Lindsay Childs
Department of Mathematics
SUNY at Albany
Albany, New York 12222
USA

Undergraduate Texts in Mathematics

AMS Subject Classification: 12-01

Library of Congress Cataloging in Publication Data

Childs, Lindsay N.
   A concrete introduction to higher algebra.

   (Undergraduate texts in mathematics)
   Bibliography: p.
   Includes index.
   1. Algebra.   I. Title.
QA155.C53   512.9   78-21870

ISBN 0-387-90333-x   Springer-Verlag   New York
ISBN 3-540-90333-x   Springer-Verlag   Berlin   Heidelberg

To Ashley and Nathan

Preface

This book is written as an introduction to higher algebra for students with a background of a year of calculus. The book developed out of a set of notes for a sophomore–junior level course at the State University of New York at Albany entitled Classical Algebra.

In the 1950s and before, it was customary for the first course in algebra to be a course in the theory of equations, consisting of a study of polynomials over the complex, real, and rational numbers, and, to a lesser extent, linear algebra from the point of view of systems of equations. Abstract algebra, that is, the study of groups, rings, and fields, usually followed such a course.

In recent years the theory of equations course has disappeared. Without it, students entering abstract algebra courses tend to lack the experience in the algebraic theory of the basic classical examples of the integers and polynomials necessary for understanding, and more importantly, for appreciating the formalism. To meet this problem, several texts have recently appeared introducing algebra through number theory.

This book combines the newer number-theoretic approach with the old theory of equations. In fact, the book contains enough of each of elementary number theory and the theory of equations that a course in either could be taught from it (see below). But the algebraic similarities of the two subjects are such that both subjects can be developed in parallel, and ideas customarily associated with one can be transferred to the other. Thus the ideas of congruence and congruence classes, normally arising in elementary number theory, can also be used with polynomials. Doing so permits passage from the study of polynomials to the study of simple field extensions, and in particular, leads to an exposition of finite fields.

There are, I feel, several advantages in beginning the study of higher algebra by studying number theory and polynomial theory.

First, the algebra is built on the student's entire mathematical experience. The study of numbers and polynomial equations dominates the precollege mathematical training. By building on this background a course in algebra will be building on the strongest possible intuitive base. And given such a base, the potential for reaching results of significance is high. I hope that this potential is realized by this book's theoretical development, numerous applications, and exercises.

Second, the dominating algebraic idea in the development of the book is that of congruence classes. The concept of quotient structure is perhaps the most difficult of the concepts of abstract algebra. The experience of seeing it in a variety of concrete contexts, and seeing worthwhile consequences of its use, should greatly aid the student when subsequently it is seen in abstract presentations. This particular feature of our approach is one which was missing from traditional theory of equations courses, and also is missing in courses in linear algebra used as background for abstract algebra.

Third, the subject matter of the book is intrinsically worth studying. Both number theory and the theory of equations have attracted the attention of the very greatest mathematicians. In particular, two of Gauss's greatest achievements, the fundamental theorem of algebra and the law of quadratic reciprocity, are important results in this book. One of the important lines of research in modern algebraic geometry stems from A. Weil's 1949 paper on solutions of equations in finite fields, a topic which is beyond the level, but very much in the tradition, of the material in this book (see Ireland and Rosen (1972) for an exposition). But even at the level of this book the subjects have attracted the notice of combinatorial analysts and computer scientists in recent decades. A surprising amount of the material in this book dates from since 1940. As I discovered only late in the writing of this book, there is considerable overlap between it and the mathematics in Chapter 4, "Arithmetic," of D. E. Knuth's fundamental treatise, *The Art of Computer Programming* (Knuth, 1969). Thus the mathematics in this book is worth learning for its own sake, apart from any value it has in preparing for more advanced mathematics.

The explicit prerequisites of the book consist for the most part only of high school algebra (in the *de facto* sense, not in the sense of Abhyankar (1976)—in his sense this *is* predominantly a high school algebra text). In various places we assume some acquaintance with calculus; however, the subject of differentiating polynomials is developed from the beginning (Chapter II-6), and the one place where integration occurs explicitly (Section II-5C) may be omitted.[1] Two-variable calculus is mentioned only in the proof of the fundamental theorem of algebra (II-3), but either the facts needed there can be taken on faith or the proof can be omitted. The

---

[1] Chapter 6 of Part II is referred to as Chapter II-6 or simply as II-6; if the reference occurs within Part II, simply as Chapter 6. II-5C refers to Section C of Chapter II-5.

use of infinite series is more substantial, however, particularly in connection with decimal expansions of fractions.

Several of the applications and a few of the theoretical sections use matrices and ideas from linear algebra. Chapter I-9 is an overly concise review of the necessary ideas. Sections C–F of I-9 should be used only for reference. If linear algebra is lacking in the student's background, the chapters particularly to avoid are II-12 and the last half of III-9. The remaining uses of linear algebra mainly involve simple matrix manipulations as described in Section I-9A, and these are quickly learned.

Exercises are scattered throughout the text as well as collected at the ends of sections. They range from routine examples to ingenious problems to extensions of theory. The most nontrivial ones are starred; comments on them are collected in the back of the book. Exercises which are mentioned subsequently either in the text or in exercises are marked with a dagger, and are indexed together with the subsequent references in the back of the book.

There is more material in this book than would be appropriate for a one semester course. For a year course it could be supplemented with a not-too-geometric introduction to linear algebra (such as Zelinsky (1973)). For a one semester course there are a variety of routes through the book.

The main development in Parts I and II is contained in

I—: 1–3, 4A, 5A, B, D, 6–8, 11, 14A;
II—: 1, 2, 3A, 4, 6, 8–10, 11A.

To get to the classification of finite fields (III-14) most efficiently, follow the main development with

III—: 1, 4, 7, 8, 10, 11, (12), 13, 14.

To get to algebraic numbers (III–18) most efficiently, follow the main development with

III—: 1, 4, 7, 8, 10, 16A, B, 18–21.

To concentrate on elementary number theory, see

I—: all except 9C–F and 4B, 4C, 5C, 9B, 10, 13, 14B, 15 as desired;
II—: 1, 2A;
III—: 1–5, 16A, B, 17; then
II—: 3A, 8–10, 11A;
III—: 7, 8, 10, 18–21.

To concentrate on theory of equations, see:

I—: 1–3, 4A, 6–8, 11, 14A;
II—: all, omitting 3C, 5, 7, 11B, 12 as desired;
III—: 1, 4, 6–8, 10, 11, 13–16.

It is probably unwise to spend too much time in Chapter I-2. Also, part of the uniqueness of this book lies in the chapters on applications, so it is hoped that any route through the book will be chosen to allow time for visits to some of the scenic wayside areas.

Finally, I wish to acknowledge with appreciation the contributions of people who in various ways influenced the book: Ed Davis, for developing and selling the idea for the course for which the book was written, and for a number of useful comments on an early version of the notes; Bill Hammond, for teaching from the notes and offering a number of improvements; Violet Larney, for teaching from the notes graciously even though her book (for a competing course) had just appeared; Morris Orzech, Paulo Ribenboim, Tony Geramita, Ted Turner, and Ivan Niven, for a variety of mathematical insights and ideas; and, especially, Malcolm Smiley, for reading through and teaching from the manuscript in its late stages and offering many substantial suggestions for improving the exposition, and David Drasin, for reading through and making many helpful comments on the nearly completed manuscript. Also I wish to thank: Michele Palleschi for typing most of the manuscript even though she didn't have to, and Mrs. Betty Turner and her staff for typing most of the manuscript even though they weren't supposed to; the Universities of Illinois (Urbana) and Oregon for their hospitality during part of the time the book was written; and Springer-Verlag, particularly Walter Kaufmann-Buehler and Joe Gannon, for their professional treatment of the manuscript. Most of all, my greatest thanks go to my wife Rhonda, for putting up with my working on the manuscript at inconvenient hours at inconvenient locations.

Fall, 1978   L. CHILDS

Contents

Contents

Contents

I. INTEGERS

This part of the book is about the natural numbers and integers. Among the highlights of this part, we show that every natural number factors uniquely into a product of primes, define congruence mod m, and invent new sets called congruence classes mod m, which for each $m \geqslant 2$ add and multiply to form a new algebraic system called $\mathbb{Z}_m$. Various related results about numbers, and applications, fill out this part.

Numbers 1

This book is about the algebraic structure of sets of numbers and poly-nomials. Here are the most familiar sets of numbers and the way we shall denote them.

$\mathbb{N}$ is the set of natural or counting numbers:

$$1, 2, 3, 4, 5 \dots .$$

$\mathbb{Z}$ is the set of integers:

$$\dots, -3, -2, -1, 0, 1, 2, 3, 4, \dots .$$

$\mathbb{Z}$ is obtained from $\mathbb{N}$ by including 0 (zero) and the negatives of the natural numbers.

$\mathbb{Q}$ is the set of rational numbers, that is, the set of all fractions a/b, where $b \neq 0$ and a are integers. The set of integers $\mathbb{Z}$ is thought of as a subset of $\mathbb{Q}$ by identifying the integer a with $a/1$. $\mathbb{Q}$ is the smallest set containing $\mathbb{Z}$ in which every nonzero integer has an inverse, or reciprocal. (See Exercise E6.)

$\mathbb{R}$ is the set of real numbers. A useful way to think of $\mathbb{R}$ is as the set of all infinite decimals. Real numbers can also be viewed geometrically, as lengths of directed line segments on a given straight line, or equivalently as coordinates on a given line (such as the x-axis used in calculus). Any rational number is a real number; the decimal expansion of a rational number a/b may be obtained by the familiar process of dividing b into a.

One way to define the real numbers is as limits of sequences (Cauchy sequences) of rational numbers. (In fact, an infinite decimal $.a_1a_2a_3 \dots$ is a limit of finite decimals $.a_1a_2 \dots a_n$, and each such is a rational number.) However, it is possible to use real numbers without having a precise understanding of how they are defined. Calculus, which uses real numbers in an essential way, was invented and used for 200 years before a rigorous definition of real numbers was achieved. So we shall adopt the 18th-century point of view and assume that the real numbers are understood, at least intuitively if not rigorously, and we shall say no more about their definition in this book.

$\mathbb{C}$ is the set of complex numbers, that is, the set of numbers of the form $a + bi$, a, b in $\mathbb{R}$, where $i = \sqrt{-1}$. You are probably less familiar with complex numbers than with real or rational numbers. Historically, they only became well understood and accepted early in the 19th century, well over a century after the development of the calculus.

We may think of $\mathbb{C}$ as the set of vectors (= directed line segments) in the plane, with $a + bi$ corresponding to the vector from the origin to the point with coordinates (a, b). If $v = a + bi$, a is called the real part of v, and b the imaginary part of v.

A convenient way to represent elements of $\mathbb{C}$ is in terms of polar coordinates (see diagram). If r is the distance from the origin to the point (a, b) (that is, $r = \sqrt{a^2 + b^2}$) and θ is the angle (measured counterclockwise) from the positive real axis to the vector v, then $a = r \cos \theta$, $b = r \sin \theta$, so $v = a + bi = r \cos \theta + ir \sin \theta$. The real number r is called the length of v and is denoted $|v|$; θ is called the argument of v and is sometimes denoted arg v. Multiplication of complex numbers when described in polar coordinates works rather neatly:

$$(r \cos \theta + ir \sin \theta)(s \cos \phi + is \sin \phi)$$
$$= rs((\cos \theta \cos \phi - \sin \theta \sin \phi) + i(\cos \theta \sin \phi + \sin \theta \cos \phi)) \quad (1)$$
$$= rs(\cos(\theta + \phi) + i \sin(\theta + \phi))$$

That is, when multiplying two complex numbers, lengths multiply and arguments add.

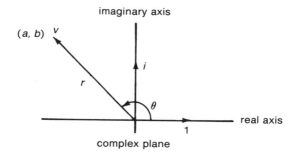

imaginary axis

(a, b) v

i

r

θ

real axis

1

complex plane

If you have had some acquaintance with infinite series, then you probably know the Taylor series for $e^x = \exp(x)$,

$$e^x = 1 + x + \frac{x^2}{2!} + \frac{x^3}{3!} + \cdots,$$

as well as for $\sin x$ and $\cos x$,

$$\sin x = x - \frac{x^3}{3!} + \frac{x^5}{5!} - \frac{x^7}{7!} + \cdots$$

and

$$\cos x = 1 - \frac{x^2}{2!} + \frac{x^4}{4!} - \frac{x^6}{6!} + \cdots,$$

which converge for all real x. Then $\cos x + i \sin x$ has a Taylor series

$$1 + ix - \frac{x^2}{2!} - \frac{ix^3}{3!} + \frac{x^4}{4!} + \frac{ix^5}{5!} - \frac{x^6}{6!} + \cdots$$

which would be the same as the Taylor series for the complex function e^{ix} if we knew what e^{ix} was. So we define e^{ix} for x real by setting $e^{ix} = \cos x + i \sin x$: that is, we define e^{ix} by replacing x by ix in the Taylor series for e^x.

Then an arbitrary complex number v can be written in polar form as

$$v = r(\cos \theta + i \sin \theta) = re^{i\theta};$$

if w is another complex number, $w = se^{i\phi}$, the multiplication of formula (1) above becomes

$$vw = re^{i\theta}se^{i\phi} = rse^{i(\theta + \phi)},$$

which is exactly what one would expect from the laws of exponents that you learned in high school.

From an algebraic point of view $\mathbb{Q}$, $\mathbb{R}$, and $\mathbb{C}$ are different from $\mathbb{Z}$. In $\mathbb{Z}$ the equation $ax = 1$ can be solved only when $a = 1$ or -1, but in $\mathbb{Q}$, $\mathbb{R}$, or $\mathbb{C}$ (see E5) every nonzero element has a reciprocal, or inverse. Thus, as we shall see, $\mathbb{Q}$, $\mathbb{R}$, and $\mathbb{C}$ are examples of fields of numbers, while $\mathbb{Z}$ is not.

We conclude this introductory chapter with a remark on logical notation. Frequently we shall make statements like "P is true iff Q is true," where P, Q are assertions. "iff" means "if and only if." The statement means: "If P is true then Q is true and if P is false then Q is false." If an exercise contains a statement to be proved which includes an "iff," there are really two statements which are to be proved. See Exercise E7 for an example.

E1. Find inverses of: $1 + i$; $3 - 2i$; $1 + 6i$.

E2. Write in polar form: $(1 + i)/2$; $(-1 - \sqrt{-3})/2$.

E3. Solve in $\mathbb{C}$ these equations: $x^3 = 1$; $x^5 = 1$; $x^n = 1$.

E4. Prove: $(\cos \theta + i \sin \theta)^n = \cos n\theta + i \sin n\theta$.

E5. What is the inverse of $re^{i\phi}(r \neq 0)$?

†E6. Let a, b, c, d, e, f be integers with b, d, $f \neq 0$. For ordered pairs (a, b) and (c, d) define the notation $(a, b) \equiv (c, d)$ by

$$(a, b) \equiv (c, d) \quad \text{iff} \quad ad = bc.$$

Show:

  (i) $(a, b) \equiv (a, b)$;
  (ii) if $(a, b) \equiv (c, d)$, then $(c, d) \equiv (a, b)$;
  (iii) if $(a, b) \equiv (c, d)$ and $(c, d) \equiv (e, f)$, then $(a, b) \equiv (e, f)$.

(These conditions are called *reflexivity*, *symmetry*, and *transitivity* respectively. If $\equiv$ is a relation such that (i), (ii), (iii) hold, then $\equiv$ is called an *equivalence relation*. The set of all pairs (c, d) which are $\equiv$ to (a, b) is called the *equivalence class of* (a, b). For $\equiv$ as defined here, the equivalence class of (a, b) is denoted a/b. Thus, $a/b = c/d$ means: the equivalence class of (a, b) is the same as the equivalence class of (c, d); this in turn means that $(a, b) \equiv (c, d)$, which in turn means that $ad = bc$. The set $\mathbb{Q}$ of rational numbers is defined to be the set of equivalence classes with respect to $\equiv$ of ordered pairs (a, b) of integers with $b \neq 0$.)

E7. Is the following fact from calculus true? If so, how is it proved? Let $f(x)$ be a function which is differentiable for all x, $a \leqslant x \leqslant b$. Then $f(x)$ is a constant function iff $f'(x) = 0$ for all x, $a \leqslant x \leqslant b$.

Induction; the Binomial Theorem 2

A. Induction

The first part of the course deals with the integers $\mathbb{Z}$ and the natural numbers $\mathbb{N}$. You know how to add and subtract, multiply and divide integers, and you know when one integer is bigger than another; we shall not review these things. One property of the integers which does need review is the principle of induction. It is the basis for most proofs involving natural numbers, for it gives a general procedure for verifying statements about *all* natural numbers (an infinite set). It comes in various formulations. Here is the first.

Induction (1). *Let $P(n)$ be a statement which makes sense for any integer $n \geqslant n_0$. If*

(a) $P(n_0)$ *is true, and*
(b) *If $P(n)$ is true, then $P(n+1)$ is true for all $n \geqslant n_0$ in $\mathbb{Z}$*
then $P(n)$ is true for all $n \geqslant n_0$ in $\mathbb{Z}$.

You have probably seen this principle used before, perhaps in evaluating sums arising in connection with the definite integral.

Here are some examples of arguments by induction.

EXAMPLE 1. $1 + 3 + 5 + \cdots + (2n - 1) = n^2$ for all $n \geqslant 1$.

PROOF. Denote the assertion by $P(n)$. The statement $P(1)$ is true, since $1 = 1^2$. Assume that for some number k, $P(k)$ is true, that is, $1 + 3 + 5 + \cdots + (2k - 1) = k^2$. We look at the left side of the assertion $P(k + 1)$:

$$1 + 3 + 5 + \cdots + (2k + 1)$$
$$= [1 + 3 + 5 + \cdots + (2k - 1)] + (2k + 2).$$

By assumption that $P(k)$ is true we can replace the expression inside the square brackets by k^2:

$$= k^2 + (2k + 1)$$
$$= (k + 1)^2.$$

Thus assuming $P(k)$ is true, it follows that $P(k + 1)$ is true. By induction $P(n)$ is true for all $n \geqslant 1$.   □

The rationale behind induction is that if statements (a) and (b) are true, then for any $n \geqslant n_0$ one can prove, in $n - n_0$ steps, that $P(n)$ is true. For example, if $P(n)$ is the equation (1) above and we wish to prove that $P(5)$ is true, we can argue logically using statements (a) and (b), as follows:

$P(1)$ is true, by statement (a).
Since $P(1)$ is true, $P(2)$ is true, by statement (b) with $n = 1$.
Since $P(2)$ is true, $P(3)$ is true, by statement (b) with $n = 2$.
Since $P(3)$ is true, $P(4)$ is true, by statement (b) with $n = 3$.
Since $P(4)$ is true, $P(5)$ is true, by statement (b) with $n = 4$.

This same kind of argument can be done, once we have shown

(a) $P(1)$ is true, and
(b) For *all* n, if $P(n)$ is true, then $P(n + 1)$ is true,

to show in n logical steps that $P(n)$ is true for any $n \geqslant 1$. The principle of induction simply asserts that given (a) and (b), and any $n \geqslant 1$, $P(n)$ can be shown true, and that therefore $P(n)$ *is* true for any n.

Here are some more examples.

EXAMPLE 2. For all $n \geqslant 1$, $2^n \geqslant 1 + n$.

PROOF. The statement is clearly true when $n = 1$. Supposing that $2^k \geqslant 1 + k$ for some $k \geqslant 1$, then it follows that $2^{k+1} = 2^k \cdot 2 \geqslant (1 + k)2 = 2k + 2 > 1 + (k + 1)$. Thus the inequality is true for all $n \geqslant 1$ by induction.   □

EXAMPLE 3. For all $n \geqslant 0$, 8 divides $3^{2n} - 1$ (that is, $3^{2n} - 1 = 8k$ for some integer k).

PROOF. The statement is true for $n = 0$ since 8 divides $3^0 - 1 = 0$. Suppose 8 divides $3^{2k} - 1$. We examine $3^{2(k+1)} - 1$:

$$3^{2(k+1)} - 1 = 3^{2k} \cdot 3^2 - 1 = 3^{2k} \cdot 3^2 - 3^2 + 3^2 - 1$$
$$= 3^2(3^{2k} - 1) + (3^2 - 1).$$

Since 8 divides $3^{2k} - 1$ and $8 = 3^2 - 1$, therefore 8 divides $3^2(3^{2k} - 1) + (3^2 - 1) = 3^{2(k+1)} - 1$. Thus the statement is true for all $n \geqslant 0$.   □

EXAMPLE 4. For all $n \geqslant -2$, $2n^3 - 3n^2 + n + 31 \geqslant 0$.

PROOF. Set $f(n) = 2n^3 - 3n^2 + n + 31$. Then $f(-2) = 1 \geqslant 0$. Suppose $f(k) \geqslant 0$ for some $k \geqslant -2$. Then

$$
\begin{aligned}
f(k+1) &= 2(k+1)^3 - 3(k+1)^2 + (k+1) + 31 \\
&= 2(k^3 + 3k^2 + 3k + 1) - 3(k^2 + 2k + 1) + (k+1) + 31 \\
&= (2k^3 - 3k^2 + k + 31) + 6k^2 + 6k + 2 - 6k - 3 + 1 \\
&= f(k) + 6k^2 \geqslant f(k) \geqslant 0.
\end{aligned}
$$

So $f(n) \geqslant 0$ for all $n \geqslant -2$. ∎

EXAMPLE 5. $dx^n/dx = nx^{n-1}$ for any integer $n \geqslant 0$.

PROOF. $n = 0$ is clear. If $dx^k/dx = kx^{k-1}$ then

$$
\lim_{\Delta x \to 0} \frac{(x + \Delta x)^k - x^k}{\Delta x} = kx^{k-1};
$$

thus

$$
\begin{aligned}
\frac{dx^{k+1}}{dx} &= \lim_{\Delta x \to 0} \frac{(x + \Delta x)^{k+1} - x^{k+1}}{\Delta x} \\
&= \lim_{\Delta x \to 0} \frac{(x + \Delta x)^k x + (x + \Delta x)^k \Delta x - x^k \cdot x}{\Delta x} \\
&= \lim_{\Delta x \to 0} \left(x \cdot \frac{(x + \Delta x)^k - x^k}{\Delta x} + \frac{(x + \Delta x)^k \Delta x}{\Delta x} \right) \\
&= xkx^{k-1} + x^k = (k+1)x^k. \quad ∎
\end{aligned}
$$

E1. Prove that for all $n \geqslant 4$, $n! \geqslant 2^n$.

E2. Prove that $1^3 + 2^3 + \cdots + n^3 = [n(n+1)/2]^2$ for all $n \geqslant 1$.

E3. Prove that for any integers x, y, $x - y$ divides $x^n - y^n$ for all $n \geqslant 1$.

E4. (a) Prove that

$$
\frac{1}{1-x} = 1 + x + x^2 + \cdots + x^{n-1} + \frac{x^n}{1-x}
$$

for any $n \geqslant 1$.

(b) Prove that $1 + 2 + 2^2 + \cdots + 2^{n-1} = 2^n - 1$ for any $n \geqslant 1$.

(c) Given any integer $n \geqslant 1$ and any integers $r, a_0, a_1, \ldots, a_{n-1}$ with $r \geqslant 2$ and $0 \leqslant a_i < r$ for all $i = 0, 1, \ldots, n-1$, prove that

$$
a_0 + a_1 r + a_2 r^2 + \cdots + a_{n-1} r^{n-1} < r^n.
$$

E5. Prove that for all $n \geqslant 1$

$$
1^4 + 2^4 + \cdots + n^4 = \frac{n(n+1)(2n+1)(3n^2 + 3n + 1)}{30}.
$$

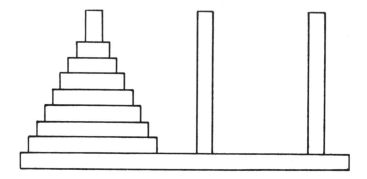

E6. Consider the ancient puzzle called the Tower of Hanoi.

The puzzle consists of *n* disks of decreasing diameter placed on a pole. There are two other poles. The problem is to move the entire pile to another pole by moving one disk at a time to any other pole, except that no disk may be placed on top of a smaller disk. Find a formula for the least number of moves needed to move *n* disks from one pole to another, and prove the formula by induction.

E7. What is wrong with the following theorem and proof?

Theorem. *All babies have the same color eyes.*

PROOF. $n = 1$ is obvious.

Suppose that in any set of *n* babies all have the same color eyes. Consider a set of $n + 1$ babies.

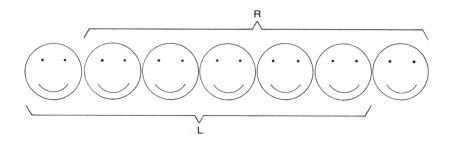

We may assume by induction that in the set *L* of the *n* babies to the left all have the same color eyes, and similarly that in the set *R* of the *n* babies to the right all have the same color eyes. But then evidently all the $n + 1$ babies have the same color eyes, for the leftmost and rightmost babies have the same color eyes as all the babies in between.

By induction, for any *n*, in every set of *n* babies all have the same color eyes. Since the set of all babies is one such set, the theorem is proved. ☐

B. Another Form of Induction

Induction (2). *Let $P(n)$ be a statement which makes sense for any integer $n \geqslant n_0$. If*

(a') *$P(n_0)$ is true, and*
(b') *For all $n > n_0$, if $P(m)$ is true for all m, $n_0 \leqslant m < n$, then $P(n)$ is true, then $P(n)$ is true for all integers $n \geqslant n_0$.*

Here is an example of the use of induction (2), an example which is not so easily done using induction (1):

EXAMPLE 6. Any natural number $n \geqslant 2$ factors into a product of prime numbers.

Recall that a natural number n is *prime* if n does not factor into the product of two natural numbers each smaller than n. We adopt the convention that a prime is a product of (one!) primes.

PROOF. Let $P(n)$ be the statement "n is a product of primes." Then $P(2)$ is true since 2 is prime (hence is a product of primes by our convention).
  Suppose $P(m)$ is true for all $m < n$. If n is prime, $P(n)$ is true. If not, $n = ab$, with $a < n$, $b < n$. So $P(a)$ is true and $P(b)$ is true, that is, a and b are both products of primes. But then $n = ab$ is also, and $P(n)$ is true. By induction (2), $P(n)$ is true for all $n \geqslant 2$.   □

Note. The statement of Example (6) is part of the fundamental theorem of arithmetic, which we will prove in full in Chapter 3.

Theorem. *If induction (1) is true then induction (2) is true.*

PROOF. Let $P(n)$ be an assertion and suppose we know of it:

(a') $P(n_0)$ is true;
(b') For all $n > n_0$, if $P(m)$ is true for all m, $n_0 \leqslant m < n$, then $P(n)$ is true.

Let $Q(n)$ be the assertion, "$P(m)$ is true for all m, $n_0 \leqslant m \leqslant n$."
  To show $P(n)$ is true for all n it suffices to show that $Q(n)$ is true for all n. We show $Q(n)$ is true by induction (1).
  We must show:

(a) $Q(n_0)$ is true;
(b) For any $n > n_0$, if $Q(n - 1)$ is true then $Q(n)$ is true.
Assertion (a) is the same as (a'), so is true. For (b), suppose $Q(n - 1)$ is true. Then $P(m)$ is true for all m, $n_0 \leqslant m \leqslant n - 1$. By (b'), $P(n)$ is true. So $P(m)$ is true for all m, $n_0 \leqslant m \leqslant n$. So $Q(n)$ is true.
By induction (1), $Q(n)$ is true for all n. That completes the proof.   □

We let you try to prove the converse.

E8. If induction (2) is true then induction (1) is true.

Exercise E8, together with the theorem, shows that if we want to prove something by induction we can use whichever version is most convenient. In (b) of the next exercise induction (2) will be more convenient.

†**E9.** *Russian peasant arithmetic*. Here is a way of multiplying attributed to Russian peasants who could only add, and multiply and divide by 2.

To multiply a and b put a and b at the top of two columns and fill in the columns below by multiplying the left number by 2 and dividing the right number by 2. Whenever division of the right number by 2 yields a new right number which is odd, subtract one from that new right number before dividing by 2 again, and put the corresponding left number in the sum column. When you reach the number 1 in the right column add the sum column to get the answer. Here is an example.

Left	Right	Sum
311	116	
622	58	
1244	29	1244
2488	14	
4976	7	4976
9952	3	9952
19904	1	19904
		36076 = 311 × 116

(a) Verify that Russian peasant arithmetic works for 218 and 195.

(b) Let $r(a, b)$ denote the result of doing Russian peasant arithmetic to a and b. Verify that

$$r(a, 1) = a,$$

$$r(a, b) = \begin{cases} r\left(2a, \dfrac{b}{2}\right) & \text{if } b \text{ is even,} \\ r\left(2a, \dfrac{b-1}{2}\right) + a & \text{if } b \text{ is odd.} \end{cases}$$

Prove by induction that $r(a, b) = a \cdot b$ for all $a, b \geq 1$.

†**E10.** The Fibonacci sequence is an example of an inductively defined sequence of numbers. It is defined as $\{a_0, a_1, a_2, \ldots\}$ where $a_0 = 0$, $a_1 = 1$, and $a_n = a_{n-1} + a_{n-2}$ for $n \geq 2$. Thus it begins

$$0 \ 1 \ 1 \ 2 \ 3 \ 5 \ 8 \ 13 \ 21 \ 34 \ 55 \ 89 \ldots.$$

(a) Prove by induction that a_n and a_{n+1} have no common factor > 0 except 1.

The *golden mean* is the ratio $b : a$ such that $b : a = (a + b) : b$. If we consider a fraction b/a where $b : a$ is the golden mean, then

$$\frac{b}{a} = \frac{a + b}{b},$$

so

$$\frac{b}{a} = \frac{a}{b} + 1, \qquad \left(\frac{b}{a}\right)^2 = 1 + \left(\frac{b}{a}\right), \quad \text{and} \quad \frac{b}{a} = \frac{1 + \sqrt{5}}{2}.$$

The golden mean $b : a$ was considered to be the most perfect ratio for the lengths of the sides of rectangular designs, such as portraits. The golden mean relates to the Fibonacci sequence by the following two facts:

(b) $a_n = \dfrac{(1 + \sqrt{5}\,)^n/2 - (1 - \sqrt{5}\,)^n/2}{\sqrt{5}}$ (prove this by induction);

(c) $\lim\limits_{n \to \infty} \dfrac{a_{n+1}}{a_n} = \dfrac{1 + \sqrt{5}}{2}$ (prove this using (b)).

It follows from (c) that the golden mean can be approximated by ratios of consecutive numbers in the Fibonacci sequence.

C. Well-ordering

Here is another version of induction.

Well-ordering principle. *Let n_0 be any fixed integer. Then any nonempty set of integers $\geqslant n_0$ has a least element.*

We prove the well-ordering principle using induction (1). To do so we prove that if there is a set of integers $\geqslant n_0$ with no least element, then it must be empty. (This approach uses the standard logical strategy for proving statements of the form "if A then B"—we prove that if B is false then A must be false. The reason that the strategy works is that the only situation under which the statement "if A then B" is false occurs when A is true and B is false. If we assume B is false and are able to show thereby that A is false, then the situation "A true and B false" cannot hold and so "if A then B" is true.)

PROOF. Let S be a set of integers $\geqslant n_0$ with no least element. Let $P(n)$ be the statement: "No number $\leqslant n$ is in S." Observe that if m is in S, then $P(m)$ is false. So by showing that $P(n)$ is true for all $n \geqslant n_0$ we will show that S is empty, which will prove the principle. Evidently $P(n_0)$ is true, for if not, n_0 is in S, and since all numbers in S are $\geqslant n_0$, therefore S would have a least element.

Suppose $P(k)$ is true for some $k \geqslant n_0$. If $P(k + 1)$ is false, then some number $\leqslant k + 1$ is in S. But $P(k)$ is true. So no such number $\leqslant k$ is in S. But then $k + 1$, the only number $\leqslant k + 1$ and not $\leqslant k$, would be in S, and would be the least element of S, which is impossible. Thus if $P(k)$ is true, so is $P(k + 1)$. By induction, $P(n)$ is true for all n, and S is empty. That finishes the proof. $\qquad\Box$

Since the well-ordering principle is true, we can use it in proofs. Here is an example.

EXAMPLE 7. Given nonnegative integers $a > 0$ and b there exist integers $q \geqslant 0$ and r with $0 \leqslant r < a$ such that $b = aq + r$.

In practice, q, r are found by long division:

q is the quotient, r the remainder.

We prove the existence of q and r using well-ordering.

PROOF. Let $S = \{b - ax | x \geqslant 0, b - ax \geqslant 0\}$. Since $b = b - a \cdot 0$ is in S, S is a nonempty set of integers $\geqslant 0$ $(n_0 = 0)$. So S has a least element, r. We show r satisfies the conditions of the theorem. Since r is in S, $r \geqslant 0$, and $r = b - aq$ for some $q \geqslant 0$. We must show that $r < a$. Now $r - a = b - qa - a = b - a(q + 1)$. So if $r \geqslant a$, then $r - a = b - a(q + 1) \geqslant 0$ would be in S. But $r - a < r$ and r was the least element of S. So $r - a$ cannot be $\geqslant 0$, so it must be that $r < a$. That completes the proof.   □

It turns out that the well-ordering principle is equivalent to induction. You can try to prove it:

*E11. (a) Prove that well-ordering implies induction (1).

(b) Prove that induction (2) is equivalent to well-ordering. (That is, prove: (i) if well-ordering is true then induction (2) is true; (ii) if induction (2) is true then well-ordering is true.)

E12. Let S be a set of integers all elements of which are less than some fixed integer N. Show that S has a maximal element.

D. The Binomial Theorem

The binomial theorem describes the coefficients when the expression $(x + y)^n$ is multiplied out.

Denote $0! = 1$, $n! = 1 \cdot 2 \cdot 3 \cdots n$ for $n > 0$.

The binomial theorem. *For* n *any integer* $\geqslant 1$,

$$(x + y)^n = \binom{n}{0}x^n + \binom{n}{1}x^{n-1}y + \cdots$$

$$+ \binom{n}{r}x^{n-r}y^r + \cdots + \binom{n}{n-1}xy^{n-1} + \binom{n}{n}y^n$$

where

$$\binom{n}{r} = \frac{n!}{r!(n-r)!} \quad \text{for} \quad 0 \leqslant r \leqslant n.$$

The proof is by induction on n, and in order to carry through the argument passing from $n - 1$ to n (the induction step) we first set up *Pascal's triangle*.

$$1$$
$$1 \quad 1$$
$$1 \quad 2 \quad 1$$
$$1 \quad 3 \quad 3 \quad 1$$
$$1 \quad 4 \quad 6 \quad 4 \quad 1$$

$$\cdots$$

We number the elements $c(n, r)$ of Pascal's triangle by the row n and the position r of the element with the row, both indices starting from 0. Thus Pascal's triangle is labeled

$$c(0, 0)$$
$$c(1, 0) \qquad c(1, 1)$$
$$c(2, 0) \qquad c(2, 1) \qquad c(2, 2)$$
$$c(3, 0) \qquad c(3, 1) \qquad c(3, 2) \qquad c(3, 3)$$

$$\cdots$$

where

$$c(0, 0) = c(n, 0) = c(n, n) = 1 \quad \text{for all } n$$

and

$$c(n, r) = c(n - 1, r - 1) + c(n - 1, r) \quad \text{for } 1 \leqslant r \leqslant n - 1.$$

That is, for each n and each r, $1 \leqslant r \leqslant n - 1$, $c(n, r)$ is the sum of the terms on its left and right shoulder:

$$c(n - 1, r - 1) + c(n - 1, r)$$
$$\|$$
$$c(n, r).$$

Fact. $c(n, r)$ *is the number of r-element subsets of a set S containing n elements.*

PROOF. We do this by induction on n, the case $n = 1$ being obvious. Let S be a set with n elements, one of them being the element y. Let S_0 be the set of all the elements of S except y. Divide the collection of all r-element subsets of S into two piles, one consisting of those subsets containing y, the other consisting of those subsets not containing y. The first pile consists of exactly those subsets of S obtained by taking an $(r - 1)$-element subset of S_0 and adjoining y. Thus there are $c(n - 1, r - 1)$ of those. The second pile consists exactly of the r-element subsets of S_0, of which there are $c(n - 1, r)$. Thus the number of r-element subsets of S is $c(n - 1, r - 1) + c(n - 1, r) = c(n, r)$, which is what we wished to show. $\square$

The elements of Pascal's triangle can be computed by the following:

Lemma.

$$c(n, r) = \binom{n}{r} = \frac{n!}{r!(n-r)!}.$$

PROOF. Induction on n. The case $n = 0$ is obvious:

$$1 = c(0, 0), \qquad \frac{0!}{0!0!} = 1.$$

Given $n > 0$ assume that for all r, $0 \leqslant r \leqslant n - 1$,

$$c(n - 1, r) = \frac{(n-1)!}{r!(n-1-r)!}.$$

Now $c(n, 0) = n!/0!(n-0)! = 1$, $c(n, n) = n!/n!(n-n)! = 1$, so the lemma is true for $c(n, r)$ when $r = 0$ or n. For $1 \leqslant r \leqslant n - 1$,

$$c(n, r) = c(n-1, r-1) + c(n-1, r)$$

$$= \frac{(n-1)!}{(r-1)!(n-r)!} + \frac{(n-1)!}{r!(n-1-r)!}$$

$$= \frac{(n-1)!}{(r-1)!(n-1-r)!}\left[\frac{1}{n-r} + \frac{1}{r}\right]$$

$$= \frac{(n-1)!}{(r-1)!(n-1-r)!} \cdot \frac{n}{(r)(n-r)}$$

$$= \frac{n!}{r!(n-r)!},$$

as was to be shown. The lemma is therefore proved by induction.   □

We therefore know that for each n, $\binom{n}{0} = \binom{n}{n} = 1$, and $\binom{n}{r} = \binom{n-1}{r}$ $+ \binom{n-1}{r-1}$ for $1 \leqslant r \leqslant n - 1$. Using these facts we can prove the binomial theorem by induction on n.

PROOF OF THE BINOMIAL THEOREM. For $n = 1$, $(x+y) = \binom{1}{0}x + \binom{1}{1}y$, so the binomial theorem is true when $n = 1$. Assume $n > 1$ and the theorem is true for $n - 1$, that is,

$$(x+y)^{n-1} = \binom{n-1}{0}x^{n-1} + \binom{n-1}{1}x^{n-2}y + \cdots$$

$$+ \binom{n-1}{r}x^{n-1-r}y^r + \cdots + \binom{n-1}{n-1}y^{n-1}.$$

We compute $(x+y)^n$ as follows:

$$(x+y)^n = (x+y) \cdot (x+y)^{n-1} = x(x+y)^{n-1} + y(x+y)^{n-1};$$

substituting for $(x + y)^{n-1}$ we get

$$\binom{n-1}{0}x^n + \binom{n-1}{1}x^{n-1}y + \cdots + \binom{n-1}{n-1}xy^{n-1}$$

$$+ \binom{n-1}{0}x^{n-1}y + \cdots + \binom{n-1}{n-2}xy^{n-1} + \binom{n-1}{n-1}y^n$$

Collecting coefficients, this becomes

$$\binom{n-1}{0}x^n + \cdots + \left[\binom{n-1}{r-1} + \binom{n-1}{r}\right]x^{n-r}y^r + \cdots + \binom{n-1}{n-1}y^n.$$

By the lemma, and the fact that $\binom{n-1}{0} = \binom{n}{0} = 1$, $\binom{n-1}{n-1} = \binom{n}{n} = 1$, this becomes

$$\binom{n}{0}x^n + \cdots + \binom{n}{r}x^{n-r}y^r + \cdots + \binom{n}{n}y^n,$$

which proves the binomial theorem by induction. □

E13. Prove that the sum of the elements of the nth row of Pascal's triangle is 2^n for each n. How many subsets of a set with n elements are there?

E14. Prove that

$$\binom{s}{s} + \binom{s+1}{s} + \cdots + \binom{N}{s} = \binom{N+1}{s+1}$$

for any s, $N \geqslant s$.

E15. Prove that

$$\binom{n}{0}^2 + \binom{n}{1}^2 + \cdots + \binom{n}{n}^2 = \binom{2n}{n}.$$

E16. Let $_n s_r$ denote the sum of the first n rth powers:

$$_n s_r = \sum_{k=1}^{n} k^r + 1^r + 2^r + \cdots + n^r.$$

You know $_n s_3$ by E2, and $_n s_4$ by E5.
(a) Find $_n s_1$.
(b) Find $_n s_2$.
*(c) Prove that for each $r \geqslant 1$,

$$(n+1)^{r+1} - (n+1) = \binom{r+1}{1}{}_n s_r + \binom{r+1}{2}{}_n s_{r-1} + \cdots$$

$$+ \binom{r+1}{r}{}_n s_1.$$

*(d) Derive from (c) that $_n s_r$ is a polynomial function of n whose degree is $r + 1$.

(e) View $_n s_r$ as the sum of the areas of a collection of rectangles lying inside the region between the x axis and the curve $y = x^r$ between $x = 1$ and $x = n + 1$. Does (d) seem reasonable when $_n s_r$ is viewed this way? (You may wish to recall the integral test if you have studied infinite series.)

*E17. *Multinomial theorem.* If you multiply out $(x_1 + \cdots + x_m)^n$ and collect coefficients you get a sum in which each term has the form

$$\binom{n}{a_1, \ldots, a_m} x_1^{a_1} x_2^{a_2} \cdots x_m^{a_m}$$

for some coefficient

$$\binom{n}{a_1, \ldots, a_m}$$

where $a_1 + a_2 + \cdots + a_m = n$, and each $a_i \geqslant 0$. Show that

$$\binom{n}{a_1, \ldots, a_m} = \frac{n!}{a_1! a_2! \cdots a_m!} \qquad \text{(where } 0! = 1)$$

(a) when $m = 3$,
(b) for any $m \geqslant 2$.
(c) Find the coefficient of x^3 in $(x^2 + 3x + 2)^4$.

Unique Factorization into Products of Primes 3

A. Euclid's Algorithm

The basic algebraic property of $\mathbb{Z}$ which we will use throughout the book is the

Division Theorem. *Given integers $a > 0$ and $b \geqslant 0$ there exist unique integers $q \geqslant 0$ and r, $0 \leqslant r < a$, such that $b = qa + r$.*

This is on first thought a pretty obvious fact about natural numbers. For it is just a restatement of long division, which is familiar from the third grade. For example, dividing 3857 by 285 looks like this:

$$
\begin{array}{r}
13 \\
285 \overline{)\,3857} \\
285 \\
\hline
1007 \\
855 \\
\hline
152\;.
\end{array}
$$

What this means is that $3857 = 285 \times 13 + 152$. So you know how to find q and r for any a and b.

But if you think about long division it is something of a marvel that it works.

E1. How do you suppose the Romans would have performed the division

$$\text{CCLXXXV} \overline{)\,\text{MMDCCCLVII}} \qquad ?$$

We shall prove the division theorem here using induction (2).

PROOF. Given the divisor $a \neq 0$ we shall prove that for every $b \geqslant 0$ there exist q, r with $0 \leqslant r < a$ such that $b = qa + r$.

First, $0 = a \cdot 0 + 0$, so when $b = 0$ we can set $q = 0$, $r = 0$.

For $b > 0$ we use induction (2).

Suppose that for all c, $0 \leqslant c < b$, there are q_0, r_0 with $0 \leqslant r_0 < a$ and $c = q_0 a + r_0$. Now consider b. We want to write b as $b = qa + r$ with $0 \leqslant r < a$.

If $b < a$ then $b = 0 \cdot a + b$; since $0 \leqslant b < a$ we can set $q = 0$, $r = b$.

If $b \geqslant a$, let $c = b - a$. Then $0 \leqslant c < b$. By induction, $c = a \cdot q_0 + r_0$ for some q_0 and r_0 with $0 \leqslant r_0 < a$. But then

$$b = a + c = a + a \cdot q_0 + r_0 = a(q_0 + 1) + r_0$$

and $0 \leqslant r_0 < a$. We can set $q = q_0 + 1$, $r = r_0$.

By induction (2), any $b > 0$ may be written as $b = a \cdot q + r$ with $0 \leqslant r < a$.

As to uniqueness, suppose we have q, r with $b = aq + r$ and $0 \leqslant r < a$, ans suppose also that $b = aq' + r'$, where q', r' are possibly different integers with $0 \leqslant r' < a$. We are to show that q' must equal q and r' must equal r.

To do this, suppose $r' \geqslant r$ and subtract $b = aq' + r'$ from $b = aq + r$ to get $a(q - q') + (r - r') = 0$, or $a(q - q') = r' - r$. Since $r' - r \geqslant 0$ and $a > 0$, $q - q' \geqslant 0$. Also, $r' - r \leqslant r' < a$, so $a(q - q') < a$, so $q - q' < 1$. Since $q - q'$ is an integer, we must have $q - q' = 0$, and $q = q'$. Hence also $r' - r = a(q - q') = 0$, and $r' = r$. That completes the proof of uniqueness, and of the division theorem. □

We proved the existence part of the division theorem by well-ordering in Chapter 2.

Here is some terminology about integers: Say that a *divides* b, if $b = aq$ for some integer q. We shall write $a|b$ if a divides b. For example, 6 divides 12, 6 doesn't divide 15. If a and b are integers, a *common divisor* of a and b is an integer e such that e divides a and e divides b. A number d is the *greatest common divisor* (g.c.d.) of a and b if: (i) $d|a$ and $d|b$; and (ii) if e is a common divisor of a and b, then $e \leqslant d$. Finally, a and b are *relatively prime* if their greatest common divisor is 1. Denote the greatest common divisor of a and b by (a, b).

EXAMPLES. $(25, 35) = 5$; $(25, 17) = 1$; $(-16, -32) = 16$.

Note. There can be only one *greatest* common divisor.

The solution of the problem of finding the greatest common divisor of two natural numbers was given by Euclid. It works as follows. Suppose the two numbers are a and b, with $b \leqslant a$. (The next two paragraphs are paraphrased from Euclid (300 B.C.).)

*If b divides a, b is a common divisor of b and a, and it is manifestly also the
greatest; for no number greater than b will divide b. But if b does not divide
a, then, the lesser of the numbers a, b being continually subtracted from the
greater, some number will be left which will divide the one before it.*

*This number which is left is the greatest common divisor of b and a, and
any common divisor of b and a divides the greatest common divisor of b and
a.*

We illustrate Euclid's method with 18 and 7. We subtract 7 from 18, to
get 11 and 7. We subtract 7 from 11, to get 7 and 4. We subtract 4 from 7,
to get 4 and 3. We subtract 3 from 4, to get 1 and 3. Now 1 divides 3, so 1
is the number which is left which divides the one before it, and 1 is the
greatest common divisor of 18 and 7.

Or consider 78 and 32. Subtract 32 from 78, to get 46 and 32. Subtract
32 from 46, to get 14 and 32. Subtract 14 from 32, to get 14 and 18.
Subtract 14 from 18, to get 4 and 14. Subtract 4 from 14, to get 10 and 4.
Subtract 4 from 10, to get 4 and 6. Subtract 4 from 6, to get 2 and 4. Now
2 divides 4, so 2 is the greatest common divisor of 78 and 32.

We can describe the algorithm somewhat more compactly by using the
division theorem. Thus for the second example we have:

$$78 = 32 \cdot 2 + 14,$$
$$32 = 14 \cdot 2 + 4,$$
$$14 = 4 \cdot 3 + 2,$$
$$4 = 2 \cdot 2 + 0.$$

It is easy to see that 2 is the greatest common divisor of 32 and 78. But we
can argue systematically as follows: First, 2 divides 4, hence divides
$4 \cdot 3 + 2 = 14$, hence $14 \cdot 2 + 4 = 32$, hence $32 \cdot 2 + 14 = 78$. So 2 is a common divisor of 32 and 78. Second, if d is a common divisor of 78 and 32, d
divides 14 (from the first equation), hence 14 and 32, hence 4 (from the
second equation), hence 4 and 14, hence 2 (from the third equation). So d
divides 2.

In mathematical symbols, here is

Euclid's algorithm. *Given natural numbers a and b, apply the division
theorem successively as follows:*

$$a = bq + r_0 \qquad (dividing\ b\ into\ a),$$
$$b = r_0 q_0 + r_1 \qquad (dividing\ r_0\ into\ b),$$
$$r_0 = r_1 q_1 + r_2 \qquad (dividing\ r_1\ into\ r_0),$$
$$r_1 = r_2 q_2 + r_3 \qquad (etc.),$$
$$\vdots$$
$$r_{n-2} = q_{n-1}r_{n-1} + r_n,$$
$$r_{n-1} = q_n r_n + 0.$$

Then r_n is the greatest common divisor of a and b.

We shall prove the last statement carefully in a few pages. You try it first.

E2. Prove the statement that r_n is the greatest common divisor of a and b by using some form of induction.

E3. What is the greatest common divisor of 1492 and 1776?

E4. Prove that the g.c.d. of a and b divides $b - a$. What is the g.c.d. of 1861 and 1865? Of your year of birth and the current year?

E5. Prove that if $d = (a, b)$, then d is the smallest natural number in the set $S = \{ax + by | x, y \text{ any integers}\}$.

E6. Let a_n and a_{n+1} be consecutive terms in the Fibonacci sequence (Chapter 2, exercise E10). Show that their g.c.d. is 1.

*E7. Try playing the *game of Euclid*. Two players play, starting with two numbers. The first player subtracts any positive multiple of the lesser of the two numbers from the greater, except that the resulting number must be nonnegative. Then the second player does the same with the two resulting numbers, then the first, etc., alternately, until one player is able to subtract a multiple of the lesser number from the greater to get 0, and then he wins. For example, the moves might start with (18, 7):

Player A	Player B
(18, 7)—(11, 7)	(11, 7)—(4, 7)
(4, 7)—(4, 3)	(4, 3)—(1, 3)
(1, 3)—(1, 0) and wins	

Try to become good at Euclid. Try the game of Euclid on successive pairs of the Fibonacci sequence.

E8. Let $N(b, a) = $ number of steps needed to get the g.c.d. of b and a $(b > a)$ using Euclid's algorithm.
  (i) Let $\{a_n\}$ be the Fibonacci sequence. What is $N(a_{n+1}, a_n)$?
  (ii) Suppose that a is an integer $< a_{n+1}$, where a_{n+1} is the $(n + 1)$st number in the Fibonacci sequence. Show that for any $b > a$, $N(b, a) \leqslant N(a_{n+1}, a_n)$. (*Hint:* Use induction.)

B. Greatest Common Divisors

We observed that the last nonzero remainder in Euclid's algorithm applied to a and b is the greatest common divisor of a and b. Thus finding the greatest common divisor is an effective computational process. Euclid's algorithm also has the following useful consequence sometimes called

Bezout's Identity. *If the greatest common divisor of a and b is d, then $d = ax + by$ for some integers x and y.*

Before we show how this is done, hide the rest of this page and try the numbers 365 and 1876. It is easy to see they are relatively prime ($365 = 5 \cdot 73$ and neither 5 nor 73 divides 1876). Try to write $1 = 365x + 1876y$ for some integers x and y.

It is not obvious how to do this!

Here is how. Do the Euclidean algorithm:

$$1876 = 365 \cdot 5 + 51,$$
$$365 = 51 \cdot 7 + 8,$$
$$51 = 8 \cdot 6 + 3,$$
$$8 = 3 \cdot 2 + 2,$$
$$3 = 2 \cdot 1 + 1;$$

so 1 is the greatest common divisor.

Now solve for the remainders (the remainders are in boldface to distinguish them from the quotients)

$$\mathbf{1} = \mathbf{3} - \mathbf{2} \cdot 1,$$
$$\mathbf{2} = \mathbf{8} - \mathbf{3} \cdot 2,$$
$$\mathbf{3} = \mathbf{51} - \mathbf{8} \cdot 6,$$
$$\mathbf{8} = \mathbf{365} - \mathbf{51} \cdot 7,$$
$$\mathbf{51} = \mathbf{1876} - \mathbf{365} \cdot 5,$$

and successively substitute the remainders into the equation $\mathbf{1} = \mathbf{3} - \mathbf{2} \cdot 1$, starting with $\mathbf{2}$:

$$\mathbf{1} = \mathbf{3} - \mathbf{2}$$
$$= \mathbf{3} - (\mathbf{8} - \mathbf{3} \cdot 2) = 3 \cdot \mathbf{3} - \mathbf{8}$$
$$= 3(\mathbf{51} - \mathbf{8} \cdot 6) - \mathbf{8} = 3 \cdot \mathbf{51} - \mathbf{8} \cdot 19$$
$$= 3 \cdot \mathbf{51} - 19(\mathbf{365} - \mathbf{51} \cdot 7) = 136 \cdot \mathbf{51} - 19 \cdot \mathbf{365}$$
$$= 136(\mathbf{1876} - 5 \cdot \mathbf{365}) - 19 \cdot \mathbf{365} = 136 \cdot \mathbf{1876} - 699 \cdot \mathbf{365}.$$

So $x = -699, y = 136$.

Another, easier, way to write $d = (a, b)$ as $d = ax + by$ is to keep track of how to write each successive remainder in Euclid's algorithm in terms of a and b, using the following layout illustrated with 1876 and 365. The first column consists of numbers obtained as remainders in Euclid's algorithm, together with a and b; the second column describes the coefficient x of 365; and the third column describes the coefficient y of 1876, when writing the number in the first column in terms of 365 and 1876.

$$r = x \cdot 365 + y \cdot 1876$$

r	x	y	
1876	0	1	
365	1	0	
$1876 =$	$365 \cdot 5$	$+51$	
1876	0	1	
$365 \cdot 5$	5	0	: subtract
51	-5	1	
$365 =$	$51 \cdot 7$	$+8$	
365	1	0	
$51 \cdot 7$	-35	7	: subtract
8	36	-7	
$51 = 8 \cdot 6 + 3$			
51	-5	1	
$8 \cdot 6$	216	-42	: subtract
3	-221	43	
$8 = 3 \cdot 2 + 2$			
8	36	-7	
$3 \cdot 2$	-442	86	: subtract
2	478	-93	
$3 = 2 \cdot 1 +$		1	
3	-221	43	
2	478	-93	: subtract
1	-699	136	

Therefore, $1 = -699 \cdot 365 + 136 \cdot 1876$.

Here is a more formal proof of the last assertion of Euclid's algorithm and of Bezout's identity using induction (1).

Theorem. *If r_N is the last nonzero remainder in Euclid's algorithm for a and b, then r_N is the greatest common divisor of a and b and $r_N = ax + by$ for some x, y.*

PROOF. If r_N is the last nonzero remainder in Euclid's algorithm for a and b, then the number of steps in the algorithm is $N + 1$.

We shall prove the theorem by induction on N. If $N = 0$, then a divides b, and the theorem is trivial. If $N = 1$, then Euclid's algorithm for a and b has the form:

$$b = aq_1 + r_1,$$
$$a = r_1 q_2 + 0.$$

Then it is easy to see that r_1 is the greatest common divisor of a and b; also $r_1 = b \cdot 1 + a \cdot (-q_1)$, so Bezout's identity holds.

Assume the theorem is true for $N = n - 1$, so that the theorem is true for any two numbers whose Euclid's algorithm takes n steps. Suppose

Euclid's algorithm takes $n + 1$ steps for a and b. If $b = aq_1 + r_1$, then the rest of the algorithm for a and b is Euclid's algorithm for r_1 and a. So Euclid's algorithm takes n steps for r_1 and a. If r_n is the last nonzero remainder for Euclid's algorithm applied to a and b, it is the last nonzero remainder for Euclid's algorithm applied to r_1 and a. By induction, r_n is the greatest common divisor of a and r_1, and $r_n = au + r_1v$. Now $b = aq_1 + r_1$, so it is very easy to see that r_n, the greatest common divisor of a and r_1, is also the greatest common divisor of b and a. Moreover, substituting for r_1 in the expression $r_n = au + r_1v$ yields $r_n = au + v(b - aq_1) = a(u - vq_1) + bv$. So Bezout's identity holds. The theorem is true by induction (1). $\square$

Corollary. *If e divides a and e divides b, then e divides (a, b).*

E9. Why is the above corollary true?

E10. Write the g.c.d. of a and b in terms of a and b as in Bezout's identity, where a and b are:

(i) 267 and 112;
(ii) 222 and 1870;
(iii) 500 and 11312;
(iv) 11312 and 11213;
(v) 135 and 216.

E11. Solve

(i) $283x + 1722y = 31$,
(ii) $365x + 72y = 18$,
(iii) $1111x + 2345y = 66$.

E12. Prove: If a divides b and a divides c then a divides $bx + cy$, for any x, y in $\mathbb{Z}$.

E13. What is the g.c.d. of a, b, c? Call it (a, b, c). Prove that $((a, b), c) = (a, (b, c)) = (a, b, c)$.

E14. What is $(a_1, \ldots, a_n)$? Prove: $(a_1, \ldots, a_n) = a_1x_1 + \cdots + a_nx_n$ for some integers $x_1, \ldots, x_n$.

E15. Prove: $(a, b) = (a, b + ax)$ for any integer x.

E16. Prove: $m(a, b) = (ma, mb)$.

E17. Prove: If $(a, r) = d$ and $(b, r) = 1$, then $(ab, r) = d$.

E18. $(13750, 12222) = ?$

E19. $(336366, 488448) = ?$

E20. Prove: If $d = (a, b)$ and $ar + bs = d$, then $(r, s) = 1$.

E21. You are given two "hour"-glasses—a 6-minute and an 11-minute hourglass—and you wish to measure 13 minutes. How do you do it?

E22. Suppose $ax_0 + by_0 = 1$. (a) Show that if $x = x_0 + nb$, $y = y_0 - na$ for some n, then $ax + by = 1$. (b) Show, conversely, if $ax_1 + by_1 = 1$, then there is some integer n such that $x_1 = x_0 + nb$ and $y_1 = y_0 - na$.

C. Unique Factorization

A natural number $p > 1$ is *prime* if the only divisor of p greater than 1 is p itself.

Note. 1 is not prime, by convention.
Primes are the building blocks of natural numbers, for

Theorem. *Any natural number > 1 is prime or factors into a product of primes.*   □

You have no doubt factored many numbers into products of primes. For example,

$$368 = 2 \cdot 2 \cdot 2 \cdot 2 \cdot 23,$$
$$369 = 3 \cdot 3 \cdot 41,$$
$$370 = 2 \cdot 5 \cdot 37,$$
$$371 = 53 \cdot 7,$$
$$372 = 2 \cdot 2 \cdot 3 \cdot 31,$$
$$373 = 373 \qquad \text{(a prime number)},$$
$$374 = 2 \cdot 11 \cdot 17.$$

The proof was given in Chapter 2 as an example of induction.

We are going to prove the fundamental theorem of arithmetic, namely, that factorization of a natural number into a product of primes is unique. What does "unique" mean?

Suppose a is a natural number. If $a = p_1 \cdots p_n$ and $a = q_1 \cdots q_m$ are factorizations of a into products of primes, we shall say that the factorizations are the same if the set of p_i's is the same as the set of q_j's (including repetitions). That is, $m = n$ and each prime occurs exactly as many times among the p_i's as it occurs among the q_j's. Thus we consider the factorizations $2 \cdot 2 \cdot 3 \cdot 31$ and $3 \cdot 2 \cdot 31 \cdot 2$ as being the same, because each prime occurs an equal number of times in each; whereas $2 \cdot 3 \cdot 3 \cdot 31$ and $2 \cdot 3 \cdot 2 \cdot 31$ are different. Factorization of a is unique if any two factorizations of a are the same.

Theorem. *Any natural number $\geqslant 2$ factors uniquely into a product of primes.*

PROOF BY INDUCTION (2). Suppose that the result is true for all numbers less than a. Suppose $a = p_1 \cdots p_n$ and $a = q_1 \cdots q_m$ are two factorizations of a. We want to show that the two factorizations are the same.

Suppose it is true that $p_1 = q_j$ for some j. Then $a/p_1 = p_2 \cdots p_n = q_1 \cdots q_{j-1}q_{j+1} \cdots q_m$. Since $a/p_1 < a$, by the induction assumption the two factorizations of a/p_1 are the same. That is, the set of primes $\{p_2, \ldots, p_n\}$ is the same as the set of primes $\{q_1, \ldots, q_{j-1}, q_{j+1}, \ldots, q_m\}$.

But since $p_1 = q_j$, the set of primes $\{p_1, p_2, \ldots, p_n\}$ is then the same as the set of primes $\{q_1, \ldots, q_{j-1}, q_j, q_{j+1}, \ldots, q_m\}$, so the two factorizations of a are the same, and the result is true for the number a. That would prove the theorem by induction (2).

We prove that if $p_1 \cdots p_n = q_1 \cdots q_m$ then $p_1 = q_i$ for some i by means of the following important

Lemma 1. *If p is a prime and p divides ab, then p divides a or p divides b.*

PROOF. Let $d = (p, a)$. If $d > 1$ then $d = p$ since p is a prime. In this case, $p|a$. If $d = 1$, by Bezout's identity we can write

$$1 = ax + py$$

for some integers x and y. Then $b = bax + bpy$. Now p divides ab and also divides p. So p divides $bax + bpy = b$, completing the proof of the lemma. □

From the lemma it follows by induction (Exercise E32) that if a prime divides a product of m numbers it must divide one of the factors. To complete the proof of uniqueness of factorization, suppose we have $p_1 p_2 \cdots p_n = q_1 \cdots q_m$. Then p_1 divides $q_1 \cdots q_m$. Since p_1 is prime, p_1 must divide one of the q's, say q_i. Since q_i is prime, q_i is divisible only by itself and 1. Since $p_1 \neq 1$, $p_1 = q_i$.

Thus the induction argument described above for proving uniqueness of factorization can always be used, and the proof of uniqueness of factorization is complete. □

The argument of Lemma 1 can be used to prove

Lemma 2. *Suppose a, b, c are natural numbers $\geqslant 1$ such that a divides bc. If the greatest common divisor of a and b is 1, then a divides c.*

E23. Prove Lemma 2.

E24. Give six counterexamples to the assertion: If a divides bc and a does not divide b, then a divides c.

The problem of factoring a natural number n into a product of primes is much harder in practice than the problem of finding the greatest common divisor of two numbers. One needs to find divisors of n, and that is a matter of trial and error. For example, to factor 3372, we first see that 2 is a divisor: $3372 = 2 \cdot 1686$. Then we look for a divisor of 1686: $1686 = 2 \cdot 843$, so $3372 = 2 \cdot 2 \cdot 843$. Then we see that 3 divides 843: $843 = 3 \cdot 281$, so $3372 = 2 \cdot 2 \cdot 3 \cdot 281$. Finally, we check that 281 is not divisible by 2, 3, 5, 7, 11 or 13, and $17 > \sqrt{281}$, so 281 must be prime by exercise E25, and we have the factorization of 3372.

There are obvious tricks for testing a number n to see whether it is divisible by 2, 3, or 5; we shall see later (Chapter 6) tests for 7, 11 and 13.

In general, however, unless n happens to be prime it is a slow process looking for divisors. So much so that it is claimed that if one had a 126-digit number N which happened to be the product of two unknown 63-digit prime numbers, and one wished to factor N, even using the best available (in 1977) methods and computers it would take about 40×10^{15} years. By comparison, finding the greatest common divisor of two 126-digit numbers would take seconds.

For a discussion of techniques for factoring composite numbers, see Knuth (1969, pp. 338–360). Some of the classical tricks are in Ore (1948).

E25. Prove that if n is not prime, n has a prime divisor $\leqslant \sqrt{n}$.

E26. One can write down all primes $\leqslant n$ for any given n by the simple device, called the *sieve of Eratosthenes*, of writing down all the numbers $\leqslant n$ except 1, then circling 2 and crossing out all proper multiples of 2, then circling the first untouched number remaining and crossing out all proper multiples of that number, then circling the first untouched number remaining and crossing out all proper multiples of that number, etc., until there are no untouched numbers $\leqslant \sqrt{n}$. All of the circled or untouched numbers $\leqslant n$ will be prime. Try this with $n = 100$. Why can you stop once you have touched all numbers $\leqslant \sqrt{n}$?

D. Exponential Notation; Least Common Multiples

Here is a convenient way of representing the prime factorization of a natural number n; we illustrate first with some examples.

$$144 = 2^4 3^2,$$

$$975 = 3 \cdot 5^2 13 = 2^0 3^1 5^2 7^0 11^0 13^1,$$

$$1000 = 2^3 5^3 = 2^3 3^0 5^3.$$

We write a natural number n as a product of powers of primes, in increasing order. We can include extra primes which do not divide n by writing the exponent as 0:

$$n = p_1^{e_1} p_2^{e_2} p_3^{e_3} \cdots p_r^{e_r}.$$

Uniqueness of factorization says there is only one way to do this, except for the inclusion of extra primes with exponent zero.

Using this notation we can interpret some of our previous ideas. Let

$$m = p_1^{a_1} p_2^{a_2} \cdots p_r^{a_r}, \qquad n = p_1^{b_1} p_2^{b_2} \cdots p_r^{b_r}$$

where $p_1, \ldots, p_r$ includes all primes which divide either m or n and some of the exponents may be zero.

E27. Complete with a statement involving a_i's and b_i's:

(i) m divides n if . . . ;
(ii) the greatest common divisor of m and n is . . . ;
(iii) the least common multiple of m and n is . . . (see E34);
(iv) m/n is an integer if . . . ;
(v) m/n is an integer divisible by p_i if . . . .

Here is an application of the fact that any natural number factors uniquely into a product of primes.

Theorem. $\sqrt{2}$ *is irrational.*

PROOF. If not, $\sqrt{2} = a/b$, a, b natural numbers. So $2a^2 = b^2$. Write $a = p_1^{e_1} \cdots p_r^{e_r}$, $b = p_1^{f_1} \cdots p_r^{f_r}$. Then in the equation $2a^2 = b^2$, 2 occurs to an even power on the right side, and to an odd power on the left side, which is impossible. □

E28. Prove: If $(a, b) = 1$ and ab is a square, then a is a square.

E29. Prove: If $d = (a, b)$, then $\{ax + by | x, y \text{ in } Z\} = \{zd | z \text{ in } Z\}$.

E30. If $(a, b) = 8$, what are the possible values of (a^3, b^4)?

E31. If $(a, b) = p^3$, p prime, what is (a^2, b^2)?

E32. Prove by induction that if a prime number p divides $a_1 \cdot a_2 \cdots a_n$, then p must divide one of the factors a_i.

E33. Show $\sqrt{1000}$ is irrational.

Definition. The *least common multiple* of two numbers a and b is the smallest number which is a common multiple of a and of b. More precisely, m is the least common multiple of a and b if

(1) $a|m$ and $b|m$, and
(2) if $a|r$ and $b|r$, then $m \leqslant r$.

The least common multiple of a and b is usually denoted by $[a, b]$.

E34. Prove that $[a, b] = ab/(a, b)$.

E35. What is $[22, 121]$; $[1001, 169]$; $[1001, 777]$?

E36. What is $[a, b, c]$? Is $[a, b, c] = abc/(a, b, c)$?

E37. What is $[a, b, c, d]$?

E38. Show: $m = [a, m]$ iff $a|m$.

*****E39.** Prove: If $(a, b) = 1$ and c is any integer, then there is some integer x so that $(a + bx, c) = 1$.

E40. Suppose $ax_0 + by_0 = r$ and $ax_1 + by_1 = r$. Show that there is some integer n so that $x_1 = x_0 + n[a, b]/a$ and $y_1 = y_0 - n[a, b]/b$

E41. You take a 12-quart jug and a 17-quart jug to the stream and want to bring back 8 quarts of water. How do you do it?

†*E42. You take a- and b-quart jugs to the stream, where a and b are integers, $a \geqslant b$. You want to bring back c quarts, where c is some integer, $0 < c < a + b$. For which c can it be done? How?

E43. Prove that $\lim_{n \to \infty}(a, b/(a^n, b)) = 1$.

E44. Prove that $(a/(a, b), b/(a, b)) = 1$.

Primes 4

A. Euclid

The last chapter showed that primes are the building blocks of natural numbers, in the sense that any number is a product of primes. For this reason mathematicians throughout history have been fascinated by primes. In this chapter we shall describe a few of the most famous results on primes.

The earliest result on primes, due to Euclid, is from Book IX, Proposition 20 of *The Elements* (300 B.C.).

Theorem. *There are infinitely many primes.*

PROOF OF EUCLID. Suppose you believe that $p_1, \ldots, p_N$ are all the primes. I shall show that you are wrong. Consider the number $p_1 p_2 \cdots p_N + 1$. If it is prime, it is a new prime. Otherwise it has a prime factor q. If q were one of the primes p_i, then q would divide $p_1 p_2 \cdots p_N$, and since q divides $p_1 p_2 \cdots p_N + 1$, q would divide the difference of these numbers, namely 1, which is impossible. So q can not be one of the p_i, and must therefore be a new prime. This completes the proof. □

E1. Use the ideas in Euclid's proof to try to prove that there are infinitely many primes of the form $4n + 3$. (*Hint*: Consider $4p_1 \cdots p_n + 3$.)

E2. Try the same for numbers of the form $6n + 5$.

*E3. Try the same for numbers of the form $4n + 1$, using $(2p_1 \cdots p_n)^2 + 1$.

In connection with these exercises we note a famous theorem of Dirichlet, which says that given any natural numbers $a \geq 2$ and b with $(a, b) = 1$, there are infinitely many primes of the form $an + b$. We give a brief introduction to some of the ideas which go into the proof of that theorem in Section B.

E4. Here is another proof of the infinitude of primes, due to G. Polya.
   *(a) Show that $2^{(2^m)} + 1$ and $2^{(2^n)} + 1$ are relatively prime if $m \neq n$.
   (b) Use (a) to show that there are at least n primes less than $2^{(2^n)} + 1$ for each n, hence there are infinitely many primes.

E5. Prove that for any n there exist n consecutive natural numbers none of which are prime. (*Hint*: Start with $n + 1! + 2$.)

E6. Prove that for any n there is a prime p with $n < p \leqslant n! + 1$.

B. Some Analytic Results

This section assumes some knowledge of calculus and of infinite series. It and the following section may be omitted without loss of continuity.

Euler provided a proof of the existence of infinitely many primes which provides the starting point for much advanced number theory. It uses the fact that the series

$$\sum_{n=1}^{\infty} \frac{1}{n} = 1 + \frac{1}{2} + \frac{1}{3} + \frac{1}{4} + \cdots$$

diverges, that is,

$$\lim_{E \to \infty} \sum_{n=1}^{E} \frac{1}{n} = \infty.$$

PROOF OF EULER. As in Euclid's proof, we assume that there are only finitely many primes, and we derive a contradiction.

If $p_1, \ldots, p_N$ are all the primes, then the product

$$\prod_{i=1}^{N} \left(\frac{1}{1 - (1/p_i)} \right)$$

is finite. Recall the geometric series: If $0 < x < 1$, then

$$\frac{1}{1 - x} = 1 + x + x^2 + \cdots + x^E + \frac{1 - x^{E+1}}{1 - x}$$

$$> \sum_{m=0}^{E} x^m.$$

Using this with $x = 1/p_i$, we get

$$\frac{1}{1 - (1/p_i)} > \sum_{m=0}^{E} \frac{1}{p_i^m}.$$

So

$$\prod_{i=1}^{N} \left(\frac{1}{1 - (1/p_i)} \right) > \prod_{i=1}^{N} \sum_{m=0}^{E} \frac{1}{p_i^m} \quad \text{for any } E > 0.$$

Multiplying out the right side,

$$\prod_{i=1}^{N} \sum_{m=0}^{E} \frac{1}{p_i^m} = \sum_{e_1=0}^{E} \sum_{e_2=0}^{E} \cdots \sum_{e_N=0}^{E} \frac{1}{p_1^{e_1} \cdot p_2^{e_2} \cdots p_N^{e_N}}. \qquad (1)$$

Now since $p_1, \ldots, p_N$ are assumed to be all the primes, and any natural number is a product of primes, each natural number n has the form $n = p_1^{e_1} \cdots p_N^{e_N}$ for some $e_1 \cdots e_N$, uniquely. So the right-hand sum of (1) is the sum of all reciprocals $1/n$ such that no prime occurs in the prime factorization of n to an exponent greater than E. In particular, if $n \leqslant E$, $1/n$ is included in the right-hand side (why?). Thus

$$\prod_{i=1}^{N} \left(\frac{1}{1 - (1/p_i)} \right) \geqslant \sum_{n=1}^{E} \frac{1}{n} \quad \text{for any } E > 0.$$

But $\lim_{E \to \infty} \sum_{n=1}^{E} (1/n)$ is infinite, and so choosing E sufficiently large, $\sum_{n=1}^{E} (1/n)$ is larger than any fixed finite number, such as

$$\prod_{i=1}^{N} \left(\frac{1}{1 - (1/p_i)} \right).$$

We get a contradiction, proving the theorem.   □

E7. Show that

$$\lim_{N \to \infty} \prod_{i=1}^{N} \left(\frac{1}{1 - (1/p_i)} \right) = \infty.$$

Euler's proof provides a starting point for a proof of Dirichlet's theorem that given any a, b with $(a, b) = 1$ there are infinitely many primes in the arithmetic progression $an + b$.

The proof of Dirichlet's theorem begins with the interesting observation, which we shall justify momentarily, that the infinite series

$$\sum_{p \text{ prime}} \frac{1}{p}$$

diverges. One wants to show that if P_b is the set of primes of the form $an + b$, then

$$\sum_{p \text{ in } P_b} \frac{1}{p}$$

diverges. This of course would mean that P_b is an infinite set. To do that one shows that in fact the ratio

$$\frac{\displaystyle \sum_{p \text{ in } P_b, p < n} \frac{1}{p}}{\displaystyle \sum_{p \text{ prime}, p < n} \frac{1}{p}}$$

approaches $1/\phi(a)$ as $n \to \infty$, where $\phi(a)$ is the number of natural numbers $< a$ which are relatively prime to a. Since the denominator diverges, that would imply that the numerator would also diverge. For example, $\phi(4) = 2$, so that

$$\frac{\displaystyle\sum_{\substack{p<n,\, p \text{ prime},\\ p=4m+3 \text{ for some } m}} \frac{1}{p}}{\displaystyle\sum_{p<n,\, \text{prime}} \frac{1}{p}} \to \frac{1}{2} \quad \text{as } n \to \infty,$$

and also

$$\frac{\displaystyle\sum_{\substack{p<n,\, p \text{ prime}\\ p=4m+1 \text{ for some } m}} \frac{1}{p}}{\displaystyle\sum_{p<n,\, p \text{ prime}} \frac{1}{p}} \to \frac{1}{2} \quad \text{as } n \to \infty.$$

Thus the proof of Dirichlet's theorem would show in particular, that not only are there infinitely many primes of each of the form $4m + 1$, $4m + 3$, but in some sense the number of primes of the form $4m + 1$ is "the same" as the number of prime of the form $4m + 3$.

We cannot prove Dirichlet's theorem in this book, but we can prove

Theorem.

$$\sum \frac{1}{p} = \frac{1}{2} + \frac{1}{3} + \frac{1}{5} + \frac{1}{7} + \frac{1}{11} + \frac{1}{13} + \frac{1}{17} + \cdots$$

diverges.

PROOF. We start from Exercise E7, a consequence of Euler's proof, that

$$\prod_{i=1}^{\infty} \left(\frac{1}{1-(1/p_i)} \right) = \infty.$$

Taking the log,

$$\infty = \log\left(\prod_{i=1}^{\infty} \left(\frac{1}{1-(1/p_i)} \right) \right) = \sum_{i=1}^{\infty} \left(-\log\left(1 - \frac{1}{p_i} \right) \right).$$

Now

$$-\log(1-x) = x + \frac{x^2}{2} + \frac{x^3}{3} + \cdots \quad \text{for } 0 < x < 1$$

(this can be shown by integrating the geometric series term by term). So

$$\sum_{i=1}^{\infty} \left(-\log\left(1 - \frac{1}{p_i} \right) \right) = \sum_{i=1}^{\infty} \sum_{n=1}^{\infty} \frac{1}{p_i^n} = \sum_{n=1}^{\infty} \sum_{i=1}^{\infty} \frac{1}{p_i^n}$$

(we can interchange the order of summation because all terms are positive)

$$= \sum_{i=1}^{\infty} \frac{1}{p_i} + \sum_{n=2}^{\infty} \sum_{i=1}^{\infty} \frac{1}{p_i^n} = \sum_{i=1}^{\infty} \frac{1}{p_i} + \sum_{i=1}^{\infty} \sum_{n=2}^{\infty} \frac{1}{p_i^n}. \tag{2}$$

Now, for any p,

$$\frac{1}{p^2} + \frac{1}{p^3} + \cdots = \frac{1}{p^2}\left(1 + \frac{1}{p} + \frac{1}{p^2} + \cdots\right);$$

using the geometric series, this equals

$$\frac{1}{p^2}\left(\frac{1}{1 - (1/p)}\right) = \frac{1}{p(p-1)}.$$

So

$$\sum_{n=2}^{\infty} \frac{1}{p^n} = \frac{1}{p(p-1)} < \frac{1}{(p-1)^2}$$

and the sum (2) is

$$\leqslant \sum_{i=1}^{\infty} \frac{1}{p_i} + \sum_{i=1}^{\infty} \frac{1}{(p_i - 1)^2} \leqslant \sum_{i=1}^{\infty} \frac{1}{p_i} + \sum_{n=1}^{\infty} \frac{1}{n^2}.$$

Since the series (2) sums to ∞, the sum of these last two series is ∞. But $\sum_{n=1}^{\infty}(1/n^2)$ converges by the integral test. Thus $\sum_{i=1}^{\infty}(1/p_i)$ must diverge. $\square$

For a proof of Dirichlet's theorem, see Apostol (1976).

C. The Prime Number Theorem

We conclude this chapter by mentioning without proof two other famous results about primes. The first concerns the function $\pi(x)$, defined to be the number of primes $\leqslant x$.

Euclid's theorem (Section A) says that as $x \to \infty$, $\pi(x) \to \infty$. A much more precise result, called the prime number theorem, says:

$$\lim_{x \to \infty} \frac{\pi(x)}{(x/\log x)} = 1.$$

This means that given any $\varepsilon > 0$ there is some number N such that for any $x > N$,

$$\left| \frac{\pi(x)}{(x/\log x)} - 1 \right| < \varepsilon,$$

or

$$\left| \pi(x) - \frac{x}{\log x} \right| < \varepsilon\left(\frac{x}{\log x}\right).$$

That is, while $\pi(x)$ and $x/\log x$ may differ by a large amount, if x is chosen large enough, that difference can be made as small a percentage of $x/\log x$ as you wish. Thus the number of primes $\leqslant x$ can be estimated rather well, no matter how large x is.

A more elementary result, which still shows how closely $\pi(x)$ and $x/\log x$ are related, is:

$$\frac{\log 2}{4} \frac{x}{\log x} \leqslant \pi(x) \leqslant 9 \log 2 \frac{x}{\log x} \quad \text{for all } x.$$

A comprehensible proof of this last result may be found in Niven and Zuckerman (1972, p. 184).

Three fairly accessible articles about the prime number theorem are those of Levinson (1969), Goldstein (1973), and Zagier (1977).

The second famous result about primes we mention concerns the gaps between consecutive primes. Exercise E6 showed that for any n, there is a prime p with $n \leqslant p < (n-1)! + 2$. There is a much better result:

Bertrand's postulate. *For any $n > 1$ there is a prime p with $n \leqslant p < 2n$.*

One consequence of Bertrand's postulate is that if $p_1 = 2$, $p_2 = 3$, $p_3, \ldots$ are the primes in increasing order, then $p_{n+1} \leqslant 2p_n + 1$.

A proof of Bertrand's postulate may be found in Niven and Zuckerman (1972, p. 189).

E8. Prove the consequence of Bertrand's postulate just described.

E9. If $p_1, p_2, \ldots$ are the primes in increasing order, prove that $p_{n+1} \leqslant p_1 + p_2 + \cdots + p_n$ for all $n > 1$.

Bases 5

A. Numbers in Base a

Our way of writing numbers is biased in favor of the number 10. When we take a number b, like $b = $ MCMLXXVI, and write it as $b = 1976$, we mean that

$$b = 1 \times 10^3 + 9 \times 10^2 + 7 \times 10 + 6;$$

we call this way of writing b the *representation of b in base* 10, or *radix* 10.

If we write a number b as a sum of powers of a ($a \geqslant 2$),

$$b = r_n a^n + r_{n-1} a^{n-1} + \cdots + r_1 a + r_0,$$

with each of $r_0, \ldots, r_n$ between 0 and $a - 1$, this is the *representation of b in base* (or *radix*) a.

There is no particular reason except convention or physiology why we have a bias towards 10. For example, we could write $b = $ MCMLXXVI in base 2:

$$b = 1 \times 2^{10} + 1 \times 2^9 + 1 \times 2^8 + 1 \times 2^7 + 1 \times 2^5 + 1 \times 2^4 + 1 \times 2^3,$$

or $b = 11110111000$. If we used base 2 with all numbers, our favorite electronic computer, which understands "on" (1) and "off" (0) very well, would be very happy and efficient.

Numbers themselves have no bias, as the following theorem shows.

Theorem. *Fix a natural number $a \geqslant 2$. We may represent any integer $b \geqslant 0$ in base a: that is, b can be written uniquely as*

$$b = r_n a^n + r_{n-1} a^{n-1} + \cdots + r_2 a^2 + r_1 a + r_0$$

with $0 \leqslant r_i < a$ for all i.

If we write b in base a we shall use the notation $b = (r_n r_{n-1} \cdots r_2 r_1 r_0)_a$. Thus $(1976)_{10} = (11110111000)_2$. We shall omit $(\ )_{10}$ in decimal notation when there is no possibility of confusion.

Here is a proof of the theorem, using induction (2).

PROOF. Suppose all numbers $< b$ may be written in base a. If a is the base, then divide a into b, using the division theorem to get $b = aq + r_0$, $0 \leqslant r_0 < a$. By induction, $q = r_1 + ar_2 + a^2 r_3 + \cdots + a^{n-1} r_n$ for unique integers $r_1, \ldots, r_n$, $0 \leqslant r_i < a$. Then

$$b = a(r_1 + ar_2 + a^2 r_3 + \cdots + a^{n-1} r_n) + r_0$$

$$= a^n r_n + a^{n-1} r_{n-1} + \cdots + a^3 r_3 + a^2 r_2 + ar_1 + r_0,$$

where $0 \leqslant r_i < a$ for all $i = 0, \ldots, n$. The expression is unique because q and r_0 are unique in the division theorem. This completes the proof. □

Notice that the proof shows how to get b in base a: we divide first b, then successive quotients, by a,

$$b = aq + r_0$$
$$q = aq_1 + r_1$$
$$q_1 = aq_2 + r_2$$
$$\vdots$$

until the last quotient is 0. The digits are the remainders: $b = (r_n r_{n-1} \cdots r_2 r_1 r_0)_a$.

Here is an example. To get 366 in base 2:

$$366 = 2 \cdot 183 + 0,$$
$$183 = 2 \cdot 91 + 1,$$
$$91 = 2 \cdot 45 + 1,$$
$$45 = 2 \cdot 22 + 1,$$
$$22 = 2 \cdot 11 + 0,$$
$$11 = 2 \cdot 5 + 1,$$
$$5 = 2 \cdot 2 + 1,$$
$$2 = 2 \cdot 1 + 0,$$
$$1 = 2 \cdot 0 + 1,$$

so $366 = (101101110)_2$.

The ancient Babylonians used a base 60 number system. To get 13056 in base 60:

$$13056 = 60 \cdot 217 + 36,$$
$$217 = 60 \cdot 3 + 37,$$

so $13056 = 3 \cdot 60^2 + 37 \cdot 60 + 36 = (3, 37, 36)_{60}$.

E1. Set $t = 10$, $e = 11$ and write 8372 in base 12.

E2. Write 8372 in base 2.

E3. Write 144 in base 6.

E4. Write $(10^{13} - 1)/3$ in base 1000.

B. Operations in Base *a*

We can add, subtract, multiply, and divide in any base. For example, multiplication in any base is done the way you learned in base 10 in grade school. The only change is that to use base *a* you must know the multiplication table in base *a*. The multiplication in base 10

$$
\begin{array}{r}
83 \\
37 \\
\hline
581 \\
249 \\
\hline
3071
\end{array}
$$

becomes in base 2

$$
\begin{array}{r}
1010011 \\
100101 \\
\hline
1010011 \\
10100110 \\
101001100 \\
\hline
101111111111 \,.
\end{array}
$$

It is very easy to remember multiplication tables in base 2!

E5. Write 176 and 398 in base 2 and multiply them.

E6. Show that in base 2, Russian peasant arithmetic (Chapter 2, E9) is identical to the usual procedure for multiplying numbers in base 2, except for the layout of the numbers on paper during the performance of the multiplication.

With long division in base 10 there is a certain amount of trial and error. For example, dividing 37 into 3071: does 37 go into 307 8 times or 9 times? You have to check. In base 2, however, all the guess work is removed, since the only possibility for each digit in the quotient is 0 or 1:

$$
\begin{array}{r}
1010011 \\
100101\overline{)101111111111} \\
100101 \\
\hline
00101011 \\
100101 \\
\hline
000110111 \\
100101 \\
\hline
100101 \\
100101 \\
\hline
0\,.
\end{array}
$$

On the other hand, in base 60 the guesswork can be horrendous! Or is it?! See Section C.

E7. Try $(1, 38)_{60}\overline{)(1, 4, 25, 46)_{60}}$; $(110110011)_2\overline{)(1100000100101)_2}$.

You may recall the square root algorithm. (If not, ask someone to show it to you.) For example, we find $\sqrt{625}$ in base 10 as follows:

$$
\begin{array}{r}
25 \\
\sqrt{6\,25} \\
4 \\
45\ \overline{\left|2\,25\right.} \\
2\,25 \\
\hline
0\,.
\end{array}
$$

In base 10 the square root algorithm involves a bit of trial and error in guessing each successive digit of the result, just as with long division. But in base 2 no guessing is needed. Here is the same problem in base 2:

$$
\begin{array}{r}
1\ \ 1\ \ 0\ \ 0\ \ 1 \\
\sqrt{10\ 01\ 11\ 00\ 01} \\
1 \\
101\quad \overline{\left|1\ 01\right.} \\
1\ 01 \\
110001\quad \overline{\left|0\ \ 11\ 00\ 01\right.} \\
11\ 00\ 01 \\
\hline
0\,.
\end{array}
$$

E8. Find $\sqrt{1010000000011}$.

Did you ever wonder why there is a square root algorithm but no cube root algorithm? Perhaps because in base 10 it is too complicated to get successive digits? If base 2 is so nice for the square root algorithm perhaps a cube root algorithm is manageable?

*E9. (a) Invent an algorithm for finding successive digits of the cube root of a number in base 2. By algorithm is meant a set of instructions which any person familiar with the instruction language can follow without thinking. (*Hint*: To do this problem it helps to really understand how and why the square root algorithm works!)

    (b) Having done so, compute $2^{1/12}$, namely, the cube root of the square root of the square root of 2, to at least three decimal places.

    (c) On the tempered musical scale, A is 440 cycles per second and B♭ is $440 \times 2^{1/12}$ cycles per second. Use (b) to compute B♭.

C. Multiple Precision Long Division

An important situation where representing numbers in a large base number is necessary is in doing highly accurate computations on a computer.

A typical computer is set up to take and work with numbers of a certain size, for example, base 2 numbers with up to 32 digits. To do precise computation with numbers bigger than 2^{32}, one has to represent the numbers in the computation in base 2^{32}, and write a program telling the computer how to add, subtract, multiply and divide such numbers, assuming that the computer has already been designed to do those operations to the digits of the base 2^{32} numbers (the digits being base 2 numbers with at most 32 binary digits.)

It is not hard to tell the computer how to add, subtract and multiply in any base, even in base 2^{32}, as we have observed in Section B. But what about division? In base 2^{32} isn't the guessing horrendous?

It turns out not to be so.

Long division in base b is a sequence of divisions by a number d, the divisor, into numbers e where $d \leqslant e < bd$. The digit of the quotient corresponding to the division of d into e is the greatest integer q such that $dq \leqslant e$.

For example, in base 10, consider

$$
\begin{array}{r}
281 \\
32\overline{)8994} \\
64 \\
\hline
259 \\
256 \\
\hline
34 \\
32 \\
\hline
2 \ .
\end{array}
$$

We have $d = 32$, $b = 10$. In doing this long division, we first divide 32 into 89 with quotient digit 2, then divide 32 into 259 with quotient digit 8, then divide 32 into 34 with quotient digit 1. Where we guess is in trying to determine these successive quotient digits. The guessing thus arises in the situation: Divide d into e with $d \le e < bd$.

In base b the standard guess is made as follows. First write d and e in base b:

$$d = d_n b^n + d_{n-1} b^{n-1} + \cdots + d_1 b + d_0 \quad \text{with } 0 \le d_i < b, d_n \ne 0,$$

$$e = e_{n+1} b^{n+1} + e_n b^n + \cdots + e_1 b + e_0 \quad \text{with } 0 \le e_i < b.$$

The standard guess is to divide d_n, the largest digit of d, into the two-digit number $e_{n+1} b + e_n$, and use the quotient as the guess. So we define our estimate q' of the correct quotient digit q by:

$$\text{if } d_n \begin{cases} = e_{n+1}, & \text{set } q' = b - 1 \\ > e_{n+1}, & \text{set } q' = \text{the largest integer such that } q' d_n \le e_{n+1} b + e_n. \end{cases}$$

$$(1)$$

E10. Show that in either case, $q' \le b - 1$ and $q' d_n \le e_{n+1} b + e_n$.

To illustrate in base 10, with q' the guess as defined by (1) and q the correct quotient (and $n = 1$):

If $d = 59$, $e = 500$, $d_n = e_{n+1}$, so $q' = 9$. Here $q = 8$.
If $d = 59$, $e = 400$, $5 \cdot 8 = 40$, so $q' = 8$. Here $q = 6$.
If $d = 19$, $e = 100$, $d_n = e_{n+1}$, so $q' = 9$. Here $q = 5$.
If $d = 19$, $e = 90$, $1 \cdot 9 = 9$, so $q' = 9$. Here $q = 4$.

In all of these cases the guess q' is too big. But observe that q' is not as bad a guess for $d = 59$ as it is for $d = 19$. This will be true in general, as the following theorem shows.

Theorem. *Let $d = (d_n d_{n-1} \cdots d_1 d_0)_b$, $e = (e_{n+1} e_n \cdots e_1 e_0)_b$ (where $d_n \ne 0$ but e_{n+1} may be 0), and suppose $d \le e < bd$. Suppose q' is the guess as defined in (1) above. Let q be the correct quotient, that is, the largest integer with $qd \le e$. Then:*

(i) $q' \ge q$—*the guess q' is always $\ge$ the correct quotient q;*
(ii) *if $d_n \ge b/2$, then $q \ge q' - 2$—the guess q' is equal either to q, to $q + 1$, or to $q + 2$.*

PROOF. (i) Since $qd \le e$, we have

$$q\big(d_n b^n + d_{n-1} b^{n-1} + \cdots + d_1 b + d_0\big)$$

$$\le e_{n+1} b^{n+1} + e_n b^n + e_{n-1} b^{n-1} + \cdots + e_1 b + e_0).$$

So

$$qd_n b^n \leqslant e_{n+1}b^{n+1} + e_n b^n + e_{n-1}b^{n-1} + \cdots + e_1 b_1 + e_0 = e.$$

Since $e_{n-1}b^{n-1} + \cdots + e_1 b + e_0 < b^n$,

$$e < e_{n+1}b^{n+1} + e_n b^n + b^n = (e_{n+1}b + e_n + 1)b^n.$$

Therefore $qd_n b^n < (e_{n+1}b + e_n + 1)b^n$, so

$$qd_n < e_{n+1}b + e_n + 1.$$

Since both sides are integers, therefore

$$qd_n \leqslant e_{n+1}b + e_n.$$

It follows that $q \leqslant q'$. For $q \leqslant b - 1$, since $qd \leqslant e < bd$, if $d_n > e_{n+1}$, then $q'd_n$ is defined to be the largest multiple of d_n which is $\leqslant e_{n+1}b + e_n$, so $q \leqslant q'$. That proves (i).

To prove (ii), suppose $d_n \geqslant b/2$. By choice of q it suffices to show that $(q' - 2)d \leqslant e$. Now

$$(q' - 2)d = (q' - 2)(d_n b^n + d_{n-1}b^{n-1} + \cdots + d_1 b + d_0). \qquad (2)$$

Since $d_{n-1}b^{n-1} + \cdots + d_1 b + d_0 < b^n$, we get from (2) that

$$(q' - 2)d < (q' - 2)(d_n + 1)b^n = [q'd_n + (q' - 2 - 2d_n)]b^n$$

$$\leqslant (be_{n+1} + e_n)b^n + (q' - 2 - 2d_n)b^n, \qquad (3)$$

by Exercise E10. Now $d_n \geqslant b/2$ and $q' \leqslant b - 1$, so $q' - 2 - 2d_n < 0$. Hence the right-hand side of (3) is

$$\leqslant (be_{n+1} + e_n)b^n$$

$$\leqslant (be_{n+1} + e_n)b^n + e_{n-1}b^{n-1} + \cdots + e_1 b + e_0 = e,$$

proving part (ii).   □

The theorem shows that in any base b, if the leading digit of the divisor is $\geqslant b/2$, the standard guess q' will never be off by more than 2: either q', $q' - 1$, or $q' - 2$ will be the correct quotient. Notice that it does not matter how large the base b is.

To insure that the leading digit of the divisor is $\geqslant b/2$, one "normalizes." Take the original divisor d and dividend e and multiply d and e by 2^s where 2^s is the largest power of 2 with $2^s d < b^{n+1}$. Then divide $2^s d$ into $2^s e$.

If $e = qd + r$, then $(2^s e) = (2^s d)q + r_0$ where $r_0 = 2^s r$. So normalizing by multiplying d and e by 2^s does not change the resulting quotient, and the remainder will be 2^s times the remainder on dividing e by d.

It is therefore quite feasible to tell a computer how to do multiple precision long division. You can tell it first to normalize, then you can tell it what to guess, and it will have no difficulty adjusting its guess by 1 or 2 if necessary to obtain the correct quotient at each stage in the long division process.

One moral of the theorem: In dividing using long division, the larger the first digit of the divisor is, the easier the division is. So, in base 10,

is easier than

$$88 \overline{\smash{\big)}\, 7135}$$

$$17 \overline{\smash{\big)}\, 7135} \,.$$

Did you know that before?

For a computer program implementing long division, together with further discussion and references, see Knuth (1969, pp. 236 ff.).

E11. In base 8
  (a) divide $(57237)_8$ into $(22457160)_8$,
  (b) divide $(16263)_8$ into $(3000012)_8$.

E12. In base 16 divide p into q by first normalizing to replace the divisor by one whose first digit is ≥ 8, where
  (a) $p = (3, 11, 2)_{16}$, $q = (1, 11, 15, 4, 0)_{16}$,
  (b) $p = (1, 6, 7)_{16}$, $q = (9, 0, 13, 1, 10)_{16}$.

E13. How big must the leading digit of the divisor be in order that the guess q' never differ from q by more than 3?

E14. (a) In base 10 if we do not normalize, how far off can q' be from the correct quotient q? (b) Same question in base b for any b.

D. Decimal Expansions

We can do expansions in base a also. In base 10 they are called decimal expansions. To begin with an example, to expand $1/7$ into a decimal, divide 7 into 1 by the division algorithm, multiply the remainder (which is 1) by 10, divide that by 7 (with quotient 1), multiply the remainder (which is 3) by 10, divide that by 7 (with quotient 4), etc. It is what you do in long division:

$$
\begin{array}{r}
14285\ldots \\
7 \overline{\smash{\big)}\, 1.00000} \\
\underline{7} \\
30 \\
\underline{28} \\
20 \\
\underline{14} \\
60 \\
\underline{56} \\
40 \\
\underline{35} \\
5\ldots\ldots
\end{array}
$$

Thus $1/7 = 0.14285\ldots\ldots$.

The base-free procedure for taking b/c and finding its expansion in base a is the the same idea: Find $b = c \cdot q + r_0$ where q is an integer and $r_0 < c$. Then

$$ar_0 = cq_1 + r_1 \quad \text{with } 0 \leqslant q_1 < a \text{ (why)? and } 0 \leqslant r_1 < c,$$
$$ar_1 = cq_2 + r_2 \quad \text{with } 0 \leqslant a_2 < a \text{ and } 0 \leqslant r_2 < c,$$
$$ar_2 = cq_3 + r_3 \quad \text{with } 0 \leqslant q_3 < a \text{ and } 0 \leqslant r_3 < c,$$

etc. Dividing these equations successively by c, ac, a^2c, ... gives

$$\frac{b}{c} = q + \frac{r_0}{c},$$
$$\frac{r_0}{c} = \frac{q_1}{a} + \frac{r_1}{ca},$$
$$\frac{r_1}{ca} = \frac{q_2}{a^2} + \frac{r_2}{ca^2},$$
$$\frac{r_2}{ca^2} = \frac{q_3}{a^3} + \frac{r_3}{ca^3},$$

etc., so, successively substituting:

$$\frac{b}{c} = q + \frac{q_1}{a} + \frac{q_2}{a^2} + \frac{q_3}{a^3} + \cdots \quad \text{with } 0 \leqslant q_i < a.$$

This is the *expansion of b/c in base a*. The quotients give the digits. We shall denote this expansion by

$$\frac{b}{c} = q + (.q_1 q_2 q_3 \ldots)_a.$$

For example, we can do fractions in the ancient Babylonian base 60 number system. To find $3/8$ in base 60, we divide as follows:

$$3 \cdot 60 = 8 \cdot 22 + 4,$$
$$4 \cdot 60 = 8 \cdot 30.$$

Then

$$\frac{3}{8} = \frac{22}{60} + \frac{30}{(60)^2}.$$

To find $137/175$ in base 60, divide as follows:

$$137 \cdot 60 = 175 \cdot 46 + 170,$$
$$170 \cdot 60 = 175 \cdot 58 + 50,$$
$$50 \cdot 60 = 175 \cdot 17 + 25,$$
$$25 \cdot 60 = 175 \cdot 8 + 100,$$
$$100 \cdot 60 = 175 \cdot 34 + 50,$$
$$50 \cdot 60 = 175 \cdot 17 + 25,$$

etc. Then

$$\frac{137}{175} = \frac{46}{60} + \frac{58}{(60)^2} + \frac{17}{(60)^3} + \frac{8}{(60)^4} + \frac{34}{(60)^5} + \frac{17}{(60)^6} + \frac{8}{(60)^7} + \cdots$$
$$= (46, 58, 17, 8, 34, 17, 8, \ldots)_{60}.$$

It repeats after the 58.

If we want the base a expansion of b/c and b and c are expressed in base a, we can do long division in base a, as we did above with $1/7$ in base 10. For example, $1/5$ in base 2 is $1/(101)_2$; its base 2 expansion is

$$
\begin{array}{r}
0.0011\overline{0011} \ldots \\
101 \overline{)\, 1.00000000 \ldots} \\
101 \\
\hline
110 \\
101 \\
\hline
1000 \\
101 \\
\hline
110 \\
101 \\
\hline
1 \\
\ddots
\end{array}
$$

We shall investigate base a expansions further in Chapters I-12 and III-2.

E15. Prove that every rational number has an eventually repeating decimal expansion (e.g., $1/7 = .14285714285714285714 \ldots$; $11/24 = .458333333 \ldots$).

†**E16.** Is Exercise E15 true for expansions of rational numbers in any base?

Definition. An expansion in base a *terminates* if all digits from some point on are equal to zero.

E17. Which rational numbers have terminating Babylonian expansions? terminating base 2 expansions? terminating base a expansions?

E18. Expand $1/31$ in base 2.

E19. If $a/b = .t_1t_2t_3t_1t_2t_3t_1t_2t_3 \ldots$ (base 10), what are the possible values of b?

E20. Find a base such that $1/11$ has a repeating base a expansion which repeats only every 10 digits.

E21. Expand $3/7$ in base 3.

E22. Expand $141/144$ in base 6.

E23. Write $1/7$ in base 1000.

E24. Use Dirichlet's theorem (I-4) to prove that given any block B of digits (such as $B = 131420$) there exist infinitely many primes P whose decimal representation includes B (for example, if $B = 37$, then 37, 137, and 373 are three such primes for B).

Congruences 6

A. Definition of Congruence

An *arithmetic progression* is a sequence of integers of the form $nk + a$, $k = \ldots, -2, -1, 0, 1, 2, \ldots$. We write the set of all integers in such a progression as $\{nk + a\}_{k \in \mathbb{Z}}$. Example: the progression $\ldots, -7, -2, 3, 8, 13, 18, \ldots$ forms the set $\{5k + 3\}_{k \in \mathbb{Z}}$.

We encountered arithmetic progressions in the chapter on primes. For example, we showed that there were infinitely many primes of the form $4k + 3$, that is, primes which when divided by 4 leave a remainder of 3. Dirichlet's theorem (Chapter 4, Sections A and B) asserts the existence of infinitely many primes in any arithmetic progression of the form $nk + a$, $k = 0, 1, 2, \ldots$, when n and a are relatively prime.

Congruence is a very convenient notation for describing whether two numbers are in the same arithmetic progression. The notation is due to Gauss.

Definition. Two integers a and b are *congruent modulo n*, written

$$a \equiv b \pmod{n},$$

if $b = a +$ (some multiple of n).

This definition is equivalent to each of the following.

(a) n divides $a - b$.
(b) b is in the arithmetic progression
  $\ldots, -3n + a, -2n + a, -n + a, a, n + a, 2n + a, 3n + a, \ldots$.
(c) The set $\{nk + a\}_{k \in \mathbb{Z}}$ is the same as the set $\{nk + b\}_{k \in \mathbb{Z}}$.

If a and b are natural numbers, it is also equivalent to:

(d) a and b leave the same remainder when divided by n.

E1. Prove that each of (a)–(d) (condition (d) when a and b are natural numbers) is equivalent to the condition that $b = a + $ (multiple of n).

Here are some numerical examples of congruences:

$$1325 \equiv 2 \pmod 9,$$
$$182 \equiv 119 \pmod 9,$$
$$3 \equiv -1 \pmod 4,$$
$$13 \equiv 0 \pmod{13},$$
$$26 \equiv 13 \pmod{13}.$$

You should check that these really are congruences.

E2. Find the smallest number $\geqslant 0$ congruent to $a \pmod m$ where a and m are
  (i) 3312 (mod 4),
  (ii) 177 (mod 8),
  (iii) 31 (mod 37),
  (iv) 111 (mod 109).

B. Basic Properties

The congruence symbol looks like an equality, and this is no accident. In fact, one can view congruence geometrically as a kind of equality. Take the real number line and lay on it a circle of circumference n. Then wind the real number line around the circle. The picture shows $n = 6$.

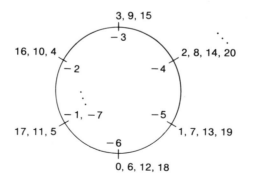

Integers which lie over the same point on the circle are congruent mod n; integers lying over different points are not.

Most of the things you normally can do with an equality you can do with congruence mod n. In particular, congruence satisfies these fundamental properties:

(i) if $a \equiv b$ (mod n) then $ka \equiv kb$ (mod n);
(ii) if $a \equiv b$ (mod n) and $b \equiv c$ (mod n), then $a \equiv c$ (mod n);
(iii) if $a \equiv b$ (mod n) and $a' \equiv b'$ (mod n), then
   (iiia) $a + a' \equiv b + b'$ (mod n), and
   (iiim) $aa' \equiv bb'$ (mod n).

All of these follow easily from the condition that $a \equiv b$ (mod n) if $a = b + nq$ for some integer q. We prove (ii) and (iiim):

PROOF OF (ii). if $a \equiv b$ (mod n) and $b \equiv c$ (mod n), then $a = b + sn$, $b = c + tn$ for some integers s and t. Substituting, we get $a = (c + tn) + sn = c +$ (multiple of n), so $a \equiv c$ (mod n). □

PROOF OF (iiim). if $a \equiv b$ (mod n) and $a' \equiv b'$ (mod n), then $a = b + sn$, $a' = b' + tn$ for some integers s and t. Then $aa' = bb' + nsb' + tbn + stn^2 = bb' +$ (multiple of n). So $aa' \equiv bb'$ (mod n). □

E3. Prove (i) and (iiia).

Property (iiim) shows particularly well the virtue of the congruence notation. If we translate $a \equiv b$ (mod n) and $a' \equiv b'$ (mod n) into divisibility notation, they read: n divides $a - b$ and n divides $a' - b'$. To conclude that n divides $aa' - bb'$ is less natural than to conclude from $a \equiv b$ (mod n) and $a' \equiv b'$ (mod n) that $aa' \equiv bb'$ (mod n).

The one thing which can be done with ordinary equality which cannot be done in general with congruences mod n is cancellation: If $ab \equiv ac$ (mod n), it does not necessarily follow that $b \equiv c$ (mod n). For example, $2 \cdot 1 \equiv 2 \cdot 3$ (mod 4), but $1 \not\equiv 3$ (mod 4). Similarly $6 \equiv 12$ (mod 6), but $1 \not\equiv 2$ (mod 12). We shall postpone for now the rule which replaces the usual rule of cancellation. Meanwhile, you try to invent a useful rule of your own. Do examples.

E4. Prove that $6 \cdot 4^n \equiv 6$ (mod 9) for all $n \geq 0$.

E5. Prove that if $n > 4$ and n is not prime, then $(n - 1)! \equiv 0$ (mod n).

C. Divisibility Tricks

There is a very old trick to check addition and multiplication: sum the digits and do the operation on the sums of the digits. For example, multiplication:

3325	$3 + 3 + 2 + 5 = 13$,	$1 + 3 = 4$
182	$1 + 8 + 2 = 11$,	$1 + 1 = 2$
605150	$6 + 0 + 5 + 1 + 5 + 0 = 17$,	$1 + 7 = 8$
		?

If the multiplication on the sums of the digits does not equal the sum of the digits of the answer, the multiplication was done incorrectly. (If the multiplication on the sums of the digits does equal the sum of the digits of the answer, the answer may still be wrong, but is likely to be correct if the person multiplying is fairly reliable.)

This trick is called "casting out 9's." You are really looking at remainders upon division by 9 when you look at the sums of the digits: $3325 = 9a + 4$ and $182 = 9b + 2$, so $3325 \cdot 182 = 9(9ab + 4b + 2a) + 8 = 9c + 8$ for some integers a, b, c. If you compute $3325 \cdot 182$ and get an answer whose remainder when divided by 9 is $\neq 8$, the answer must be wrong.

Here is why it works: For any $n > 0$,

$$10^n - 1 = (999 \ldots 9)_{10} \qquad (n \text{ times})$$
$$= 9 \cdot (111 \ldots 1)_{10} \qquad (n \text{ times}),$$

so $10^n = 1 + $ (multiple of 9). Multiplying by any number a gives $a \cdot 10^n = a + $ (multiple of 9). Thus

$$3325 = 3 \cdot 10^3 + 3 \cdot 10^2 + 2 \cdot 10 + 5 = 3 + 3 + 2 + 5 + \text{(multiples of 9)}.$$

So 3325 differs from the sum of its digits by a multiple of 9.

Using the notation and properties of congruence mod n we can conveniently describe some tests for deciding when a number expressed in base 10 is divisible by a certain number. (Recall from Chapter 1 that "iff" means "if and only if.")

Let $a = (a_n a_{n-1} \ldots a_1 a_0)_{10} = a_n 10^n + a_{n-1} 10^{n-1} + \cdots + a_0$.

Fact. 9 *divides a if 9 divides the sum of its digits.*

PROOF. We just did this. We redo it using congruence notation. Since $10^r \equiv 1 \pmod 9$ for all $r \geq 0$, it follows from the properties (i)–(iii) of congruence that

$$a = a_n 10^n + \cdots + a_1 10 + a_0 \equiv a_n + a_{n-1} + \cdots + a_0 \pmod 9.$$

That is, a is congruent mod 9 to the sum of its digits. Thus 9 divides a iff $a \equiv 0 \pmod 9$ iff $a_n + a_{n-1} + \cdots + a_0 \equiv 0 \pmod 9$ iff 9 divides the sum of the digits of a. $\qquad \square$

Fact. 3 *divides a if 3 divides the sum of its digits.*

PROOF. Since 9 divides $10^r - 1$, so does 3. So the proof is the same as for 9. $\qquad \square$

Fact. 2 (*resp.* 5) *divides a if 2 (resp. 5) divides a_0.*

PROOF. $10^r \equiv 0 \pmod 2$ and $10^r \equiv 0 \pmod 5$ for all $r \geq 1$. So $a \equiv a_0 \pmod 2$ and $a \equiv a_0 \pmod 5$. $\qquad \square$

Fact. 11 *divides a if 11 divides $a_0 - a_1 + a_2 - \cdots + (-1)^n a_n$.*

PROOF. $10 \equiv -1 \pmod{11}$, so $10^r \equiv (-1)^r \pmod{11}$. $\qquad \square$

Fact. 7 (*resp.* 11, 13) *divides a if* 7 (*resp.* 11, 13) *divides* $(a_2a_1a_0)_{10} - (a_5a_4a_3)_{10} + (a_8a_7a_6)_{10} - \cdots$.

PROOF. Write a in base 1000. If a is in base 10 this is very easy! Then $a = b_m 1000^m + b_{m-1} 1000^{m-1} + \cdots + b_1 1000 + b_0$. Now $7 \cdot 11 \cdot 13 = 1000 + 1$. So $1000 \equiv -1 \pmod{7}$, and also $\pmod{11}$ and $\pmod{13}$. Thus $b_m 1000^m + b_{m-1} 1000^{m-1} + \cdots + b_1 1000 + b_0 \equiv (-1)^m b_m + (-1)^{m-1} b_{m-1} + \cdots + b_2 - b_1 + b_0 \pmod{c}$, where $c = 7$, 11, or 13. □

E6. If we are doing base 12 arithmetic can we check it by casting out 11's? Why? How?

E7. Does casting out 9's work with respect to long division? How?

E8. What about casting out 1's in base 2? How could we check arithmetic in base 2?

E9. Find nice tests for divisibility of numbers in base 34 by each of 2, 3, 5, 7, 11, 13, and 17.

E10. Is $(332, 587, 399, 817, 443)_{12}$ divisible by $(11)_{10}$?

D. More Properties of Congruence

The properties (i)–(iii) show that congruence to a fixed modulus m is much like equality, except for cancelling; in this section we list properties which relate congruences to different moduli, and describe how to cancel.

(iv) *If* $a \equiv b \pmod{m}$ *and d divides m, then* $a \equiv b \pmod{d}$.

We already used this property in discussing the divisibility test for 7, 11, and 13, for we observed that if $a \equiv b \pmod{1001}$ then $a \equiv b \pmod{7}$ since 7 divides 1001.

(v) *If* $a \equiv b \pmod{r}$ *and* $a \equiv b \pmod{s}$, *then* $a \equiv b \pmod{[r, s]}$.

Thus if $a \equiv b \pmod{7}$ and $a \equiv b \pmod{11}$ and $a \equiv b \pmod{13}$, then $a \equiv b \pmod{[7, 11, 13]}$, that is, $a \equiv b \pmod{1001}$.

The cancellation properties of congruences are summed up by the following. Here $r \neq 0$.

(vi) *If* $ra \equiv rb \pmod{m}$ *then*

$$a \equiv b \left(\mathrm{mod}\ \frac{m}{(r, m)} \right).$$

For example, $4 \cdot 3 \equiv 4 \cdot 8 \pmod{10}$, so

$$3 \equiv 8 \left(\mathrm{mod}\ \frac{10}{(10, 4)} \right),$$

that is, $3 \equiv 8 \pmod{5}$.

Two special cases are when $r|m$ or when $(m, r) = 1$:

(vii) *If* $ra \equiv rb$ *(mod* rm*), then* $a \equiv b$ *(mod* m*).*

PROOF. if $ra \equiv rb$ (mod rm), then $ra - rb = rmc$ for some c; cancelling r gives $a - b = mc$, so $a \equiv b$ (mod m).   □

(viii) *If* $ra \equiv rb$ *(mod* m*) and* $(r, m) = 1$*, then* $a \equiv b$ *(mod* m*).*

Thus, if $2 \cdot r \equiv 2 \cdot s$ (mod 11), we may conclude that $r \equiv s$ (mod 11) since $(2, 11) = 1$.

Proofs of all these results are exercises.

E11. Prove Property (iv).

E12. Prove property (v).

E13. Prove property (viii) without assuming Property (vi).

E14. Prove property (vi).

E15. Prove that $a \equiv a$ (mod m), and that if $a \equiv b$ (mod m), then $b \equiv a$ (mod m). These properties and property (ii) show that $\equiv$ (mod m) is an equivalence relation. See Chapter 1, Exercise E6.

E16. Prove: For $x, y \geqslant 1$, if $x \equiv y$ (mod m), then $(m, x) = (m, y)$.

E. Congruence Problems

We begin considering the problem of solving congruences containing unknowns.

The simplest such congruences are

(α) $x + c \equiv d$ (mod m), and
(β) $ax \equiv b$ (mod m).

Congruence (α) is trivial to solve: add $-c$ to both sides to get $x \equiv d - c$ (mod m).

Congruence (β) is already interesting. Here are some examples to try:

(i) $10x \equiv 14$ (mod 15);

(ii) $10x \equiv 14$ (mod 18);

(iii) $10x \equiv 14$ (mod 21).

The first of these is not solvable. For if it were, there would exist integers x and y such that $10x + 15y = 14$. Since 5 divides $10x + 15y$ for any x and y, but 5 does not divide 14, no such x and y can exist.

The second congruence is solvable. Using property (viii) above with $r = 2$, we get

(ii') $5x \equiv 7 \pmod 9$.

By trial and error we get $x = 5$ as a solution.

The third congruence is solvable. Since $(10, 21) = 1$ we can find integers u, v with $1 = 10u + 21v$ by Bezout's identity (or by inspection: $v = 1$, $u = -2$). Then

$$10u \equiv 1 \pmod{21}$$

so

$$10 \cdot 14u \equiv 14 \pmod{21}$$

and $x = 14u$ is a solution. Bezout's identity would work also for congruence (ii) since we can solve $10u + 18v = 2$ and 14 is divisible by 2.

E17. Decide whether each of the following congruences has a solution. If so, find the least nonnegative solution:
   (a) $12x \equiv 7 \pmod{21}$;
   (b) $12x \equiv 7 \pmod{84}$;
   (c) $12x \equiv 7 \pmod{73}$;
   (d) $12x \equiv 7 \pmod{46}$;
   (e) $12x \equiv 7 \pmod{35}$.

E18. Prove: $ax \equiv b \pmod m$ has a solution iff b is divisible by (a, m).

E19. How many solutions x of $ax \equiv b \pmod m$ are there with $0 < x < m$?

E20. Solve $18x \equiv 1 \pmod{25}$.

E21. Solve $12x \equiv 33 \pmod{57}$.

When trying to solve quadratic congruences, that is, congruences of the form

$$ax^2 + bx + c \equiv 0 \pmod m, \qquad (1)$$

already the theory becomes very subtle. The simplest case is

$$x^2 \equiv a \pmod m. \qquad (2)$$

The general case (1) reduces to the form (2) by multiplying everything by $4a$:

$$4a^2x^2 + 4abx + 4ac \equiv 0 \pmod{4am}$$

or

$$(2ax + b)^2 \equiv b^2 - 4ac \pmod{4am},$$

which is of the form (2).

Gauss (1801) was the first to give a complete treatment of the solution of (1), by means of his famous law of quadratic reciprocity. We will examine this law in Chapter III-16.

For aspects of the solution of polynomials of degree > 2, a much harder problem, see Chapter II-12 and Gerst and Brillhart (1971).

E22. Prove: If x, y are odd integers, then $x^2 + y^2$ is not the square of an integer.

E23. Prove: given a modulus m and numbers a, b, r with $(a, m) = 1$, there exists x with $ax + b \equiv r \pmod{m}$.

*E24. Let a, b, n be natural numbers. By the binomial theorem,

$$(a + b)^n = a^n + \binom{n}{1}a^{n-1}b + \cdots + \binom{n}{k}a^{n-k}b^k$$
$$+ \binom{n}{k+1}a^{n-k-1}b^{k+1} + \cdots + b^n.$$

Prove that there are two adjacent terms in this expansion which are equal iff $n \equiv -1 \pmod{a + b/(a, b)}$.

E25. A person wishes to buy a \$4900 car. He has a \$10000 check and three \$100 bills. The dealer has as possible change only six \$1000 bills and seven \$500 bills. Can the sale be completed?

F. Round Robin Tournaments

Congruences can be applied to the design of round robin tournaments. We need the following easily proved

Lemma. *For any integer a and n ($n \geq 2$), there is a **unique** x with $1 \leq x \leq n$ such that $x \equiv a \pmod{n}$.*

E26. Prove this, paying special attention to the case $a < 0$, and to uniqueness.

Suppose we have n players, labeled $1, 2, 3, \ldots, n$, where n is even. If there are an odd number of players, then add a "bye," an extra, vacuous player, to make an even number. Each player is to play all $n - 1$ other players, so there are $n - 1$ rounds. For round r the opponent for player x_0, $x_0 < n$, is given by the rule: x_0 plays y if y is the (unique) solution, $1 \leq y \leq n - 1$, of the congruence $y \equiv r - x_0 \pmod{n - 1}$ provided $x_0 \neq y$. If the congruence $y \equiv r - x_0 \pmod{n - 1}$ is solved by $y = x_0$, so that $2x_0 \equiv r \pmod{n - 1}$, then x_0 plays n. Player n plays x in round r if x is the unique solution, $1 \leq x \leq n - 1$, of the equation $2x \equiv r \pmod{n - 1}$.

E27. If n is even, show that for each r there is exactly one solution of $2x \equiv r \pmod{n - 1}$ with $1 \leq x \leq n - 1$.

E28. Describe how the round robin scheme works when $n = 8$ by filling in the table below, which will describe whom player x plays in round r.

$x \setminus r$	1	2	3	4	5	6	7
1							
2							
3							
4							
5							
6							
7							
8							

For more subtle designs, see III-17.

7 Congruence Classes

There is a way of interpreting divisibility and congruences which leads to the invention of new "number" systems.

The new systems we invent consist of things we shall call *congruence classes*. Each congruence class is a set consisting of an arithmetic progression, hence each congruence class contains infinitely many integers.

We begin with a simple example of such a number system; we shall call it $\mathbb{Z}_2$. It consists of two elements, two congruence classes, which we shall call $[0]_2$ and $[1]_2$:

$$[0]_2 = \{ \dots, -4, -2, 0, 2, 4, 6, \dots \}, \quad \text{the } even \text{ integers,}$$

and

$$[1]_2 = \{ \dots, -3, -1, 1, 3, 5, \dots \}, \quad \text{the } odd \text{ integers.}$$

Thus $[0]_2$ is the set of integers in the arithmetic progression $2k + 0$, $k = \dots -2, -1, 0, 1, 2, \dots$, and $[1]_2$ is the set of integers in the progression $2k + 1$, $k = \dots -2, -1, 0, 1, \dots$. Any two elements of $[0]_2$ (or of $[1]_2$) are congruent mod 2, and no element of $[0]_2$ is congruent to any element of $[1]_2$ mod 2.

We define arithmetic operations with $[0]_2$ and $[1]_2$ by recalling that:

even integer + odd integer = odd integer;
odd integer + odd integer = even integer;
even integer + even integer = even integer;
even integer · even integer = even integer;
even integer · odd integer = even integer;
odd integer · odd integer = odd integer.

Thus we define addition and multiplication in $\mathbb{Z}_2 = \{[0]_2, [1]_2\}$ according to the following table.

+	$[0]_2$	$[1]_2$
$[0]_2$	$[0]_2$	$[1]_2$
$[1]_2$	$[1]_2$	$[0]_2$

$\cdot$	$[0]_2$	$[1]_2$
$[0]_2$	$[0]_2$	$[0]_2$
$[1]_2$	$[0]_2$	$[1]_2$

That is,

$$[0]_2 + [1]_2 = [1]_2,$$
$$[1]_2 + [1]_2 = [0]_2,$$
$$\vdots$$

$$[0]_2 \cdot [1]_2 = [0]_2,$$
$$[1]_2 \cdot [1]_2 = [1]_2;$$
$$\vdots$$

These equations give the sum or product of any two elements of $\mathbb{Z}_2$, in such a way that $\mathbb{Z}_2$ satisfies the formal properties required of something called a field. (Examples of fields which you already know are the rational numbers, the real numbers, and the complex numbers.) We shall define a field in the next chapter.

Another example of congruence classes arises with time. If you get to work at 9 o'clock and work 8 hours, the time when you stop working is 5 o'clock. Thus $9 + 8 = 5$ on the clock. This addition makes sense because a clock divides time into congruence classes. Two hours t_1 and t_2 read the same on the clock iff t_1 and t_2 differ by a multiple of 12 hours iff $t_1 \equiv t_2$ (mod 12). The congruence class mod 12 of the time t_0 consists of all times t so that the clock reads the same at t_0 as it does at t.

In particular, if you go to work 9 hours after midnight and work 8 hours, you quit 17 hours after midnight. The clock reads the same 17 hours after midnight as it does 5 hours after midnight. So if we denote the congruence class of t by $[t]_{12}$, then $[9 + 8]_{12} = [5]_{12}$.

Now we describe congruence classes in general.

Pick and fix a natural number $m \geqslant 2$.

We know that $a \equiv b$ (mod m) means $b = a +$ (multiple of m).

We define the *congruence class of a* (mod m), written $[a]_m$, to be the set of numbers of the form $km + a$. For all k, that is, $[a]_m = \{km + a\}_{k \in \mathbb{Z}}$.

The picture in Chapter 6 in which we wrapped the real number line about a circle of circumference m gives a very good view of congruence classes: The congruence class mod m of the number a is the set of all integers which lie at the same point as a on the circle.

From that picture it is clear that two congruence classes are either equal or disjoint (have no elements in common).

The congruence classes $[a]_m$ and $[b]_m$ are equal exactly when $a \equiv b$ (mod m). This is obvious from the picture in Chapter 6. It can be proved using property (ii), Section 6B, as follows: Suppose $a \equiv b$ (mod m). If c is in $[a]_m$, then $a \equiv c$ (mod m); since $a \equiv b$ (mod m), then $c \equiv b$ (mod m) by property (ii), Section 6B, so c is in $[b]_m$. A similar argument shows that if c

is in $[b]_m$, then c is in $[a]_m$. On the other hand, if $a \not\equiv b \pmod{m}$, then b is not in the congruence class of a, and so $[a]_m \neq [b]_m$.

There are exactly m congruence classes mod m. This is obvious from the picture in Chapter 6. To prove it without a picture, note that every element of $\mathbb{Z}$ is congruent to one of the numbers $0, 1, 2, \ldots, m-2, m-1$. You can see this for any a by the division theorem: If $a = qm + r$ where $0 \leqslant r < m$, then $a \equiv r \pmod{m}$, so $[a]_m = [r]_m$. So every congruence class mod m is equal to one of $[0]_m, [1]_m, \ldots, [m-1]_m$. Since these classes are all different, there are exactly m classes mod m.

For example, there are 2 congruence classes mod 2: $[0]_2$, the congruence class consisting of the even integers, and $[1]_2$, the congruence class consisting of the odd integers. There are 3 congruence classes mod 3, which we can call $[0]_3, [1]_3, [2]_3$:

$$[0]_3 = \{ \ldots, -9, -6, -3, 0, 3, 6, 9, 12, \ldots, 99354, \ldots \};$$
$$[1]_3 = \{ \ldots, -20, \ldots, -8, -5, -2, 1, 4, 7, \ldots 82, \ldots 93547, \ldots \};$$
$$[2]_3 = \{ \ldots, -334, \ldots, -1, 2, 5, \ldots, 811211, \ldots \}.$$

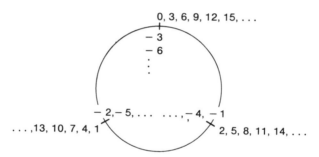

E1. In which congruence class in $\mathbb{Z}_3$ is 3124? -8337? 8391?

In general we shall call the set of congruence classes mod m by $\mathbb{Z}_m$. Then

$$\mathbb{Z}_2 = \{[0]_2, [1]_2\},$$
$$\mathbb{Z}_3 = \{[0]_3, [1]_3, [2]_3\}$$

and, in general,

$$\mathbb{Z}_m = \{[0]_m, [1]_m, [2]_m, \ldots, [m-1]_m\}.$$

Any element b of a congruence class $[r]_m$ is called a *representative* of the congruence class $[r]_m$. If b is in a congruence class $[r]_m$, then $[r]_m = [b]_m$: any representative of a congruence class completely determines the class, and a congruence class can be labeled by any representative. Thus $[1]_3 = [4]_3 = [901]_3 = [-8]_3$. The tendency is to label a congruence class by its smallest positive or nonnegative representative. For example, a clock labels hours with the representatives $1, 2, 3, \ldots, 12$. But one can do otherwise. Thus we could describe $\mathbb{Z}_3$ as $\{[30]_3, [-2]_3, [11]_3\}$ if we wished.

It is common to abbreviate the notation by writing $[a]_m$ as a when it is clear we are working with elements of $\mathbb{Z}_m$, i.e., "mod m." This abbreviation yields statements which would be absurd if a were taken as an integer, not as an integer mod m or as the representative of a congruence class in $\mathbb{Z}_m$. For example, the statement about the integers 1 and 3,

$$1 = 3,$$

is absurd, but

$$[1]_2 = [3]_2$$

is true, so as an abbreviation for this last equality

$$1 = 3$$

is true. We shall usually give explicit warnings when we are using abbreviated notation for congruence classes.

In $\mathbb{Z}_2$ above we defined addition and multiplication. We do the same for $\mathbb{Z}_m$, in the most natural way possible: If a and b are integers,

$$[a]_m + [b]_m = [a+b]_m,$$
$$-[a]_m = [-a]_m,$$
$$[a]_m \cdot [b]_m = [a \cdot b]_m.$$

We shall show shortly that these operations defined on congruence classes make sense. First, here is how $+$ and $\cdot$ work for some examples. We did $\mathbb{Z}_2$ above.

$\mathbb{Z}_3$: Here we abbreviate—$0 = [0]_3$, $1 = [1]_3$, $2 = [2]_3$.

+	0	1	2
0	0	1	2
1	1	2	0
2	2	0	1

·	0	1	2
0	0	0	0
1	0	1	2
2	0	2	1

$\mathbb{Z}_4$: Here $0 = [0]_4$, $1 = [1]_4$, $2 = [2]_4$, $3 = [3]_4$.

+	0	1	2	3
0	0	1	2	3
1	1	2	3	0
2	2	3	0	1
3	3	0	1	2

·	0	1	2	3
0	0	0	0	0
1	0	1	2	3
2	0	2	0	2
3	0	3	2	1

E2. Write down the tables for $\mathbb{Z}_5$, $\mathbb{Z}_6$.

E3. What time is it on the clock:
   (i) 34 hours after 7 o'clock?
   (ii) 19 hours after 3 o'clock?
   (iii) 29 hours before 1 o'clock?

We define the operations in $\mathbb{Z}_m$ by $[a]_m + [b]_m = [a + b]_m$, $[a]_m \cdot [b]_m = [a \cdot b]_m$, $-[a]_m = [-a]_m$. Notice that the operations seem to depend on the representatives used to label the congruence classes of $\mathbb{Z}_m$. In fact, they do not.

We have to show that if we label congruence classes by different representatives it will not affect the outcome of addition, negation, or multiplication. That is,

if $[a]_m = [a']_m$ *and* $[b]_m = [b']_m$, *then* $[a + b]_m = [a' + b']_m$, $[-b]_m = [-b']_m$, *and* $[a \cdot b]_m = [a' \cdot b']_m$.

(Think about this with odd and even integers and $m = 2$.) To prove these facts, we translate into congruence notation. The first becomes

if $a \equiv a'$ (mod m) *and* $b \equiv b'$ (mod m), *then* $a + a' \equiv b + b'$ (mod m).

This is property (iiia) of congruences. Similarly, *if* $b \equiv b'$ (mod m) *then* $-b \equiv -b'$ (mod m). This is property (i) of Section 6B with $k = -1$. You try proving the assertion about multiplication:

E4. Prove that if $[a]_m = [a']_m$, $[b]_m = [b']_m$ then $[a \cdot b]_m = [a' \cdot b']_m$.

The classes $[0]_m$ and $[1]_m$ are a bit special, and satisfy

$$[0]_m + [b]_m = [b]_m \quad \text{for all } b,$$
$$[0]_m \cdot [b]_m = [0]_m \quad \text{for all } b,$$
$$[1]_m \cdot [b]_m = [b]_m \quad \text{for all } b.$$

They act just like 0 and 1 do in $\mathbb{Z}$, which is hardly surprising in view of how we defined addition and multiplication in $\mathbb{Z}_m$.

Casting out 9's can be rephrased in the new symbolism using congruence classes in $\mathbb{Z}_9$: Let $(a_n a_{n-1} \ldots a_1 a_0)_{10}$ be a number expressed on ordinary base 10. Then using the definition of addition and multiplication of congruence classes, we have

$$\begin{aligned}
[(a_n a_{n-1} \ldots a_1 a_0)_{10}]_9 &= [a_n 10^n + a_{n-1} 10^{n-1} + \cdots + a_1 10 + a_0]_9 \\
&= [a_n 10^n]_9 + \cdots + [a_1 10]_9 + [a_0]_9 \\
&= [a_n]_9 [10^n]_9 + \cdots + [a_1]_9 [10]_9 + [a_0]_9 \\
&= [a_n]_9 [1]_9 + \cdots + [a_1]_9 [1]_9 + [a_0]_9 \\
&= [a_n]_9 + \cdots + [a_1]_9 + [a_0]_9 \\
&= [a_n + a_{n-1} + \cdots + a_1 + a_0]_9
\end{aligned}$$

since $[10^n]_9 = [1]_9$ for any n. Thus any number differs from the sum of its digits by a multiple of 9.

We emphasize that $\equiv$ (mod m) for integers is the same as equality for congruence classes in $\mathbb{Z}_m$: that is why $\equiv$ (mod m) shares so many properties of equality.

We conclude this chapter by observing that we now have several different notations for the same idea. Given integers a and b, and $m > 1$, $a = b +$ (multiple of m) iff m divides $b - a$ iff $a \equiv b$ (mod m) iff $[a]_m = [b]_m$. The reason for introducing this last notation is that it enables us to conveniently introduce the new systems $\mathbb{Z}_m$ of "numbers" which are quite analogous to the usual systems, like $\mathbb{R}$, and which we can work with in similar ways. The value of having these new systems available will, we hope, be made clear by some of the applications which we shall study later.

E5. Using the notation of congruence classes mod 11, derive the test for divisibility of a number by 11.

If $[a]_m[b]_m = [1]_m$ we call $[b]_m$ an *inverse* for $[a]_m$ and write it as $[a]_m^{-1}$.

E6. Solve $3x \equiv 1$ (mod 25). That is, find $[3]_{25}^{-1}$.

E7. Which elements of $\mathbb{Z}_4$ have inverses? of $\mathbb{Z}_5$? of $\mathbb{Z}_6$?

E8. Can you solve $14x \equiv 1$ (mod 77)? Can you find $[14]_{77}^{-1}$?

E9. (a) Find all solutions of $[2142]_{238}[x]_{238} = [442]_{238}$.
   (b) Find all solutions of $[14]_{77}[x]_{77} = [21]_{77}$.
   (c) Find all solutions of $[1239]_{154}[x]_{154} = [6]_{154}$.

Call r the *least nonnegative residue* of a (mod m) if $0 \leqslant r < m$ and $a \equiv r$ (mod m).

E10. Find the least nonnegative residue:
   (a) of 3^{17} (mod 7);
   (b) of 81^{119} (mod 13);
   (c) of 310^{71} (mod 12);
   (d) of 13^{216} (mod 19).

A set of numbers $\{r_1, \ldots, r_m\}$ which are not congruent to each other and such that every integer is congruent to exactly one of the numbers $r_1, \ldots, r_m$ is called a *complete set of representatives of* $\mathbb{Z}_m$. Thus $\{0, 1, 2, \ldots, m - 2, m - 1\}$ is a complete set of representatives of $\mathbb{Z}_m$.

E11. Is $\{80, 11, 211, 32, -5, 994, 5\}$ a complete set of representatives of $\mathbb{Z}_7$? Is $\{0, 92, 118, 324, -32, 153, 9\}$?

*E12. Let a, b be relatively prime integers with $a > b > 0$. Define the sequence of numbers $s_1, s_2, \ldots, s_k$, by:
$$s_1 = a$$
$$s_2 = a - b$$
$$s_{k+1} = \begin{cases} s_k + a & \text{if } s_k < b \\ s_k - b & \text{if } s_k \geqslant b. \end{cases}$$

Prove that the integers $\{s_1, s_2, \ldots, s_{a+b}\}$ form a complete set of representatives mod $a + b$ (cf. Chapter 3, Exercise E40.) What if $(a, b) > 1$?

E13. (i) Find a reasonable definition of congruence mod m for $m = 0$. What would $\mathbb{Z}_0$ be? (ii) Same question for $m = 1$.

8 Rings and Fields

A. Axioms

In this chapter we describe a number of axioms for sets on which addition and multiplication are defined. These axioms were originally found by isolating the basic properties of addition and multiplication which are common to all or most of the examples we already know: $\mathbb{Z}$, $\mathbb{Q}$, $\mathbb{R}$, $\mathbb{C}$. A ring, a commutative ring, or a field will be defined as a set with addition and multiplication satisfying certain of the axioms.

The value of these definitions is twofold. First, having the notion of field, for example, gives us a way of examining new sets with addition and multiplication. We shall ask: Is $\mathbb{Z}_m$ a field? We shall look for other examples of fields. By the end of the book we shall have found a vast array of new fields. Second, since a commutative ring or a field shares, by definition, many of the properties of $\mathbb{Z}$, $\mathbb{Q}$, $\mathbb{R}$, etc., examples we already know, then, if we find a new example of a field or a commutative ring, we shall know we can manipulate its elements in the same way we would manipulate numbers—rearranging sums, rearranging products, collecting common factors of sums, etc. In Chapter 9 we shall observe that much of the elementary linear algebra you may (or may not) have learned works when the scalars come from any commutative ring with identity, and in some later chapters, when we use linear algebra it will be with scalars in some of the new fields and rings we are about to discover.

We begin with the most general concept.

Definition. A *ring* (*with identity*) is a set R with three operations, $+$, $\cdot$, and $-$, and two special elements, 0 and 1, which satisfy the various properties listed below as axioms (i)–(vii). The operations $+$ and $\cdot$ may each be thought of as functions from $R \times R$ (ordered pairs of elements of the set R) to R, so that for any pair (a, b), a, b in R, $a + b$ is an element of R, and $a \cdot b$ is an element of R. The operation $-$ similarly is a function from R to R which takes a in R to $-a$ in R.

The set R, together with the operations $+$, $\cdot$ and $-$ and special elements 0, 1, is a *ring with identity* if the following axioms hold:

(i) for any a, b, c in R, $(a + b) + c = a + (b + c)$ (*associativity of addition*);
(ii) for any a, b, in R, $a + b = b + a$ (*commutativity of addition*);
(iii) for all a in R, $a + 0 = a$ (0 is a *zero element*);
(iv) for any a in R, $a + (-a) = 0$ ($-a$ is the *negative* of a);
(v) for any a, b, c in R, $(a \cdot b) \cdot c = a \cdot (b \cdot c)$ (*associativity of multiplication*);
(vi) for all a in R, $a \cdot 1 = 1 \cdot a = a$ (1 is an *identity element*).
(vii) for any a, b, c in R, $a \cdot (b + c) = (a \cdot b) + (a \cdot c)$ and $(a + b) \cdot c = (a \cdot c) + (b \cdot c)$ (*distributive laws*).

These basic axioms are all satisfied by $\mathbb{Z}$, $\mathbb{Q}$, $\mathbb{R}$, $\mathbb{C}$, $\mathbb{Z}_m$, and the set of $n \times n$ real matrices (see Section 9B).

Some examples of sets which are not rings are:

the set of natural numbers $\mathbb{N}$;
the set of nonnegative real numbers $\mathbb{R}_+$, with the usual $+$ and $\cdot$;
the set $\mathbb{Z} - \{3\}$ of all integers except 3.

The set $\mathbb{Z} - \{3\}$ is not a ring with respect to the usual addition and multiplication in $\mathbb{Z}$ for the reason that if a, b are in $\mathbb{Z} - \{3\}$ then $a + b$ need not be in $\mathbb{Z} - \{3\}$. For example, $1 + 2 = 3$: 1 and 2 are in $\mathbb{Z} - \{3\}$ but 3 is not. When this kind of thing occurs we say that $\mathbb{Z} - \{3\}$ is not *closed under addition*, by which we mean that $\mathbb{Z} - \{3\}$ is a subset of a bigger set, $\mathbb{Z}$, on which addition and multiplication always makes sense, and if we take the sum in $\mathbb{Z}$ of two elements of $\mathbb{Z} - \{3\}$ the result in some cases is not in $\mathbb{Z} - \{3\}$. The notion of (not) *closed under multiplication* is similar.

We now state some axioms which are satisfied by special types of rings.

(viii) For all a, b in R, $a \cdot b = b \cdot a$ (*commutativity of multiplication*).

A ring (like $\mathbb{Z}$, $\mathbb{Q}$, $\mathbb{R}$, $\mathbb{C}$, $\mathbb{Z}_m$) which satisfies this axiom is called a *commutative* ring.

(ix) For all a, b in R, if $a \cdot b = 0$, then $a = 0$ or $b = 0$.

A ring satisfying this axiom is said to *have no zero divisors*. Examples are $\mathbb{Z}$, $\mathbb{Q}$, $\mathbb{R}$, $\mathbb{C}$, and $\mathbb{Z}_m$ for some but not all m.

E1. Give examples of m_1, $m_2 > 6$ so that $\mathbb{Z}_{m_1}$ has zero divisors and $\mathbb{Z}_{m_2}$ does not.

Definition. An element a of a ring with identity R is called a *unit* of R if there exists some b in R with $ab = ba = 1$.

EXAMPLES. In $\mathbb{Z}$ only 1 and -1 are units. In $\mathbb{Q}$ every nonzero rational number is a unit.

If $ab = ba = 1$, b is unique (Exercise E4(ii)) and is called the *inverse* of a; it is usually denoted by a^{-1}.

Here are the last two axioms.

(x) Each $a \neq 0$ in R is a unit.

(xi) $1 \neq 0$ in R.

E2. Prove that if axiom (xi) does not hold in R, a ring with identity, then R must be a set with only one element.

A commutative ring satisfying axioms (x) and (xi) is called a *field*.

EXAMPLES. $\mathbb{Q}$, $\mathbb{R}$, $\mathbb{C}$ are fields; $\mathbb{Z}_m$ is for some but not all m. $\mathbb{Z}$ does not satisfy axiom (x) and so is not a field.

E3. Show that a field has no zero divisors.

E4. Show that in a commutative ring R:
  (i) if $a + b = d$ and $a + c = d$, then $b = c$;
  (ii) if $a \cdot b = b \cdot a = 1$ and $a \cdot c = 1$, then $b = c$;
  (iii) if a has an inverse in R, then there is a unique solution in R to the equation $ax = d$.

E5. If m is odd, does $[2]_m$ have an inverse in $\mathbb{Z}_m$?

E6. Determine which axioms hold and which fail for each of the nonrings $\mathbb{N}$ and $\mathbb{R}_+$ mentioned above.

E7. Show that if R is a ring and a is any element of R, then $a \cdot 0 = 0$.

E8. Let R be a ring with identity. Prove:
  (i) for all a, b in R, if $a + b = 0$, then $b = -a$;
  (ii) $(-1) \cdot (-1) = 1$;
  (iii) $-(-a) = a$ for all a in R;
  (iv) for all a, b in R, $(-a) \cdot b = -(a \cdot b)$.

E9. For a/b and c/d rational numbers, say $a/b \equiv c/d$ (mod 1) if $(a/d) - (c/d)$ is an integer. Call the set of congruence classes mod 1, $\mathbb{Q}/\mathbb{Z}$.
  (i) Show that every rational number is congruent (mod 1) to a rational number a/b with $0 \leq a < b$.
  (ii) Is $\mathbb{Q}/\mathbb{Z}$ a commutative ring? a field? Check the axioms.

E10. Let $\mathbb{R}^3$ be the set of vectors in real 3-space with the usual addition (parallelogram law) and with multiplication given by the cross or vector product. Which axioms for a ring hold and which do not for $\mathbb{R}^3$?

B. $\mathbb{Z}_m$

We now check these axioms on $\mathbb{Z}_m$. We shall write $[a]_m$ as $[a]$ if we think no confusion will arise in doing so.

Theorem. $\mathbb{Z}_m$ *is a commutative ring with unity for any m.*

We defined addition, multiplication, and subtraction in $\mathbb{Z}_m$ by $[a] + [b] = [a + b]$, $-[a] = [-a]$, $[a] \cdot [b] = [a \cdot b]$. Set $1 = [1]$, $0 = [0]$. With these definitions it is easy to show that if $\mathbb{Z}$ is a commutative ring (which it is), then so is $\mathbb{Z}_m$. For example, the verification that the associativity law for multiplication holds goes as follows: for any a, b, c, in $\mathbb{Z}$, $[a] \cdot ([b] \cdot [c]) = [a] \cdot [b \cdot c] = [a(b \cdot c)] = [(a \cdot b) \cdot c] = [a \cdot b] \cdot [c] = ([a] \cdot [b]) \cdot [c]$.

E11. Verify a distributive law.

The only axioms which might not hold in $\mathbb{Z}_m$ are the axioms (ix) (no zero divisors) and (x) (existence of multiplicative inverses). For example, check back to the multiplication tables for $\mathbb{Z}_3$ and $\mathbb{Z}_4$. If you do, you will see that $\mathbb{Z}_3$ is in fact a field, whereas in $\mathbb{Z}_4$ neither axiom holds because $[2]_4 \cdot [2]_4 = 0$.

Theorem. *If m is not prime, $\mathbb{Z}_m$ is not a field.*

PROOF. Write $m = a \cdot b$ with $a, b < m$. Then $[a]_m [b]_m = [m]_m = [0]_m$, but $[a]_m \neq [0]_m$, $[b]_m \neq [0]_m$. $\qquad\square$

Theorem. *In $\mathbb{Z}_m$ $[a]_m$ has an inverse if and only if the greatest common divisor of a and m is 1.*

PROOF. If $[a]_m$ has an inverse then there is some b in $\mathbb{Z}$ with $[a]_m \cdot [b]_m = [1]_m$. This means that $a \cdot b = 1 + r \cdot m$ for some r, so the greatest common divisor of a and m is 1. On the other hand, if the greatest common divisor of a and m is 1, then we can use the Euclidean algorithm to find b, r in $\mathbb{Z}$ so that $a \cdot b + r \cdot m = 1$. Then in $\mathbb{Z}_m$, $[a]_m \cdot [b]_m = [1]_m$, so $[b]_m$ is the inverse of $[a]_m$. $\qquad\square$

The point should be emphasized that if $[a]$ has a inverse in $\mathbb{Z}_m$ the Euclidean algorithm allows us to actually find it.

EXAMPLES. In $\mathbb{Z}_{13}$, $[9]^{-1} = ?$ The Euclidean algorithm or some intuition reveals that $1 = 9 \cdot 3 - 13 \cdot 2$, so $[9]^{-1} = [3]$.
    In $\mathbb{Z}_{14}$, $[9]^{-1} = ?$ Since $1 = 14 \cdot 2 - 9 \cdot 3$, $[1] = [9] \cdot [-3]$, so $[9]^{-1} = [-3] = [11]$.

E12. In $\mathbb{Z}_{26}$, what is $[9]^{-1}$? $[11]^{-1}$? $[17]^{-1}$? $[22]^{-1}$?

Theorem. $\mathbb{Z}_m$ *is a field if and only if* m *is a prime.*

E13. Prove this.

Using notions of congruence classes we redo an exercise about congruence mod m:

Proposition. *Given a congruence* $ax \equiv ay$ (mod m) *one is free to cancel a and conclude* $x \equiv y$ (mod m) *iff* $(a, m) = 1$.

PROOF. $[ax]_m = [ay]_m$ iff $[a]_m[x - y]_m = 0$. We are allowed to cancel $[a]_m$ iff $[a]_m$ is not a zero divisor in $\mathbb{Z}_m$.

Claim. $[a]_m$ *is a nonzero divisor in* $\mathbb{Z}_m$ *iff* $(a, m) = 1$.

PROOF. If $(a, m) = 1$, then $[a]_m$ has an inverse in $\mathbb{Z}_m$, $[b]_m$. Then $[a]_m[r]_m = [0]_m$ implies $[b]_m[a]_m[r]_m = [r]_m = [0]_m$, so $[a]_m$ is a nonzero divisor. Conversely, if $(a, m) = d > 1$, then $m = de$, $a = df$ with $1 < e, f < m$. Then $[a]_m[e]_m = [f]_m[d]_m[e]_m = [f]_m[m]_m = 0$ but $[e]_m \neq 0$. So $[a]_m$ is a zero divisor.   □

Of course the complete story with respect to cancellation in congruences is:

$$ax \equiv ay \ (\text{mod } m) \quad \text{iff} \quad x \equiv y \left(\text{mod} \frac{m}{(a, m)}\right).$$

The proposition covers the case when $(a, m) = 1$.

E14. How many solutions x are there of $6x \equiv 6$ (mod 18) with $0 \leqslant x < 18$?

E15. Find $[6]_{13}^{-1}$.

E16. Find the smallest positive solution of $34x \equiv 12$ (mod 23).

E17. Find an element α in $\mathbb{Z}_{34}$ so that every invertible element of $\mathbb{Z}_{34}$ is a power of α.

E18. In the ring of 2×2 matrices with entries in $\mathbb{Z}_8$, find the inverse of $\begin{pmatrix} 3 & 5 \\ 1 & 0 \end{pmatrix}$.

†**E19.** Here is a new example of a field. In $\mathbb{Z}_3$, just as in $\mathbb{R}$, one cannot solve the equation $x^2 + 1 = 0$. So let i be a solution of $x^2 + 1 = 0$, and consider the set $\mathbb{Z}_3[i]$ of numbers $a + bi$ with a, b in $\mathbb{Z}_3$. (This field is also known as $GF(9)$, which denotes "Galois field of 9 elements"—Galois was a famous French mathematician of the Napoleonic era.) Operate on elements of $\mathbb{Z}_3[i]$ as though they were complex numbers. (1) Write down all nine elements of $\mathbb{Z}_3[i]$. (2) Show that every nonzero element of $\mathbb{Z}_3[i]$ has an inverse.

E20. Consider, as in E19, the set of numbers of the form $a + bi$ with a, b in $\mathbb{Z}_2$ and $i^2 = 1 = -1$. Which elements of $\mathbb{Z}_2[i]$ have inverses?

E21. Let $\mathbb{Q}[i]$ denote the set of complex numbers of the form $a + bi$, where a, b are rational numbers. Show that $\mathbb{Q}[i]$ is a field.

E22. Let $\mathbb{Q}[\sqrt{2}\,]$ denote the set of real numbers of the form $a + b\sqrt{2}$, where a, b are rational numbers. Show that $\mathbb{Q}[\sqrt{2}\,]$ is a field.

E23. For which n, r can one solve the congruence $2x \equiv r \pmod{n}$?

E24. If p is prime, and $\alpha^2 = \alpha$ in $\mathbb{Z}_p$, show that α must be 0 or 1. Find all α such that $\alpha^2 = \alpha$ in $\mathbb{Z}_{34}$; in $\mathbb{Z}_{30}$.

9 Matrices and Vectors

This chapter is written to give a brief survey of those aspects of vectors and matrices which we wish to use elsewhere in the book. Except for Section B, which gives an example of a noncommutative ring, the chapter is intended as reference and not as an introduction to linear algebra. The rest of the book has been written so that those few places where linear algebra is used may be omitted without loss of continuity. However, linear algebra plays a central role in mathematics. So when you get to a place where some linear algebra is used, it is better that you not skip it but, rather, try to understand it, with the aid, if needed, of this chapter or some introductory text in linear algebra.

We assume R is some commutative ring, like $\mathbb{Q}$, $\mathbb{Z}$, $\mathbb{Z}_n$, or $\mathbb{R}$.

A. Matrix Multiplication

This section reviews the most basic properties of matrices.

A column vector is a column of elements of R, viz.,

$$\begin{bmatrix} a_1 \\ \vdots \\ a_n \end{bmatrix}.$$

A row vector is a row of elements of R, viz., $(a_1, \ldots, a_n)$.

An $m \times n$ matrix is a rectangular array of mn elements of R, viz.,

$$\begin{bmatrix} a_{11} & a_{12} & \cdots & a_{1n} \\ a_{21} & a_{22} & \cdots & a_{2n} \\ \vdots & \vdots & & \vdots \\ a_{m1} & a_{m2} & \cdots & a_{mn} \end{bmatrix},$$

which can be thought of as a collection of row vectors placed in a column, or as a collection of column vectors laid out in a row. When we say that a matrix is $m \times n$, the first number, m, is the number of rows, the second, n, is the number of columns.

Given a row vector with n elements (placed on the left) and a column vector with the same number of elements (placed on the right), we may multiply them to get an element of the ring R;

$$(a_1, \ldots, a_n) \begin{bmatrix} b_1 \\ \vdots \\ b_n \end{bmatrix} = a_1 b_1 + a_2 b_2 + \cdots + a_n b_n.$$

Examples where $R = \mathbb{Z}$ are:

$$(325) \begin{bmatrix} 1 \\ -2 \\ 1 \end{bmatrix} = 4, \ (123) \begin{bmatrix} 2 \\ 2 \\ -2 \end{bmatrix} = 0, \ (12) \begin{pmatrix} 2 \\ 0 \end{pmatrix} = 2.$$

Given an $m \times n$ matrix A, we can multiply the matrix (placed on the left) with an n-element column vector X (placed on the right) by thinking of the matrix as a collection of m n-element row vectors and doing m multiplications of the row vectors of A with X. The result, AX, is a column of m elements:

$$\begin{bmatrix} 1 & 2 \\ 2 & 4 \\ 2 & 3 \end{bmatrix} \cdot \begin{pmatrix} -1 \\ 2 \end{pmatrix} = \begin{bmatrix} 3 \\ 6 \\ 4 \end{bmatrix}, \quad \begin{pmatrix} 1 & 2 & 3 \\ 0 & 0 & 1 \end{pmatrix} \cdot \begin{bmatrix} 2 \\ 0 \\ 1 \end{bmatrix} = \begin{pmatrix} 5 \\ 1 \end{pmatrix}.$$

Given an $m \times n$ matrix A (on the left) and an $n \times p$ matrix B (on the right) we can multiply them by thinking of A as a collection of n-element rows and B as a collection of n-element columns. The result, AB, is an $m \times p$ matrix whose element in the ith row and jth column is obtained by multiplying the ith row of A and the jth column of B. Thus in the example

$$\begin{pmatrix} 1 & 2 & 1 \\ 2 & 3 & 0 \end{pmatrix} \begin{bmatrix} 2 & 1 \\ 0 & 1 \\ 1 & 3 \end{bmatrix} = \begin{pmatrix} 3 & 6 \\ 4 & 5 \end{pmatrix},$$

the 3 comes from multiplying

$$(1 \quad 2 \quad 1) \begin{bmatrix} 2 \\ 0 \\ 1 \end{bmatrix};$$

the 6 from

$$(1 \quad 2 \quad 1) \begin{bmatrix} 1 \\ 1 \\ 3 \end{bmatrix},$$

etc. Other examples:

$$\begin{pmatrix} 1 \\ 2 \end{pmatrix}(3 \quad 4) = \begin{pmatrix} 3 & 4 \\ 6 & 8 \end{pmatrix}, \qquad \begin{bmatrix} 2 & 1 \\ 0 & 1 \\ 1 & 2 \end{bmatrix}\begin{pmatrix} 1 & 2 & 1 \\ 2 & 3 & 0 \end{pmatrix} = \begin{bmatrix} 4 & 7 & 2 \\ 2 & 3 & 0 \\ 5 & 8 & 1 \end{bmatrix},$$

$$\begin{bmatrix} 2 & 1 \\ 0 & 1 \\ 1 & 2 \end{bmatrix}\begin{pmatrix} 0 & 1 \\ 1 & 1 \end{pmatrix} = \begin{bmatrix} 1 & 3 \\ 1 & 1 \\ 2 & 3 \end{bmatrix}.$$

Note that the order in which the matrices are multiplied (i.e., which matrix is on the left and which is on the right) is very important. In the last example,

$$\begin{pmatrix} 0 & 1 \\ 1 & 1 \end{pmatrix}\begin{bmatrix} 2 & 1 \\ 0 & 1 \\ 1 & 2 \end{bmatrix}$$

makes no sense, because it requires multiplying row vectors and column vectors with differing numbers of elements. Even when it makes sense to multiply in either order the results may be different:

$$\begin{pmatrix} 1 \\ 2 \end{pmatrix}(3 \quad 1) = \begin{pmatrix} 3 & 1 \\ 6 & 2 \end{pmatrix}; \quad (3 \quad 1)\begin{pmatrix} 1 \\ 2 \end{pmatrix} = (5);$$

also

$$\begin{pmatrix} 1 & 0 \\ 0 & 0 \end{pmatrix}\begin{pmatrix} 0 & 1 \\ 0 & 0 \end{pmatrix} = \begin{pmatrix} 0 & 1 \\ 0 & 0 \end{pmatrix},$$

whereas

$$\begin{pmatrix} 0 & 1 \\ 0 & 0 \end{pmatrix}\begin{pmatrix} 1 & 0 \\ 0 & 0 \end{pmatrix} = \begin{pmatrix} 0 & 0 \\ 0 & 0 \end{pmatrix}.$$

B. The Ring of $n \times n$ Matrices

In this section we show that the set of square matrices of a given size satisfies many of the axioms introduced in Chapter 8.

Let $M_n(R)$ be the set of $n \times n$ (square) matrices. If $\mathbf{A}$, $\mathbf{B}$ are in $M_n(R)$, then $\mathbf{A}$ and $\mathbf{B}$ can be multiplied to get $\mathbf{AB}$, another $n \times n$ matrix. Thus $M_n(R)$ is closed under multiplication, in the sense that given any pair $(\mathbf{A}, \mathbf{B})$ of $n \times n$ matrices, their product $\mathbf{AB}$ is defined and in $M_n(R)$.

We can also define addition of vectors and matrices, first for column vectors,

$$\begin{bmatrix} a_1 \\ \vdots \\ a_n \end{bmatrix} + \begin{bmatrix} b_1 \\ \vdots \\ b_n \end{bmatrix} = \begin{bmatrix} a_1 + b_1 \\ \vdots \\ a_n + b_n \end{bmatrix},$$

then for matrices of the same size by thinking of them as rows of column vectors:

$$\begin{bmatrix} a_{11} & \cdots & a_{1n} \\ \vdots & & \vdots \\ a_{m1} & \cdots & a_{mn} \end{bmatrix} + \begin{bmatrix} b_{11} & \cdots & b_{1n} \\ \vdots & & \vdots \\ b_{m1} & \cdots & b_{mn} \end{bmatrix} = \begin{bmatrix} a_{11}+b_{11} & \cdots & a_{1n}+b_{1n} \\ \vdots & & \vdots \\ a_{m1}+b_{m1} & \cdots & a_{mn}+b_{mn} \end{bmatrix}.$$

Note that $\mathbf{A}+\mathbf{B}$ makes sense only if $\mathbf{A}$ and $\mathbf{B}$ have the same shape. In particular, if $\mathbf{A}, \mathbf{B}$ are square $n \times n$ matrices, $\mathbf{A}+\mathbf{B}$ is defined.

Thus $M_n(R)$ is a set with two operations, $+$ and $\cdot$.

Theorem. *If R is a commutative ring with identity, then $M_n(R)$ is a ring with identity.*

To prove this we have to check a number of axioms. We leave the checking as exercises.

Those for addition follow almost immediately from the fact that R satisfies the same axioms.

Associativity of addition: $\mathbf{A}+(\mathbf{B}+\mathbf{C})=(\mathbf{A}+\mathbf{B})+\mathbf{C}$.

E1. Check this.

Additive identity. Let $\mathbf{0}$ be the $n \times n$ matrix consisting of all zeros. Then
$\mathbf{0}+\mathbf{A}=\mathbf{A}+\mathbf{0}=\mathbf{A}$ for any $n \times n$ matrix $\mathbf{A}$.

Inverse. If

$$\mathbf{A} = \begin{bmatrix} a_{11} & \cdots & a_{1n} \\ \vdots & & \vdots \\ a_{n1} & \cdots & a_{nn} \end{bmatrix}$$

let

$$-\mathbf{A} = \begin{bmatrix} -a_{11} & \cdots & -a_{1n} \\ \vdots & & \vdots \\ -a_n 1 & \cdots & -a_{nn} \end{bmatrix}.$$

Then $\mathbf{A}+(-\mathbf{A})=\mathbf{0}$.

Commutativity of addition: $\mathbf{A}+\mathbf{B}=\mathbf{B}+\mathbf{A}$.

E2. Check this.

Associativity of multiplication: $\mathbf{A}(\mathbf{B}\cdot\mathbf{C})=(\mathbf{A}\cdot\mathbf{B})\mathbf{C}$.

E3. Check this. (It is a mess!)

Identity. Let

$$
\mathbf{I} = \begin{bmatrix} 1 & 0 & \cdots & 0 \\ 0 & 1 & & \\ \vdots & & \ddots & \vdots \\ 0 & & \cdots & 1 \end{bmatrix}.
$$

Then $\mathbf{AI} = \mathbf{IA} = \mathbf{A}$ for any $n \times n$ matrix $\mathbf{A}$.

E4. Verify this.

Distributivity: $(\mathbf{A}(\mathbf{B} + \mathbf{C}) = \mathbf{A} \cdot \mathbf{B} + \mathbf{A} \cdot \mathbf{C}; (\mathbf{A} + \mathbf{B})\mathbf{C} = \mathbf{A} \cdot \mathbf{C} + \mathbf{B} \cdot \mathbf{C}.$

E5. Check distributivity.

Thus $M_n(R)$ is a ring with identity.

The example concluding Section A shows that $M_2(R)$ does not have a commutative multiplication, and also that $M_2(R)$ has zero divisors.

E6. Show that $M_n(R)$ does not have a commutative multiplication and also has zero divisors, for any $n \geqslant 2$.

E7. For any $n \geqslant 2$, find a nonzero $n \times n$ matrix without an inverse.

E8. Find the elements of $M_2(\mathbb{Z}_2)$ which are units, i.e. have multiplicative inverses. Write down the multiplication table for the units of $M_2(\mathbb{Z}_2)$.

We conclude this section by mentioning one other operation involving matrices, namely scalar multiplication. If $\mathbf{A}$ is a matrix of any size (in particular, a column or row vector) and s is an element of R, that is, a scalar, then define the matrix $s\mathbf{A}$ to be the matrix in which each element of $\mathbf{A}$ is multiplied by s. That is,

$$
s\mathbf{A} = s \begin{bmatrix} a_{11} & \cdots & a_{1n} \\ \vdots & & \vdots \\ a_{m1} & \cdots & a_{mn} \end{bmatrix} = \begin{bmatrix} sa_{11} & \cdots & sa_{1n} \\ \vdots & & \vdots \\ sa_{m1} & \cdots & sa_{mn} \end{bmatrix}.
$$

EXAMPLES.

$$
3\begin{pmatrix} 1 & 2 \\ 3 & 1 \end{pmatrix} = \begin{pmatrix} 3 & 6 \\ 9 & 3 \end{pmatrix}, \qquad -2\begin{bmatrix} 1 \\ 2 \\ 3 \end{bmatrix} = \begin{bmatrix} -2 \\ -4 \\ -6 \end{bmatrix}, \qquad x\begin{bmatrix} a_1 \\ a_2 \\ a_3 \end{bmatrix} = \begin{bmatrix} xa_1 \\ xa_2 \\ xa_3 \end{bmatrix}.
$$

C. Linear Equations

Matrices and vectors are a convenient way to describe systems of linear equations.

Suppose given a system of m equations in n unknowns:

$$
\begin{aligned}
a_{11}x_1 + a_{12}x_2 + \cdots + a_{1n}x_n &= b_1, \\
a_{21}x_1 + a_{22}x_2 + \cdots + a_{2n}x_n &= b_2, \\
&\;\;\vdots \\
a_{m1}x_1 + a_{m2}x_2 + \cdots + a_{mn}x_n &= b_m.
\end{aligned}
$$

We call such a system *homogeneous* if $b_1 = b_2 = \cdots = b_m = 0$, and *nonhomogeneous* otherwise.

We can make the two sides into column vectors and write the system as an equality of column vectors,

$$
\begin{bmatrix}
a_{11}x_1 + \cdots + a_{1n}x_n \\
\vdots \\
a_{m1}x_1 + \cdots + a_{mn}x_n
\end{bmatrix}
=
\begin{bmatrix}
b_1 \\
\vdots \\
b_m
\end{bmatrix},
\tag{1}
$$

because two column vectors are equal precisely when their respective components are equal.

We can rewrite (1) in either of two ways. On the one hand we can use the definition of addition and scalar multiplication of column vectors ($= m \times 1$ matrices) to write equation (1) as

$$
x_1
\begin{bmatrix}
a_{11} \\
\vdots \\
a_{m1}
\end{bmatrix}
+ x_2
\begin{bmatrix}
a_{12} \\
\vdots \\
a_{m2}
\end{bmatrix}
+ \cdots + x_n
\begin{bmatrix}
a_{1n} \\
\vdots \\
a_{mn}
\end{bmatrix}
=
\begin{bmatrix}
b_1 \\
\vdots \\
b_m
\end{bmatrix}.
$$

This says that to solve the original system is the same as to write the vector

$$
\begin{bmatrix}
b_1 \\
\vdots \\
b_m
\end{bmatrix}
$$

as a linear combination (i.e., a sum of scalar multiples) of the column vectors

$$
\begin{bmatrix}
a_{11} \\
\vdots \\
a_{m1}
\end{bmatrix}
\cdots
\begin{bmatrix}
a_{1n} \\
\vdots \\
a_{mn}
\end{bmatrix}.
$$

On the other hand, we can write down the $m \times n$ matrix whose columns are the vectors we just wrote down, and observe that the left side of (1) is the product of that matrix, called the matrix of coefficients of the original system, with a column vector of the x_i's:

$$\begin{bmatrix} a_{11} & \cdots & a_{1n} \\ \vdots & & \\ a_{m1} & \cdots & a_{mn} \end{bmatrix} \begin{bmatrix} x_1 \\ \vdots \\ x_n \end{bmatrix}.$$

If we set

$$\mathbf{A} = \begin{bmatrix} a_{11} & \cdots & a_{1n} \\ \vdots & & \\ a_{m1} & \cdots & a_{mn} \end{bmatrix}, \quad \mathbf{X} = \begin{bmatrix} x_1 \\ \vdots \\ x_n \end{bmatrix}, \quad \mathbf{B} = \begin{bmatrix} b_1 \\ \vdots \\ b_m \end{bmatrix}.$$

then the set of equations can be written in the form $\mathbf{AX} = \mathbf{B}$.

EXAMPLE. The set of equations

$$3x_1 - 2x_2 + x_3 = 4,$$
$$x_1 + x_2 - x_3 = 2,$$
$$x_1 \qquad + 3x_3 = 1$$

may be written

$$x_1 \begin{bmatrix} 3 \\ 1 \\ 1 \end{bmatrix} + x_2 \begin{bmatrix} -2 \\ 1 \\ 0 \end{bmatrix} + x_3 \begin{bmatrix} 1 \\ -1 \\ 3 \end{bmatrix} = \begin{bmatrix} 4 \\ 2 \\ 1 \end{bmatrix}$$

or as

$$\begin{bmatrix} 3 & -2 & 1 \\ 1 & 1 & -1 \\ 1 & 0 & 3 \end{bmatrix} \begin{bmatrix} x_1 \\ x_2 \\ x_3 \end{bmatrix} = \begin{bmatrix} 4 \\ 2 \\ 1 \end{bmatrix}.$$

Suppose there were an $n \times m$ matrix $\mathbf{C}$ such that $\mathbf{CA} = \mathbf{I}$. If we could find such a $\mathbf{C}$, then $\mathbf{CB} = \mathbf{CAX} = \mathbf{IX} = \mathbf{X}$ would be a solution of the equations. Thus solving equations is closely related to finding inverses of matrices.

For example,

$$\begin{bmatrix} 3 & -2 & 1 \\ 1 & 1 & -1 \\ 1 & 0 & 3 \end{bmatrix}$$

turns out to have the inverse

$$\begin{bmatrix} 3/16 & 6/16 & 1/16 \\ -1/4 & 1/2 & 1/4 \\ -1/16 & -1/8 & 5/16 \end{bmatrix},$$

so

$$\begin{pmatrix} x_1 \\ x_2 \\ x_3 \end{pmatrix} = \begin{bmatrix} 3/16 & 6/16 & 1/16 \\ -1/4 & 1/2 & 1/4 \\ -1/16 & -1/8 & 5/16 \end{bmatrix} \begin{pmatrix} 4 \\ 2 \\ 1 \end{pmatrix} = \begin{pmatrix} 25/16 \\ 1/4 \\ -3/16 \end{pmatrix}.$$

D. Determinants and Inverses

If A is an $n \times n$ (square) matrix, the determinant of A is defined and is an element of R. For 1×1, 2×2 and 3×3 matrices, the determinant of A is defined as follows:

$$\det(a) = a; \qquad \det\begin{pmatrix} ab \\ cd \end{pmatrix} = ad - bc;$$

$$\det\begin{bmatrix} a_1 & b_1 & c_1 \\ a_2 & b_2 & c_2 \\ a_3 & b_3 & c_3 \end{bmatrix} = a_1 b_2 c_3 + b_1 c_2 a_3 + c_1 a_2 b_3 - a_3 b_2 c_1 -$$

$$a_2 b_1 c_3 - a_1 b_3 c_2.$$

If A is a triangular matrix, that is, a matrix of the form

$$\begin{bmatrix} a_{11} & 0 & 0 & \cdots & 0 \\ a_{21} & a_{22} & 0 & & \\ a_{31} & a_{32} & a_{33} & 0 & \vdots \\ \vdots & & & \ddots & 0 \\ a_{n1} & a_{n2} & \cdots & & a_{nn} \end{bmatrix},$$

then $\det(A) = a_{11} a_{22} \cdots a_{nn}$.

For nontriangular 4×4 or larger matrices the explicit formula for the determinant is too complicated to write down and will not be needed in this book.

If A is an $n \times n$ matrix, sometimes A has an inverse, an $n \times n$ matrix B such that $AB = BA = I$. There is a theorem which says that such a B exists iff the determinant of A is an invertible element of R. In particular, for 2×2 matrices, if the determinant of A is invertible, the inverse of A can be found as follows: When

$$A = \begin{pmatrix} a & b \\ c & d \end{pmatrix}$$

with $1/(ad - bc)$ in R

$$B = \begin{bmatrix} \dfrac{d}{ad - bc} & \dfrac{-b}{ad - bc} \\ \dfrac{-c}{ad - bc} & \dfrac{a}{ad - bc} \end{bmatrix},$$

as is easily checked. For **A** 3×3 or bigger, the formula for getting the inverse of **A** using determinants is too complicated to state here. We can find the inverse of **A** using row operations.

E. Row Operations

The material from linear algebra surveyed in the remainder of this chapter is somewhat less elementary than what has been discussed above. It will be needed only in a very few subsequent chapters, and should be referred to only when needed. The ideas presented may be found in most introductory textbooks on linear algebra, such as Zelinsky (1973).

A most useful computational technique for matrices is the use of *row operations*, known also as Gaussian elimination. There are three such operations:

(1) interchanging two rows of a matrix;
(2) multiplying a row by an invertible scalar;
(3) replacing a row by itself plus a multiple of another row.

These correspond to things which can be done to sets of simultaneous equations without affecting the solutions of these equations. Using them, it is possible, if R is a field, to take an $m \times n$ matrix **A** and, by using operation (3) to introduce zero entries into the matrix, transform **A** into its *row reduced echelon form*. This is a matrix of the following forms, with "steps":

$$\begin{bmatrix} \ldots & 1 & \ldots & 0 & & 0 & & 0 \\ & & & 1 & \ldots & 0 & & 0 \\ & & & & & 1 & \ldots & 0 \\ & & & & & & & 1 \ldots \\ & & 0 & & & & & \\ & & & & & & & \ddots \end{bmatrix}$$

Examples are:

$$\begin{bmatrix} 1 & 2 & 0 & 0 & 3 & 0 \\ 0 & 0 & 1 & -1 & 0 & 0 \\ 0 & 0 & 0 & 0 & 0 & 1 \end{bmatrix}; \quad \begin{bmatrix} 1 & 0 & 2 \\ 0 & 1 & 2 \\ 0 & 0 & 0 \end{bmatrix}; \quad \begin{bmatrix} 1 & 0 & 0 \\ 0 & 1 & 0 \\ 0 & 0 & 1 \end{bmatrix};$$

$$\begin{pmatrix} 1 & 2 & 3 \\ 0 & 0 & 0 \end{pmatrix}; \quad \begin{bmatrix} 0 & 0 & 1 & 0 \\ 0 & 0 & 0 & 1 \\ 0 & 0 & 0 & 0 \end{bmatrix}.$$

To illustrate the technique, to put

$$\begin{bmatrix} 2 & 1 & 4 & 2 \\ 1 & 2 & 3 & 1 \\ 3 & -3 & 3 & 3 \end{bmatrix}$$

into row reduced echelon form we would proceed as follows, with each arrow representing one row operation:

$$\begin{pmatrix} 2 & 1 & 4 & 2 \\ 1 & 2 & 3 & 1 \\ 3 & -3 & 3 & 3 \end{pmatrix} \rightarrow \begin{pmatrix} 1 & 2 & 3 & 1 \\ 2 & 1 & 4 & 2 \\ 3 & -3 & 3 & 3 \end{pmatrix} \rightarrow \begin{pmatrix} 1 & 2 & 3 & 1 \\ 2 & 1 & 4 & 2 \\ 1 & -1 & 1 & 1 \end{pmatrix}$$

$$\rightarrow \begin{pmatrix} 1 & 2 & 3 & 1 \\ 0 & -3 & -2 & 0 \\ 1 & -1 & 1 & 1 \end{pmatrix} \rightarrow \begin{pmatrix} 1 & 2 & 3 & 1 \\ 0 & -3 & -2 & 0 \\ 0 & -3 & -2 & 0 \end{pmatrix}$$

$$\rightarrow \begin{pmatrix} 1 & 2 & 3 & 1 \\ 0 & -3 & -2 & 0 \\ 0 & 0 & 0 & 0 \end{pmatrix}$$

$$\rightarrow \begin{pmatrix} 1 & 2 & 3 & 1 \\ 0 & 1 & 2/3 & 0 \\ 0 & 0 & 0 & 0 \end{pmatrix} \rightarrow \begin{pmatrix} 1 & 0 & 5/3 & 1 \\ 0 & 1 & 2/3 & 0 \\ 0 & 0 & 0 & 0 \end{pmatrix}.$$

(There are of course other choices of row operations which would achieve the same result.)

If we wished to solve

$$2x_1 + x_2 + 4x_3 + 2x_4 = 0,$$
$$x_1 + 2x_2 + 3x_3 + x_4 = 0, \qquad (1)$$
$$3x_1 - 3x_2 + 3x_3 + 3x_4 = 0,$$

or

$$\begin{pmatrix} 2 & 1 & 4 & 2 \\ 1 & 2 & 3 & 1 \\ 3 & -3 & 3 & 3 \end{pmatrix} \begin{pmatrix} x_1 \\ x_2 \\ x_3 \\ x_4 \end{pmatrix} = \begin{pmatrix} 0 \\ 0 \\ 0 \end{pmatrix},$$

the row operations done above would correspond to manipulations on the equations which would yield new equations having the same solutions as the original set. Thus the solution of equations (1) is the same as that of

$$\begin{pmatrix} 1 & 0 & 5/3 & 1 \\ 0 & 1 & 2/3 & 0 \\ 0 & 0 & 0 & 0 \end{pmatrix} \begin{pmatrix} x_1 \\ x_2 \\ x_3 \\ x_4 \end{pmatrix} = \begin{pmatrix} 0 \\ 0 \\ 0 \end{pmatrix} \qquad (2)$$

or

$$x_1 + \tfrac{5}{3}x_3 + x_4 = 0,$$
$$x_2 + \tfrac{2}{3}x_3 = 0,$$
$$0 = 0.$$

The solutions to (2) are all obtained by assigning arbitrarily chosen values to x_3 and x_4. In fact, the vector of solutions can be written as

$$
\begin{bmatrix} x_1 \\ x_2 \\ x_3 \\ x_4 \end{bmatrix} = \begin{bmatrix} -5/3x_3 - x_4 \\ -2/3x_3 \\ x_3 \\ x_4 \end{bmatrix} = x_3 \begin{bmatrix} -5/3 \\ -2/3 \\ 1 \\ 0 \end{bmatrix} + x_4 \begin{bmatrix} -1 \\ 0 \\ 0 \\ 1 \end{bmatrix}, \tag{2'}
$$

where x_3 and x_4 may be chosen at will.

If we wish to solve a nonhomogeneous system of equations, one in the form $\mathbf{AX} = \mathbf{Y}$, with $\mathbf{Y}$ not a vector of all zeros, we can do row operations to the so-called augmented matrix. For example, suppose we wish to solve

$$
\begin{align}
2x_1 + x_2 + 4x_3 &= 2, \\
x_1 + 2x_2 + 3x_3 &= 1, \tag{3} \\
3x_1 - 3x_2 + 3x_3 &= 3,
\end{align}
$$

or

$$
\begin{bmatrix} 2 & 1 & 4 \\ 1 & 2 & 3 \\ 3 & -3 & 3 \end{bmatrix} \begin{bmatrix} x_1 \\ x_2 \\ x_3 \end{bmatrix} = \begin{bmatrix} 2 \\ 1 \\ 3 \end{bmatrix}.
$$

This equation is the same as the equation (1) we just solved, except that we require that $x_4 = -1$:

$$
\begin{bmatrix} 2 & 1 & 4 & 2 \\ 1 & 2 & 3 & 1 \\ 3 & -3 & 3 & 3 \end{bmatrix} \begin{bmatrix} x_1 \\ x_2 \\ x_3 \\ -1 \end{bmatrix} = \begin{bmatrix} 0 \\ 0 \\ 0 \end{bmatrix}. \tag{3'}
$$

Thus to solve $\mathbf{AX} = \mathbf{Y}$, where $\mathbf{A}$ is $m \times n$, we take the $m \times (n+1)$ matrix with $\mathbf{A}$ in the leftmost n columns and $\mathbf{Y}$ in the rightmost column, and reduce it to row reduced echelon form. If the resulting leftmost n columns are the $m \times n$ matrix $\mathbf{E}$ and the rightmost column is $\mathbf{Z}$, then solving $\mathbf{AX} = \mathbf{Y}$ is the same as solving $\mathbf{EX} = \mathbf{Z}$. Doing this with equation (3'), we get

$$
\begin{bmatrix} 1 & 0 & 5/3 & 1 \\ 0 & 1 & 2/3 & 0 \\ 0 & 0 & 0 & 0 \end{bmatrix} \begin{bmatrix} x_1 \\ x_2 \\ x_3 \\ -1 \end{bmatrix} = \begin{bmatrix} 0 \\ 0 \\ 0 \end{bmatrix}
$$

or

$$
\begin{align}
x_1 + 5/3x_3 - 1 &= 0, \\
x_2 + 2/3x_3 &= 0, \\
0 &= 0,
\end{align}
$$

which has solutions

$$\begin{bmatrix} x_1 \\ x_2 \\ x_3 \end{bmatrix} = \begin{bmatrix} -5/3x_3 + 1 \\ -2/3x_3 \\ x_3 \end{bmatrix} = \begin{bmatrix} 1 \\ 0 \\ 0 \end{bmatrix} + x_3 \begin{bmatrix} -5/3 \\ -2/3 \\ 1 \end{bmatrix}$$

with x_3 arbitrary.

Sometimes we cannot solve the equation $\mathbf{AX} = \mathbf{Y}$. Here is an example. If we consider

$$\begin{bmatrix} 2 & 1 & 4 \\ 1 & 2 & 3 \\ 3 & -3 & 3 \end{bmatrix} \begin{bmatrix} x_1 \\ x_2 \\ x_3 \end{bmatrix} = \begin{bmatrix} 2 \\ 1 \\ 6 \end{bmatrix},$$

the augmented matrix would be

$$\begin{bmatrix} 2 & 1 & 4 & 2 \\ 1 & 2 & 3 & 1 \\ 3 & -3 & 3 & 6 \end{bmatrix},$$

whose row reduced echelon form is

$$\begin{bmatrix} 1 & 0 & 5/3 & 0 \\ 0 & 1 & 2/3 & 0 \\ 0 & 0 & 0 & 1 \end{bmatrix},$$

which yields the equations

$$x_1 + 5/3x_3 = 0,$$
$$x_2 + 2/3x_3 = 0,$$
$$-1 = 0,$$

which are clearly unsolvable!

In an equation of the form

$$\mathbf{AX} = \mathbf{Y}, \quad \mathbf{X} = \begin{bmatrix} x_1 \\ \vdots \\ x_n \end{bmatrix},$$

the variable x_i has as coefficients the entries in the ith column of $\mathbf{A}$. When we reduce $\mathbf{A}$ to row reduced echelon form, certain columns contain steps and certain columns do not. The variables $x_1, \ldots, x_n$ then divide up into those corresponding to columns with steps, and those corresponding to the columns without steps. Reducing the matrix of coefficients of a set of equations to row reduced echelon form always gives instant solutions for the variables corresponding to the columns with steps, in terms of the other variables.

We can find the inverse of an $n \times n$ matrix $\mathbf{A}$, if it exists, by solving for $\mathbf{X}$ in the equation

$$\mathbf{AX} = \mathbf{Y} = \mathbf{IY} \quad \text{or} \quad (\mathbf{A}, \mathbf{I})\begin{pmatrix} \mathbf{X} \\ -\mathbf{Y} \end{pmatrix} = 0,$$

where $(\mathbf{A}, \mathbf{I})$ is the $n \times 2n$ matrix whose left side is $\mathbf{A}$ and whose right side is the $n \times n$ identity matrix $\mathbf{I}$, and where $\mathbf{X}, \mathbf{Y}$ are column vectors of unknowns. If the row reduced echelon form for the $n \times 2n$ matrix $(\mathbf{A}, \mathbf{I})$ is in the form $(\mathbf{I}, \mathbf{B})$, that is, if the leftmost n columns of the row reduced echelon form of the $n \times 2n$ matrix $(\mathbf{A}, \mathbf{I})$ form the $n \times n$ identity matrix $\mathbf{I}$, then

$$(\mathbf{A}, \mathbf{I})\begin{pmatrix} \mathbf{X} \\ -\mathbf{Y} \end{pmatrix} = \mathbf{0} \quad \text{iff} \quad (\mathbf{I}, \mathbf{B})\begin{pmatrix} \mathbf{X} \\ -\mathbf{Y} \end{pmatrix} = \mathbf{0} \quad \text{iff} \quad \mathbf{X} = \mathbf{BY}$$

iff $\mathbf{B}$ is the inverse of $\mathbf{A}$. So write, side by side, $(\mathbf{A}, \mathbf{I})$, and do row operations on the whole matrix to try to reduce $\mathbf{A}$ to $\mathbf{I}$. If this succeeds, then by the same operations $\mathbf{I}$ becomes transformed into the inverse of $\mathbf{A}$.

For example, to find the inverse of $\begin{pmatrix} 2 & 1 \\ 1 & 3 \end{pmatrix}$, proceed with row operations as follows:

$$\left[\begin{array}{cc|cc} 2 & 1 & 1 & 0 \\ 1 & 3 & 0 & 1 \end{array}\right] \rightarrow \left[\begin{array}{cc|cc} 1 & 3 & 0 & 1 \\ 2 & 1 & 1 & 0 \end{array}\right] \rightarrow \left[\begin{array}{cc|cc} 1 & 3 & 0 & 1 \\ 0 & -5 & 1 & -2 \end{array}\right]$$

$$\rightarrow \left[\begin{array}{cc|cc} 1 & 3 & 0 & 1 \\ 0 & 1 & -\frac{1}{5} & \frac{2}{5} \end{array}\right] \rightarrow \left[\begin{array}{cc|cc} 1 & 0 & \frac{3}{5} & -\frac{1}{5} \\ 0 & 1 & -\frac{1}{5} & \frac{2}{5} \end{array}\right];$$

then

$$\begin{pmatrix} \frac{3}{5} & -\frac{1}{5} \\ -\frac{1}{5} & \frac{2}{5} \end{pmatrix}\begin{pmatrix} 2 & 1 \\ 1 & 3 \end{pmatrix} = \begin{pmatrix} 1 & 0 \\ 0 & 1 \end{pmatrix}.$$

E9. Find the inverse of

$$\begin{pmatrix} 3 & 7 \\ 1 & 2 \end{pmatrix}; \quad \begin{pmatrix} 1 & 3 & 2 \\ 2 & 1 & 2 \\ 3 & 4 & 3 \end{pmatrix}$$

(entries in $\mathbb{Q}$).

E10. Solve

$$\begin{aligned} x + 3y + 2z &= 2, \\ 2x + y + 2z &= 1, \\ 3x + 4y + 3z &= 1, \end{aligned}$$

for x, y, z in $\mathbb{Q}$.

F. Subspaces, Bases, Dimension

For F a field, denote by F^n the set of all column vectors with n entries from F. Thus $F^2 = \left\{ \begin{pmatrix} a \\ b \end{pmatrix} \right\}$ where a, b are elements of F. The set F^n of all n-element column vectors ("n-tuples") has operations of addition and scalar multiplication as follows:

$$\begin{pmatrix} a_1 \\ \vdots \\ a_n \end{pmatrix} + \begin{pmatrix} b_1 \\ \vdots \\ b_n \end{pmatrix} = \begin{pmatrix} a_1 + b_1 \\ \vdots \\ a_n + b_n \end{pmatrix},$$

$$r \begin{pmatrix} a_1 \\ \vdots \\ a_n \end{pmatrix} = \begin{pmatrix} ra_1 \\ \vdots \\ ra_n \end{pmatrix} \qquad (r \text{ in } F);$$

these operations make F^n into what is called a vector space.

A subset S of F^n is called a *subspace* if whenever $\mathbf{X}$, $\mathbf{Y}$ are in S, so is any linear combination, that is, so is $r\mathbf{X} + s\mathbf{Y}$ for any r, s in F.

The relevant examples of subspaces in this work are null spaces and row spaces of $m \times n$ matrices.

Let $\mathbf{A}$ be an $m \times n$ matrix. The *null space* of $\mathbf{A}$ is the set S of all $\mathbf{X}$ in F^n such that $\mathbf{AX} = \mathbf{0}$. If we interpret $\mathbf{A}$ as a matrix of coefficients of a set of homogeneous equations, then the null space of $\mathbf{A}$ is the set of solutions to the corresponding set of equations. The set S is a subspace of F^n. For if $\mathbf{X}$, $\mathbf{Y}$ are solutions, so is any linear combination: If r, s are any elements of F, then

$$\mathbf{A}(r\mathbf{X} + s\mathbf{Y}) = r(\mathbf{AX}) + s(\mathbf{AY}) = r\mathbf{0} + s\mathbf{0} = \mathbf{0}.$$

(On the other hand, the set of $\mathbf{X}$ such that $\mathbf{AX} = \mathbf{B}$, for fixed $\mathbf{B} \neq \mathbf{0}$, is not a subspace, for if $\mathbf{AX} = \mathbf{B}$, then $\mathbf{A}(\mathbf{X} + \mathbf{X}) = 2\mathbf{B}$, not $\mathbf{B}$.)

A *basis* of a vector space S is a set $\mathbf{X}_1, \ldots, \mathbf{X}_r$ of vectors in S such that any vector $\mathbf{W}$ in S can be written as a linear combination of $\mathbf{X}_1, \ldots, \mathbf{X}_r$ in exactly one way. This condition can be described as a combination of two conditions. We say that a set $\mathbf{X}_1, \ldots, \mathbf{X}_r$ of vectors of S *spans* S if any vector in S can be written as a linear combination of $\mathbf{X}_1, \ldots, \mathbf{X}_r$ (perhaps in many ways), and we say that $\mathbf{X}_1, \ldots, \mathbf{X}_r$ are *linearly independent* if the only solution to the equation

$$c_1\mathbf{X}_1 + c_2\mathbf{X}_2 + \cdots + c_r\mathbf{X}_r = 0$$

is the solution $c_1 = 0, c_2 = 0, \ldots, c_r = 0$. Then a basis of S is a set of linearly independent vectors which spans S. It is not hard to see that this last description of basis is equivalent to the first.

Here are some examples of bases.

In F^2, a basis is $\begin{pmatrix} 1 \\ 0 \end{pmatrix}$ and $\begin{pmatrix} 0 \\ 1 \end{pmatrix}$. For given any vector $\begin{pmatrix} a \\ b \end{pmatrix}$ in F^2, we can write

$$\begin{pmatrix} a \\ b \end{pmatrix} = a \begin{pmatrix} 1 \\ 0 \end{pmatrix} + b \begin{pmatrix} 0 \\ 1 \end{pmatrix},$$

and it is very easy to see that $\begin{pmatrix} 1 \\ 0 \end{pmatrix}$ and $\begin{pmatrix} 0 \\ 1 \end{pmatrix}$ are linearly independent.

If N is the null space of the matrix of coefficients of the set of equations (1) of the last section, then the vectors in N are described by (2′) as arbitrary linear combinations of

$$\begin{bmatrix} -5/3 \\ -2/3 \\ 1 \\ 0 \end{bmatrix} \quad \text{and} \quad \begin{bmatrix} -1 \\ 0 \\ 0 \\ 1 \end{bmatrix}.$$

Thus

$$\begin{bmatrix} -5/3 \\ -2/3 \\ 1 \\ 0 \end{bmatrix} \quad \text{and} \quad \begin{bmatrix} -1 \\ 0 \\ 0 \\ 1 \end{bmatrix}$$

span the null space N, and since those two vectors are linearly independent, they are a basis of the null space N.

In general, to find a basis of the null space N of an $m \times n$ matrix $\mathbf{A}$, solve $\mathbf{AX} = \mathbf{0}$ by putting $\mathbf{A}$ into row reduced echelon form. Then as in equation (2′) of Section E, spreading out the resulting solution vector describes the solutions as linear combinations of a basis of N.

A space S generally has many different bases. But it is a theorem that all bases of a space S contain the same number of vectors. The *dimension* of a space is equal to the number of vectors in any basis of the space. Thus the dimension of F^n is n, and the dimension of the space N of solutions to (1), Section E, is 2.

The *row space* of an $m \times n$ matrix $\mathbf{A}$ is the set of all n-tuples (row vectors) which are linear combinations of the rows of $\mathbf{A}$. It is not hard to see that if you do a row operation to $\mathbf{A}$, the row space of the new matrix is the same as the row space of $\mathbf{A}$. Thus, since the row reduced echelon form $\mathbf{E}$ of $\mathbf{A}$ is obtained by a sequence of row operations, the row space of $\mathbf{E}$ is the same as the row space of $\mathbf{A}$.

For example, if $\mathbf{A}$ is the matrix of coefficients of the equations (1) of the last section, then the row space is the set of vectors of the form

$$\mathbf{X} = a_1(2 \ 1 \ 4 \ 2) + a_2(1 \ 2 \ 3 \ 1) + a_3(3 \ -3 \ 3 \ 3),$$

where a_1, a_2, a_3 are arbitrary scalars. Any such vector $\mathbf{X}$ can be written as a linear combination of the vectors in the row reduced echelon form of $\mathbf{A}$. In fact, the vector $\mathbf{X}$ may be written

$$\mathbf{X} = (2a_1 + a_2 + 3a_3)\left(1 \ 0 \ \tfrac{5}{3} \ 1\right) + (a_1 + 2a_2 - 3a_3)\left(0 \ 1 \ \tfrac{2}{3} \ 0\right).$$

Conversely, if

$$\mathbf{Y} = b_1\left(1 \ 0 \ \tfrac{5}{3} \ 1\right) + b_2\left(0 \ 1 \ \tfrac{2}{3} \ 0\right)$$

is a typical vector in the row space of the echelon form $\mathbf{E}$, then

$$\mathbf{Y} = \left(\tfrac{2}{3}b_1 - \tfrac{1}{3}b_2\right)(2 \ 1 \ 4 \ 2) + \left(\tfrac{1}{3}b_1 + \tfrac{2}{3}b_2\right)(1 \ 2 \ 3 \ 1) + 0(3 \ -3 \ 3 \ 3),$$

so is in the row space of $\mathbf{A}$. Thus the row spaces of $\mathbf{A}$ and of $\mathbf{E}$ are the same.

It is easy to check that the nonzero rows of the row reduced echelon form of a matrix **A** are a basis of the row space of **A**. Thus the number of nonzero rows in the row reduced echelon form of **A** is the dimension of the row space of **A**. That dimension is called the (row) *rank* of **A**.

It is a theorem that for **A** an $m \times n$ matrix, the rank of **A** + the dimension of the null space of **A** = the number of columns of **A**. If you interpret the rank as the number of columns with steps in the row reduced echelon form of **A**, and the dimension of the null space as the number of columns without steps, the theorem becomes very reasonable.

E11. What is the rank of an invertible $n \times n$ matrix?

E12. Find the rank and the dimension of the null space for A when
   (i)

$$A = \begin{pmatrix} 2 & 3 & 1 & 6 \\ 3 & 1 & -1 & 2 \\ 1 & 2 & 1 & 4 \end{pmatrix},$$

   (ii)

$$A = \begin{pmatrix} 2 & 2 & 3 & 1 & 4 \\ 1 & 2 & 1 & 1 & 3 \end{pmatrix},$$

   (iii)

$$A = \begin{pmatrix} 1 & 2 & 7 \\ 1 & -2 & 1 \\ -1 & 4 & 2 \end{pmatrix}.$$

10 Secret Codes, I

This chapter may be omitted without loss of continuity.

One interesting use of $\mathbb{Z}_n$ is in designing ciphers, or codes designed to preserve the secrecy of messages being communicated. In this chapter, we describe a way of creating ciphers based on the use of matrices with entries in $\mathbb{Z}_{26}$. This approach is based on an article by Hill (1931) that is now somewhat dated, but still provides interesting practice in doing arithmetic mod n. (See Chapter 15 for a more up-to-date example of ciphers.)

We shall assume some of the facts about matrices and vectors reviewed in Sections A, C, and D of Chapter 9.

The messages we start with will be in English, so the first step is to change letters into numbers.

We let the letters of the alphabet correspond 1–1 with the integers from 1 to 26, which we view as representatives of the congruence classes in $\mathbb{Z}_{26}$: if a is an integer, we shall write $[a]_{26}$ as a in this section. Thus all algebraic operations on integers will be mod 26. We can correspond letters to elements of $\mathbb{Z}_{26}$ in any way we wish; for simplicity we shall use the following.

$$\begin{cases} A & B & C & D & \ldots & J & \ldots & O & \ldots & T & Z \\ 1 & 2 & 3 & 4 & & 10 & & 15 & & 20 & 26 \end{cases} \quad (1)$$

Since $\mathbb{Z}_{26}$ is a commutative ring we can form column vectors of elements of $\mathbb{Z}_{26}$ and $n \times n$ matrices with entries in $\mathbb{Z}_{26}$, and we can multiply vectors by matrices and multiply matrices of appropriate sizes. We can even sometimes find the inverse of an $n \times n$ matrix: we note the useful fact (Chapter 9D) that an $n \times n$ matrix $\mathbf{A}$ with entries in a commutative ring R has an inverse $\mathbf{B}$ with entries in R (so that $\mathbf{AB} = \mathbf{BA} = \mathbf{I}_n$) if and only if the determinant of $\mathbf{A}$ has an inverse in R.

The ciphers we can form involve matrices of various sizes. We shall illustrate several different sizes by coding and decoding the message:

ATTACKXATXDAWN

where we put the X into separate words. We write the message as a sequence of elements of $\mathbb{Z}_{26}$ using the correspondence (1)

1, 20, 20, 1, 3, 11, 24, 1, 20, 24, 4, 1, 23, 14

or which is easier (since we're thinking of these numbers as representing classes in $\mathbb{Z}_{26}$),

$$1, -6, -6, 1, 3, 11, -2, 1, -6, -2, 4, 1, -3, 14. \tag{2}$$

Codes of size 1×1

Here is how we code. Take an invertible element of $\mathbb{Z}_{26}$, such as 5: $5 \cdot (-5) \equiv 1$. Multiply each number in the message (2) by 5 (mod 26): we get

5, -30, -30, 5, 15, 55, -10, 5, -30, -10, 20, 5, -15, 70

or, choosing different representatives (mod 26),

$$5, 22, 22, 5, 15, 3, 16, 5, 22, 16, 20, 5, 11, 18. \tag{3}$$

Using the correspondence (1) to change back to letters, we get

EVVEOCPEVPTEKR.

The receiver decodes by translating the received message using (1) back into numbers, getting (3), and multiplying each number of (3) by -5 (mod 26). Since $-5 \cdot 5 \equiv 1$ (mod 26), the receiver ends up with the sequence of numbers corresponding to the original message.

E1. Encode and decode

HAPPYXBIRTHDAY

using this 1×1 code.

This code is pretty easy to crack, if messages of some length are sent. Since each letter, such as A (or T), always corresponds to the same letter in code, here E (or V), cryptanalysts can note letter frequencies in the code, and, using known information about frequencies of occurrences of letters in English tests, decipher the message without knowing the code. This is sufficiently easy to do that 1×1 coded messages are given as puzzles in some magazines.

So we shall try to make more complicated codes.

Codes of size 2×2

Break the enumerated message as in (2) up into a sequence of 2-tuples:

$$\begin{pmatrix} 1 \\ -6 \end{pmatrix}, \begin{pmatrix} -6 \\ 1 \end{pmatrix}, \begin{pmatrix} 3 \\ 11 \end{pmatrix}, \begin{pmatrix} -2 \\ 1 \end{pmatrix}, \begin{pmatrix} -6 \\ -2 \end{pmatrix}, \begin{pmatrix} 4 \\ 1 \end{pmatrix}, \begin{pmatrix} -3 \\ 14 \end{pmatrix}. \tag{4}$$

To code, multiply each vector in (4) by some invertible 2×2 matrix $\mathbf{A}$.

EXAMPLE 1. Let

$$\mathbf{A} = \begin{pmatrix} 8 & 13 \\ -5 & -8 \end{pmatrix}.$$

Then det $\mathbf{A} = 1$, and

$$\mathbf{A}^{-1} = \begin{pmatrix} -8 & -13 \\ 5 & 8 \end{pmatrix}.$$

The coded message is the sequence of 2-tuples:

$$\begin{pmatrix} 8 \\ 17 \end{pmatrix}, \begin{pmatrix} 17 \\ 22 \end{pmatrix}, \begin{pmatrix} 11 \\ 1 \end{pmatrix}, \begin{pmatrix} 23 \\ 8 \end{pmatrix}, \begin{pmatrix} 4 \\ 20 \end{pmatrix}, \begin{pmatrix} 19 \\ 24 \end{pmatrix}, \begin{pmatrix} 2 \\ 23 \end{pmatrix}, \tag{5}$$

obtained by multiplying each 2-tuple of (4) by $\mathbf{A}$. In letters, the message in code is

HQQVKAWHDTSXBW.

Notice that now A in the original message is replaced by H, V, or X, depending on its location in the original message. Now, only pairs of letters are set to the same thing, and then only if they both begin at an odd, or both at an even location in the message.

The receiver would take the coded message, put it back into a sequence of 2-tuples (5), and multiply each 2-tuple by $\mathbf{A}^{-1}$. Since $\mathbf{A}^{-1} \cdot \mathbf{A} = \mathbf{I}$, she will end up with the original set of 2-tuples (4) and finally, using (1), the original message.

EXAMPLE 2. We could use a matrix like

$$\mathbf{A} = \begin{pmatrix} 2 & -3 \\ 3 & 1 \end{pmatrix},$$

whose determinant is 11, an invertible element of $\mathbb{Z}_{26}$. Then

$$\mathbf{A}^{-1} = \begin{pmatrix} -7 & 5 \\ -5 & 12 \end{pmatrix}.$$

Applying this matrix $\mathbf{A}$ to the 2-tuples (4), the coded message will then be

TWKIYTSUTFEMDE,

which can be deciphered using $\mathbf{A}^{-1}$.

How did we get $\mathbf{A}^{-1}$? recall that if $D = ad - bc$ is invertible, $DE = 1$, then

$$\begin{pmatrix} a & b \\ c & d \end{pmatrix}^{-1} = \begin{pmatrix} dE & -bE \\ -cE & aE \end{pmatrix}.$$

In $\mathbb{Z}_{26}$, if $\mathbf{A} = \begin{pmatrix} 2 & -3 \\ 3 & 1 \end{pmatrix}$, $D = \det(\mathbf{A}) = 2 + 9 = 11$. Now 11 has an inverse (mod 26) since 11 and 26 are relatively prime; using the Euclidean algorithm or Euler's theorem (Chapter 12) we get $11 \cdot 19 \equiv 1$ (mod 26), and $E = 19$. So

$$\mathbf{A}^{-1} = \begin{pmatrix} 1 \cdot 19 & 3 \cdot 19 \\ -3 \cdot 19 & 2 \cdot 19 \end{pmatrix} = \begin{pmatrix} 19 & 57 \\ -57 & 38 \end{pmatrix} = \begin{pmatrix} -7 & 5 \\ -5 & 12 \end{pmatrix}.$$

E2. Encode and decode HAPPY BIRTHDAY using the 2×2 code of Example 2.

E3. Decode MXGWGCCCUKMQNGRC using the code of Example 2.

If higher security is needed we can use larger matrices.

Codes of size 3×3

Break up the message into words of length 3, using extra dummy X's at the end to fill out a word:

$$\begin{array}{ccccc} \text{ATT} & \text{ACK} & \text{XAT} & \text{XDA} & \text{WNX} \\ \begin{bmatrix} 1 \\ -6 \\ -6 \end{bmatrix} & \begin{bmatrix} 1 \\ 3 \\ 11 \end{bmatrix} & \begin{bmatrix} -2 \\ 1 \\ -6 \end{bmatrix} & \begin{bmatrix} -2 \\ 4 \\ 1 \end{bmatrix} & \begin{bmatrix} -3 \\ 14 \\ -2 \end{bmatrix} \end{array}$$

Find a 3×3 matrix with invertible determinant, like

$$\mathbf{A} = \begin{bmatrix} 2 & 3 & 5 \\ 5 & 11 & 2 \\ 1 & 2 & 2 \end{bmatrix},$$

$$\det \mathbf{A} = 44 + 6 + 50 + -8 - 55 - 30 = 100 - 93 = 7.$$

Since $(26, 7) = 1$, $\mathbf{A}$ is invertible. Its inverse turns out to be

$$\mathbf{A}^{-1} = \begin{bmatrix} 10 & 8 & -7 \\ 10 & 11 & 3 \\ 11 & 11 & 1 \end{bmatrix}.$$

E4. (a) Verify that $\mathbf{A}^{-1}$ is as claimed. (b) Encode and decode HAPPY BIRTH-DAY using this 3×3 code.

Codes of size 5×5

Break up the message into words of length 5, using extra dummy X's at the end to fill out a word:

$$\text{ATTAC KXATX DAWNX.}$$

Translate into vectors using (1):

$$\begin{bmatrix} 1 \\ -6 \\ -6 \\ 1 \\ 3 \end{bmatrix} \quad \begin{bmatrix} 11 \\ -2 \\ 1 \\ -6 \\ -2 \end{bmatrix} \quad \begin{bmatrix} 4 \\ 1 \\ -3 \\ 14 \\ -2 \end{bmatrix}$$

Find a 5×5 invertible matrix, like

$$\mathbf{A} = \begin{bmatrix} 1 & 0 & 0 & 0 & 0 \\ 2 & 3 & 0 & 0 & 0 \\ 8 & 0 & 7 & 0 & 0 \\ 0 & 2 & 1 & 9 & 0 \\ -10 & -6 & 8 & 0 & 3 \end{bmatrix},$$

which has inverse in $\mathbb{Z}_{26}$:

$$\mathbf{A}^{-1} = \begin{bmatrix} 1 & 0 & 0 & 0 & 0 \\ 8 & 9 & 0 & 0 & 0 \\ 10 & 0 & 15 & 0 & 0 \\ 0 & -2 & 7 & 3 & 0 \\ 10 & -8 & 12 & 0 & 9 \end{bmatrix}.$$

E5. Verify that $\mathbf{A}^{-1}\mathbf{A} = I$. (I knew I could find $\mathbf{A}^{-1}$ because $\det(\mathbf{A}) = 1 \cdot 3 \cdot 7 \cdot 9 \cdot 3$, a product of invertible elements of $\mathbb{Z}_{26}$.)

Encode, by multiplying the 5-tuples by $\mathbf{A}$, to get

$$\begin{bmatrix} 1 \\ 10 \\ 18 \\ 17 \\ 13 \end{bmatrix} \quad \begin{bmatrix} 11 \\ 16 \\ 11 \\ 21 \\ 4 \end{bmatrix} \quad \begin{bmatrix} 4 \\ 11 \\ 11 \\ 21 \\ 2 \end{bmatrix} \tag{6}$$

or AJRQM KPKUD DKKUB.

Decode, by multiplying the received 5-tuples (6) by $\mathbf{A}^{-1}$.

Twisted codes

If even more security is desired, add a twist: Use a matrix $\mathbf{A}$ which has entries which are functions of a variable t, but whose determinant is an invertible constant in $\mathbb{Z}_{26}$, so that the inverse of $\mathbf{A}$ can be found for any t. Then vary t by some rule, such as $t = n$ if the nth "tuple" is being coded (or decoded).

For example, in the 2×2 case, we could use

$$\mathbf{A} = \begin{pmatrix} 5 & -3 + t \\ 15 & 2 + 3t \end{pmatrix}$$

with inverse

$$\mathbf{A}^{-1} = \begin{pmatrix} 18 + t & 1 - 9t \\ -5 & -7 \end{pmatrix}.$$

Or suppose there were some q so that $t^q = t$ for all t in $\mathbb{Z}_{26}$. Then we could use a matrix like

$$\mathbf{A} = \begin{pmatrix} 2 + 3t & 9 + 2t^{q-1} \\ 3 - t & 2 + 3t^{q-1} \end{pmatrix}$$

whose determinant is 3 for any t in $\mathbb{Z}_{26}$ and so whose inverse can be found.

E6. Note that if $a^q = a$ in $\mathbb{Z}_{26}$ it does *not* necessarily follow that $a^{q-1} = 1$. Example?

†**E7.** Is there some $q > 1$ such that $t^q = t$ for all t in $\mathbb{Z}_{26}$? If so, what is the smallest q which works?

Another possible strategy is to use nonsquare matrices. If $\mathbf{A}$ is an $m \times n$ matrix, $m > n$, sometimes there is an $n \times m$ matrix $\mathbf{B}$ with $\mathbf{BA} = \mathbf{I}$. For example:

$$\begin{pmatrix} -2 & -2 & 3 \\ 3 & -10 & 9 \end{pmatrix} \begin{bmatrix} 1 & 3 \\ 3 & 0 \\ 3 & 2 \end{bmatrix} = \begin{pmatrix} 1 & 0 \\ 0 & 1 \end{pmatrix}.$$

So we could encode ATTACKXATXDAWN by multiplying the 2-tuples

$$\begin{pmatrix} 1 \\ -6 \end{pmatrix} \begin{pmatrix} -6 \\ 1 \end{pmatrix} \begin{pmatrix} 3 \\ 11 \end{pmatrix},$$

etc., by

$$\mathbf{A} = \begin{bmatrix} 1 & 3 \\ 3 & 0 \\ 3 & 2 \end{bmatrix},$$

to get

$$\begin{bmatrix} 9 \\ 3 \\ -9 \end{bmatrix} \begin{bmatrix} -3 \\ -18 \\ -16 \end{bmatrix} \begin{bmatrix} 36 \\ 9 \\ 31 \end{bmatrix},$$

etc., or ICQ WHJ JIE, etc.

If we did this we would be sending redundant information, and that might be useful, for example, if the radio signal being used to send the messages were full of static. We shall consider that problem later!

E8. Complete the encoding of ATTACKXATXDAWN. Check your answer by decoding the result.

There is nothing special about $\mathbb{Z}_{26}$ in all we have done. We might prefer to add three symbols to our alphabet and formulate all our secret codes in $\mathbb{Z}_{29}$, a field. Then we could use (1) to translate from letters to elements of $\mathbb{Z}_{29}$, and let 27, 28, 29 denote ".", "?" and "-" (space). So "attack at dawn" is set up as ATTACK-AT-DAWN.-, and numerically as

1, 20, 20, 1, 3, 11, 29, 1, 20, 29, 4, 1, 23, 14, 27, 29.

E9. Do Example 2 (2×2 case) with the same matrix $\mathbf{A}$, except think of $\mathbf{A}$ as having entries in $\mathbb{Z}_{29}$, and illustrate the example by encoding and decoding the message ATTACK-AT-DAWN.-. Note: $\mathbf{A}^{-1}$ will be different.

11 Fermat's Theorem, I: Abelian Groups

A. Fermat's Theorem

A famous and important theorem of number theory due to Fermat (c. 1640) gives, among other uses, a different way to find the inverse of a nonzero element of $\mathbb{Z}_p$, p prime. We have already seen how to find the inverse of $[a]_p$ by solving $ax + py = 1$, using Euclid's algorithm and Bezout's identity.

Fermat's theorem. *If p is a prime and a is an integer not divisible by p, then $a^{p-1} \equiv 1 \pmod{p}$.*

Corollary. *In $\mathbb{Z}_p$, $[a]_p^{-1} = [a^{p-2}]_p$.* □

As an example, notice that the corollary is true in $\mathbb{Z}_7$. In that case, $p - 2 = 5$:

$1^5 = 1$,	and $1 \cdot 1 \equiv 1 \pmod 7$
$2^5 = 32 \equiv 4 \pmod 7$,	and $2 \cdot 4 \equiv 1 \pmod 7$
$3^5 = 243 \equiv 5 \pmod 7$,	and $5 \cdot 3 \equiv 1 \pmod 7$
$4^5 = (-3)^5 \equiv -5 \equiv 2 \pmod 7$,	and $2 \cdot 4 \equiv 1 \pmod 7$
$5^5 = (-2)^5 \equiv -4 \equiv 3 \pmod 7$,	and $5 \cdot 3 \equiv 1 \pmod 7$
$6^5 = (-1)^5 \equiv -1 \equiv 6 \pmod 7$,	and $6 \cdot 6 \equiv -1 \cdot -1 = 1 \pmod 7$.

E1. Verify the corollary in $\mathbb{Z}_{11}$.

The following proof of Fermat's theorem goes back at least to 1803.

PROOF. Let a be a number relatively prime to p. We want to show that $[a^{p-1}]_p = [1]_p$. Let $\mathbb{Z}_p^*$ be the set $\{[1]_p, [2]_p, \ldots, [p-1]_p\}$ consisting of all the invertible elements of $\mathbb{Z}_p$. Consider the set

$$V = \{[a]_p, [2a]_p, [3a]_p, \ldots, [(p-1)a]_p\}.$$

We observe that V is the same as the set $\mathbb{Z}_p^*$. For since a and p are relatively prime, there is some integer b with $[b]_p[a]_p = [1]_p$. If $[m]_p$ is in $\mathbb{Z}_p^*$, $[m]_p = [(mb)a]_p$, which is in V. Thus the set $\mathbb{Z}_p^*$ is a subset of the set V. But V has at most $p-1$ distinct elements in it, while $\mathbb{Z}_p^*$ has exactly $p-1$ elements which are all distinct. So $\mathbb{Z}_p^*$ must be all of V.

It follows that when we take the product of all the elements of $\mathbb{Z}_p^*$ it is the same as the product of all the elements of V. Thus

$$[1 \cdot 2 \cdot 3 \cdots p-1]_p = [a \cdot 2a \cdot 3a \cdots (p-1)a]_p.$$

(The right side is the product of the elements of V.) There are $p-1$ factors of a in the right side. Collecting them together, we get

$$[(p-1)!]_p = [a^{p-1} \cdot (p-1)!]_p = [a^{p-1}]_p \cdot [(p-1)!]_p.$$

Now $(p-1)!$ is relatively prime to p (why?) so we can cancel $[(p-1)!]_p$ to get

$$[1]_p = [a^{p-1}]_p$$

so that $a^{p-1} \equiv 1 \pmod{p}$. That completes the proof. $\qquad\square$

We shall give an entirely different proof of Fermat's theorem in Chapter II-10.

E2. Prove that if a and b are relatively prime integers, and a and c are relatively prime, then a and bc are relatively prime.

E3. Show that if p is prime, then p is relatively prime to $(p-1)!$.

E4. Show that n^5 and n have the same last digit (in base 10).

E5. Show: If 7 does not divide n then 7 divides $n^{12} - 1$.

E6. Show $n^{13} - n$ is divisible by 2, 3, 5, 7 and 13 for all n.

E7. Show that $n^5/5 + n^3/3 + 7n/15$ is an integer for any n.

E8. Prove that for any integer n, $n^9 + 2n^7 + 3n^3 + 2n$ is divisible by 5.

B. Abelian Groups

The proof we gave of Fermat's theorem in Section A depends on the fact that the set $\mathbb{Z}_p^*$ of nonzero elements of $\mathbb{Z}_p$ forms an abelian group, which we now define.

Definition. A *group* G is a set $\{a, b, c, \ldots, e\}$ together with an operation $*$ (so that for any a, b in G, $a * b$ is an element of G), which satisfies the following three axioms:

(i) $(a * b) * c = a * (b * c)$ for all a, b, c in G (*associativity*);

(ii) there is an element e in G (the *identity*) such that for all a in G, $a * e = e * a = a$;

(iii) for each a in G, there is some b in G (the *inverse* of a) such that $a * b = b * a = e$.

The group G is *abelian* if

(iv) for all a, b in G, $a * b = b * a$ (*commutativity*).

E9. Prove that in a group G, the cancellation law holds: if $a * b = a * c$, then $b = c$.

Here are some examples of groups:

(1) any ring R, where the operation $*$ is $+$ and $e = 0$;
(2) the set of all elements except 0 in a field F, where the operation $*$ is $\cdot$ and $e = 1$;
(3) all $[r]_m$ in $\mathbb{Z}_m$ with r relatively prime to m, where the operation $*$ is $\cdot$, and $e = [1]_m$;
(4) all real numbers > 0, where the operation $*$ is $\cdot$ and $e = 1$;
(5) the real numbers 1 and -1, where the operation $*$ is $\cdot$ and $e = 1$;
(6) all complex numbers on the unit circle in the complex plane, where the operation $*$ is $\cdot$ and $e = 1$;
(7) any vector space V, where the operation $*$ is addition of vectors and e is the zero vector;
(8) the set of all invertible $n \times n$ real matrices, where the operation $*$ is multiplication and $e = \mathbf{I}_n$, the $n \times n$ identity matrix;
(9) the set of all $n \times n$ real matrices with determinant $= 1$, with the operation $*$ being multiplication and $e = \mathbf{I}_n$.

All but the last two examples are abelian groups.

Fermat's theorem as we have proved it is really a theorem about abelian groups. We show that by sketching an abstraction of our proof of Fermat's theorem. Here it is:

Abstract Fermat Theorem. *If an abelian group G consists of n elements, then for any a in G, $a * a * a * \cdots * a = e$ (where a appears n times on the left side).*

SKETCH OF PROOF. Let $a_1, a_2, \ldots, a_n$ be the elements of G. Then $a * a_1$, $a * a_2, \ldots, a * a_n$ are n elements of G which are all distinct, for if $a * a_i = a * a_j$, then $a_i = a_j$ by cancellation (Exercise E9). Thus the set $\{a_1, a_2, \ldots, a_n\}$ is the same as the set $\{a * a_1, a * a_2, \ldots, a * a_n\}$.

To complete the proof along the lines of Fermat's theorem we need two consequences of the axioms:

(i') *Generalized associativity.* If A is a group, so that $a(bc) = (ab)c$ for any a, b, c in A, then for any $n \geqslant 3$ all possible ways of associating the product of any n elements of A are equal.

For example, with $n = 5$,

$$(a(bc))(de) = ((ab)(cd))e = (a(b(c(de)))) = \cdots .$$

This consequence means that when we see a product $abcde$ we are free to associate it in any way. The resulting product will not depend on how we did it. For this reason we can without confusion omit parentheses entirely.

(iv') *Generalized commutativity.* If A is an abelian group, so that $ab = ba$ for any a, b in A, then for any $n \geqslant 2$ all possible ways of multiplying n elements of A, regardless of order, give the same result.

For example, $abcde = ebcda = acedb = \cdots$ (where we have omitted parentheses by assuming generalized associativity).

The proofs of both (i') and (iv') can be done by induction: see Exercises E11 and E12.

These two facts permit the manipulations in the remainder of the proof of the abstract Fermat theorem.

Since the set $\{a_1, \ldots, a_n\}$ is the same as the set $\{a * a_1, \ldots, a * a_n\}$, the products of all the elements in the two sets are the same (by generalized commutativity):

$$a_1 * a_2 * a_3 * \cdots * a_n = a * a_1 * a * a_2 * \cdots * a * a_n$$

(where we are allowed to ignore parentheses by generalized associativity). We rearrange the right side:

$$(a * a * \cdots * a) * a_1 * a_2 * \cdots * a_n$$

(where we are allowed to rearrange by commutativity). By cancellation

$$e = a * a * \cdots * a \qquad (n \text{ times}),$$

which was to be proved. $\qquad\qquad\qquad\qquad\qquad\qquad\qquad\qquad\square$

Corollary. *Fermat's theorem.*

Apply the abstract theorem to Example (2), with $F = \mathbb{Z}_p$, or to Example (3) with $m = p$, a prime. $\qquad\qquad\qquad\qquad\qquad\qquad\qquad\qquad\square$

For a different proof of the abstract Fermat theorem not depending on generalized associativity and commutativity, see Section G.

E10. Write down all possible ways of associating the product $abcd$, where a, b, c, d are elements of a group A, and show, using the associative law, that the products are all equal.

E11. Assuming generalized associativity, prove generalized commutativity in an abelian group by induction on n.

E12. Prove generalized associativity for $a_1, a_2, \ldots, a_n$ by showing that any product u of $a_1, \ldots, a_n$, associated in any way, equals

$$a_1(a_2(a_3(a_4(\ldots (a_{n-1}a_n) \ldots))))$$

by induction on the number of parentheses to the left of a_1 in the product u, and assuming that the result is true for any product of $n - 1$ elements.

C. Euler's Theorem

Euler's theorem is a generalization of Fermat's theorem obtained by applying the abstract theorem to Example (3) for any m:

Euler's theorem. *Let G be the set of units of $\mathbb{Z}_m$, that is, the set of congruence classes in $\mathbb{Z}_m$ represented by natural numbers m which are relatively prime to m. If $\phi(m)$ is the number of elements in G, then $[a^{\phi(m)}] = [1]$.*

E13. Prove that G is closed under multiplication.

In congruence notation Euler's theorem reads:

If $(a, m) = 1$, then $a^{\phi(m)} \equiv 1 \pmod{m}$.

EXAMPLE. In $\mathbb{Z}_{26}$, G consists of $[a]$ where $a = 1, 3, 5, 7, 9, 11, 15, 17, 19, 21, 23,$ and 25. Thus $\phi(26) = 12$. So $3^{12} \equiv 1 \pmod{26}$, $11^{12} \equiv 1 \pmod{26}$, etc.

Of course, to apply Euler's theorem one must know $\phi(m)$, the number of natural numbers $< m$ which are relatively prime to m. The function ϕ is called *Euler's phi function*. If m can be factored, then $\phi(m)$ can be computed by use of the following facts, which you can try to prove:

*E14. Prove that

(a) $\phi(p) = p - 1$ if p is prime,
(b) $\phi(p^n) = p^n - p^{n-1}$ for p a prime,
(c) if a and b are relatively prime, $\phi(ab) = \phi(a)\phi(b)$. What then is $\phi(8772)$?

Euler's theorem gives an alternate way to find the inverse of an invertible element of $\mathbb{Z}_m$. If $[a]$ is invertible in $\mathbb{Z}_m$, that is, $(a, m) = 1$, then $[a]^{-1} = [a^{\phi(m)-1}]$.

E15. Verify that in $\mathbb{Z}_{26}$, $[3]^{12} = 1$. Find the remainder r upon dividing 3^{11} by 26 and verify that $[3r] = 1$ in $\mathbb{Z}_{26}$. Do the same for 11^{11}.

†E16. If p is a prime and $p \neq 2, 5$ show p divides one of the numbers in the set $\{1; 11; 111; 1111; 11111; 111111; \ldots \}$.

D. Finding High Powers mod *m*

The use of Fermat's and Euler's theorems is facilitated by the following base 2 trick. We will demonstrate it by looking at the problem of finding n^{100} (mod *m*).

Begin by writing 100, the exponent, in base 2: $100 = 64 + 32 + 4 = (1100100)_2$. This is a seven-digit base 2 number, so write down $7 - 1 = 6$ copies of *S*:

$$S\ S\ S\ S\ S\ S.$$

Think of the six copies of *S* as separators of the digits of (1100100). Insert *X* wherever the digit is 1, leave blank (or put a dot) when the digit is 0. We get

$$XSXS.S.SXS.S.\quad \text{or}\quad XSXSSSXSS.$$

Now, starting with the number 1, view *X* and *S*, from left to right, as the following operations: *X* means multiply by *n* and reduce mod *m*; *S* means square and reduce mod *m*.

If we do not reduce mod *m* at each step, we get:

$$1 \xrightarrow{X} n \xrightarrow{S} n^2 \xrightarrow{X} n^3 \xrightarrow{S} n^6 \xrightarrow{S} n^{12} \xrightarrow{S} n^{24} \xrightarrow{X} n^{25} \xrightarrow{S} n^{50} \xrightarrow{S} n^{100}.$$

If we do reduce mod *m* at each step, then after each step we get a number *m* which is congruent mod *m* to the corresponding power.

For example, we do 12^{100} (mod 34), $n = 12$, $m = 34$.

Power	Operation	Number 12 to that power, mod 34
0		1
	X	
1		12
	S	
2		$144 \equiv 8$
	X	
3		$96 \equiv -6$
	S	
6		$36 \equiv 2$
	S	
12		4
	S	
24		16
	X	
25		$192 \equiv 22 \equiv -12$
	S	
50		$144 \equiv 8$
	S	
100		$64 \equiv 30$

Thus $12^{100} \equiv 30$ (mod 34).

This process is based on the way one gets 100 in base 2: successively divide 100 by 2, just as in Russian peasant arithmetic (Chapter 2, Exercise E9)—

$$
\begin{aligned}
100 &= 50 \cdot 2, \\
50 &= 25 \cdot 2, \\
25 &= 12 \cdot 2 + 1, \\
12 &= 6 \cdot 2, \\
6 &= 3 \cdot 2, \\
3 &= 1 \cdot 2 + 1, \\
1 &= 0 \cdot 2 + 1.
\end{aligned}
$$

We know that the remainders, reading from bottom to top, are the digits, from left to right, in the representation of 100 in base 2. But each of these equations tells how to get the successive powers of n, starting with n^0: for example, given n^{12}, $n^{25} = n^{12 \cdot 2 + 1} = (n^{12})^2 \cdot n$: to get n^{25}, square n^{12} and then multiply by n.

E17. Find 2^{47} (mod 23). Note: do not forget Fermat's theorem!

E18. Find 2^{47} (mod 30).

E19. What is the last digit of 7^{126} in base 10?

E20. What is the last digit in the base 12 representation of 5^{400}? of 3^{400}?

E21. Use Fermat's theorem to find a number $r < 29$ such that $17r \equiv 1$ (mod 29).

E22. Find a number $r < 41$ such that $12r \equiv 1$ (mod 41).

E. The Order of an Element

We know that in $\mathbb{Z}_{26}$, $[11]^{12} = [1]$ by Euler's theorem. What other powers of $[11]$ can equal $[1]$? The following two results describe the possibilities.

Let G be an abelian group with n elements. For a in G write $a \cdot a \cdots a$ (m times) as a^m; $a^0 = e$, the identity of G. The usual laws of exponents apply. For a in G let $S = \{m \geqslant 1 | a^m = e\}$. Since $a^n = e$, S is nonempty and so has a least element, n_0, called the order of a.

Proposition 1. *If n_0 is the order of a in G, then n_0 divides n, the number of elements of G.*

Since $a^n = e$ by the abstract Fermat theorem, this is a special case of:

Proposition 2. *If n_0 is the order of a in G and $a^m = e$, then n_0 divides m.*

PROOF OF PROPOSITION 2. Suppose $a^m = e$. Let $m = n_0 q + r$ with $0 \leqslant r < n_0$. If $r = 0$ then n_0 divides m. So suppose $r > 0$. Then $a^m = a^{n_0 q + r} = (a^{n_0})^q a^r$.

But $a^m = a^{n_0} = e$, so $a^r = e$. Since $1 \leqslant r < n_0$ and n_0 was the least number such that $a^{n_0} = e$, we have a contradiction. Thus $r = 0$, completing the proof. □

E23. What is the order of 11 in $\mathbb{Z}_{26}$? Given Proposition 2, which powers of 11 need to be checked? How about the order of 11 in $\mathbb{Z}_{23}$? the order of 11 in $\mathbb{Z}_{19}$?

†**E24.** What are the possible orders of nonzero elements of $\mathbb{Z}_{13}$? Find the order of each element of $\mathbb{Z}_{13}$. Do the same for $\mathbb{Z}_{11}$, $\mathbb{Z}_{17}$. Find the orders of the invertible elements of $\mathbb{Z}_{20}$, $\mathbb{Z}_{15}$.

E25. If G is a group, α is in G, and α has order d, show that the order of α^r is $d/(r, d)$.

†**E26.** Find an element α of $\mathbb{Z}_{23}$ such that every nonzero element of $\mathbb{Z}_{23}$ is a power of α.

F. About Finite Fields

Here is one more application of our abstract version of Fermat's theorem.

Proposition 3. *If F is a field with n elements, and $a \neq 0$ is any element of F, then $a^{n-1} = 1$.*

PROOF. The nonzero elements of F form an abelian group of order $n - 1$ under multiplication. This is Example (2). □

To show that there really exist fields with a finite number of elements besides $\mathbb{Z}_p$, so that Proposition 3 is really more general than the original Fermat's theorem, we examine $GF(9)$, a field containing 9 elements which appeared in Chapter 8, Exercise E19. Note that $\mathbb{Z}_9$ is not a field.

The set $GF(9)$ is constructed from $\mathbb{Z}_3$ in exactly the same way that the complex numbers are constructed from the real numbers.

In $\mathbb{Z}_3$, just as in $\mathbb{R}$, we cannot solve the equation $x^2 = -1$. So we let i be a "formal" solution, $i^2 = -1$, and let $GF(9)$ be the set of numbers $a + bi$ where a and b are in $\mathbb{Z}_3$. Denote the elements of $\mathbb{Z}_3$ by $[0]_3 = 0$, $[1]_3 = 1$, and $[-1]_3 = [2]_3 = 2$. Then $GF(9)$ consists of the set of all linear combinations of 1 and i with coefficients in $\mathbb{Z}_3$, namely:

$$0 = 0 \cdot 1 + 0 \cdot i \qquad i = 0 \cdot 1 + 1 \cdot i$$
$$1 = 1 \cdot 1 + 0 \cdot i \qquad 1 + i = 1 \cdot 1 + 1 \cdot i$$
$$2 = 2 \cdot 1 + 0 \cdot i \qquad 2 + i = 2 \cdot 1 + 1 \cdot i$$
$$2i = 0 \cdot 1 + 2 \cdot i$$
$$1 + 2i = 1 \cdot 1 + 2 \cdot i$$
$$2 + 2i = 2 \cdot 1 + 2 \cdot i$$

Multiplication of these elements is defined by the distributive law and the condition that $i \cdot i = [-1]_3 = 2$. For example, $(2 + i)(1 + i) = 2 + 2i + i + i^2 = 2 + 2i + i + 2 = 1$.

One way of thinking about $GF(9)$ is that it is the set of congruence classes mod 3 of the set of complex numbers of the form $a + bi$, where a and b are integers.

We will show in a later chapter that $GF(9)$ is a field. In it, Proposition 3 holds. For example, in $GF(9)$, $(1 + 2i)^8 = 1$. In fact, the order of $(1 + 2i)$ is 8, and every nonzero element of $GF(9)$ is a power of $1 + 2i$:

$$(1 + 2i)^1 = 1 + 2i,$$

$$(1 + 2i)^2 = i,$$

$$(1 + 2i)^3 = 1 + i,$$

$$(1 + 2i)^4 = 2,$$

$$(1 + 2i)^5 = 2 + i,$$

$$(1 + 2i)^6 = 2i,$$

$$(1 + 2i)^7 = 1 + 2i,$$

$$(1 + 2i)^8 = 1.$$

It follows easily that if x is any nonzero element of $GF(9)$, then $x^8 = 1$. For if $x = (1 + 2i)^r$, then $x^8 = (1 + 2i)^{8r} = 1^r = 1$.

We shall discover many more examples of finite fields like $GF(9)$ in Part III.

E27. What are the orders of the nonzero elements of $GF(9)$?

E28. Given $n > 1$, what are the elements of order n in $\mathbb{R}$? in $\mathbb{C}$?

E29. Prove that if F is a field with m elements and α is in F with $\alpha^r = 1$, then $\alpha^d = 1$ with $d = (r, m - 1)$.

G. Nonabelian Groups

The abstract version of Fermat's theorem generalizes to nonabelian groups. The proof we gave in Section B does not work in this case, however, so we present here another one.

Theorem. *If G is a (not necessarily abelian) group with n elements, and if g is in G, then $g^n = 1$.*

To do the proof we need to develop for groups a notion analogous to congruence classes.

Definition. Let g be an element of G. Say x, y in G are *right congruent* (mod g) if $x = yg^r$ for some integer r. Write $x \equiv y$ if x and y are right congruent (mod g).

Proposition 1. *For any x, y, z in G: $x \equiv x$ (reflexivity); if $x \equiv y$ then $y \equiv x$ (symmetry); if $x \equiv y$ and $y \equiv z$ then $x \equiv z$ (transitivity).*

E30. Prove Proposition 1.

Recall that congruence (mod m) of natural numbers satisfied the properties listed in proposition 1.

Definition. A relation $\equiv$ which is reflexive, symmetric and transitive is called an *equivalence relation*.

Fact. *If $\equiv$ is an equivalence relation on a set S, then $\equiv$ divides up S into a collection of mutually disjoint subsets as follows: Define x, y to be in the same subset iff $x \equiv y$. The collection of subsets is denoted $S/\equiv$.*

Definition. Let $\langle g \rangle$ denote the set of all positive and negative powers of g. Then the collection of disjoint subsets of G under right congruence mod g is called the set of *left cosets* of $\langle g \rangle$. The left coset containing x is denoted $x\langle g \rangle$. Then $x\langle g \rangle = y\langle g \rangle$ iff $x = yg^r$ for some integer r iff $y^{-1}x$ is in $\langle g \rangle$.

Proposition 2. *The function ϕ_x which takes g^r in $\langle g \rangle$ to xg^r in $x\langle g \rangle$ is a 1-1 correspondence. Thus any two left cosets have the same number of elements.*

E31. Show that ϕ_x is 1-1 and maps onto $x\langle g \rangle$.

Proposition 3. *If g has order r in G and $\langle g \rangle$ has s cosets, then G has rs elements.*

PROOF. If g has order r, then $\langle g \rangle$ has r elements. So $x\langle g \rangle$ has r elements for each x. Since each element of G is in exactly one coset, if there are r elements in each coset and s cosets, there are rs elements in G. □

PROOF OF THEOREM. The theorem follows, for if G has n elements and g has order r, then $n = rs$ for some s, so $g^n = g^{rs} = e$. □

E32. If $G = \mathbb{Z}$ with $* =$ addition, and $g = m$, what is $\langle g \rangle$? What are the left cosets of $\langle g \rangle$?

E33. Let $GL_2(\mathbb{Z}_3)$ be the set of invertible 2×2 matrices with entries in $\mathbb{Z}_3$, a group under multiplication. Write down all the elements of $GL_2(\mathbb{Z}_3)$ and verify that the order of each element divides the number of elements of the group.

E34. Prove the fact about equivalence relations cited above. (Compare Chapter 7, the definition of congruence class, and Chapter 1, Exercise E6.)

E35. (a) Write down the multiplication table for the group $GL_2(\mathbb{Z}_2)$ of invertible 2×2 matrices with entries in $\mathbb{Z}_2$. Verify that the order of each element divides the order of the group.

(b) The group $GL_2(\mathbb{Z}_2)$ may be thought of as permutations of the vertices of a triangle. Think of 0, 1 as being the elements of $\mathbb{Z}_2$. Consider the set $T = \left\{ \begin{pmatrix} 0 \\ 1 \end{pmatrix}, \begin{pmatrix} 1 \\ 0 \end{pmatrix}, \begin{pmatrix} 1 \\ 1 \end{pmatrix} \right\}$, which may be thought of as labels for the vertices of a triangle. The six invertible 2×2 matrices with entries in $\mathbb{Z}_2$ give six ways of permuting the elements of T, as follows: If $\mathbf{A}$ is an invertible 2×2 matrix, then $\begin{pmatrix} 0 \\ 1 \end{pmatrix} \rightarrow \mathbf{A} \begin{pmatrix} 0 \\ 1 \end{pmatrix}, \begin{pmatrix} 1 \\ 0 \end{pmatrix} \rightarrow \mathbf{A} \begin{pmatrix} 1 \\ 0 \end{pmatrix}, \begin{pmatrix} 1 \\ 1 \end{pmatrix} \rightarrow \mathbf{A} \begin{pmatrix} 1 \\ 1 \end{pmatrix}$ is a permutation of T. Show that each permutation of T arises from some matrix $\mathbf{A}$.

Repeating Decimals, I   12

In Chapter 5 we introduced base a expansions of fractions. It turns out, as we shall see in this chapter, that Euler's theorem gives information about those expansions.

We begin by solving Exercises E15 and E16 of Chapter 5.

Theorem 1. *Any rational number has an eventually repeating base a expansion, for any base a.*

EXAMPLES. Let $a = 10$. Then $1/7 = .1428571428\ldots$ is repeating with period of length 6; $1/12 = .08333\ldots$ is repeating starting with the third digit with a period of length 1; $3/14 = .21428571428\ldots$ is repeating starting with the second digit with a period of length 6; $1/5 = .2000\ldots$ is repeating starting with the second digit with a period of length 1. We shall denote these eventually repeating decimal expansions by

$$1/7 = (.\,\overline{142857}\,)_{10},$$

$$1/12 = (.08\overline{3})_{10},$$

$$3/14 = (.2\,\overline{142857}\,)_{10},$$

$$1/5 = (.2\overline{0})_{10}.$$

We shall say that $1/7$ is *strictly repeating* since it has the form $(.a_1 \ldots a_d)$.

Define the *period* of u/t in base a to be the smallest $d \geqslant 1$ such that $u/t = (.r_1 r_2 \ldots r_m r_{m+1} \ldots r_{m+d})_a$ for some m.

PROOF OF THEOREM 1. Let the fraction be r/s, which we may as well assume is reduced to lowest terms, so that $(r, s) = 1$. To get the base a

expansion we successively divide by s as follows:

$$r = sq_0 + r_0 \quad \text{with} \quad r_0 < s,$$
$$ar_0 = sq_1 + r_1, r_1 < s,$$
$$ar_1 = sq_2 + r_2, r_2 < s,$$

$$\vdots$$

Then

$$\frac{r}{s} = q_0 + \frac{q_1}{a} + \frac{q_2}{a_2} + \frac{q_3}{a_3} + \cdots .$$

Consider the sequence of digits

$$\{ r_0, r_1, r_2, \ldots, r_{s-1}, r_s, \ldots \}.$$

They are all numbers $\geqslant 0$ and $< s$, so at least two of the numbers in the set $\{ r_0, r_1, \ldots, r_s \}$ must be equal. Suppose $r_m = r_{m+d}$, where $d > 0$, $0 \leqslant m < m + d \leqslant s$. Then dividing ar_{m+d} by s yields the same quotient and remainder as dividing ar_m by s. Thus $q_{m+1} = q_{m+d+1}$ and $r_{m+1} = r_{m+d+1}$. The same argument then shows that $q_{m+2} = q_{m+d+2}$ and $r_{m+2} = r_{m+d+2}$, etc. Thus $r/s = (q_0.q_1q_2 \cdots q_m\overline{q_{m+1} \cdots q_{m+d}})_a$, as was to be shown.   □

Note that the period $d \leqslant s$.

E1. Show that in fact, $d \leqslant s - 1$. *Hint:* What if some $r_i = 0$?

In any base a, r/s has an eventually repeating expansion of period $\leqslant s - 1$. Can we be more precise about the possibilities for the period? It turns out that Euler's theorem and Proposition 1 of the last section relate to this question. In order to simplify the explanation, we first note the following.

E2. Let r/s be a fraction. Show that given any $a \geqslant 2$ there is some $n \geqslant 0$ such that $a^n r/s$, when reduced to lowest terms $= u/t$ with $(u, t) = 1$ and with $(t, a) = 1$. (Example: with $a = 10$, $r/s = 3/14$, $10 \cdot 3/14 = 15/7$, where $(7, 10) = 1$.)

Given exercise E2, any fraction r/s may be written as $r/s = (1/a^n)(u/t)$ with $(u, t) = 1$ and with $(t, a) = 1$. We shall show in our next result that u/t has a strictly repeating base a expansion; since $r/s = (1/a^n)(u/t)$ it is clear that the periods of the fractions r/s and u/t are equal.

Theorem 2. *Let $a \geqslant 2$ be a base and let u/t be a fraction with $u < t$, reduced to lowest terms (i.e., $(u, t) = 1$), such that t and a are relatively prime. Then u/t has a strictly repeating base a expansion, and the period is the smallest integer $d > 0$ such that $a^d \equiv 1 \pmod{t}$. Hence that length divides $\phi(t)$, the number of natural numbers $< t$ which are relatively prime to t.*

We have called the smallest integer $d > 0$ so that $a^d \equiv 1 \pmod{t}$ the *order* of a mod t.

Notice that in the theorem the numerator is irrelevant, provided only that it is relatively prime to the denominator.

For examples, $[10]_7$ has order 6 in $\mathbb{Z}_7$, so $1/7$ has a repeating decimal expansion of period 6. So does $2/7, 3/7, 4/7, 5/7, 6/7$.

$[10]_{101}$ has order 4 in $\mathbb{Z}_{101}$ so $13/101$ has a repeating decimal expansion of period 4. In fact, $13/101 = .\overline{1287})_{10}$.

PROOF OF THEOREM 2.   The proof involves geometric series. Given u/t reduced to lowest terms and with $(a, t) = 1$, suppose d is some number > 0 such that $a^d \equiv 1 \pmod{t}$. We show u/t repeats every d digits, as follows. If $a^d \equiv 1 \pmod{t}$, $a^d - 1 = tw$ for some w. Then

$$\frac{u}{t} = \frac{uw}{tw} = \frac{uw}{a^d - 1}$$

with $uw < a^d - 1$ (since we assumed $u < t$).

Now

$$\frac{1}{a^d - 1} = \frac{1}{a^d(1 - (1/a^d))} = \frac{1}{a^d}\left(\frac{1}{1 - (1/a^d)}\right).$$

The factor in parentheses has a geometric series expansion, valid for $a^d > 1$:

$$\frac{1}{1 - (1/a^d)} = 1 + \frac{1}{a^d} + \frac{1}{a^{2d}} + \frac{1}{a^{3d}} + \cdots = \sum_{i=0}^{\infty} \frac{1}{a^{di}}.$$

Thus

$$\frac{uw}{a^d - 1} = \frac{uw}{a^d}\left(\sum_{i=0}^{\infty} \frac{1}{a^{di}}\right).$$

Now uw has a base a representation as

$$uw = r_0 + r_1 a + r_2 a^2 + \cdots + r_{d-1} a^{d-1}$$

(since $uw < a^d$), with $0 \leqslant r_i < a$. Thus

$$\frac{u}{t} = \frac{uw}{a^d - 1} = \sum_{i=0}^{\infty} \left(\frac{r_{d-1}a^{d-1} + r_{d-2}a^{d-2} + \cdots + r_2 a^2 + r_1 a + r_0}{a^d}\right)\frac{1}{a^{di}}$$

$$= \left(\frac{r_{d-1}}{a} + \frac{r_{d-2}}{a^2} + \cdots + \frac{r_1}{a^{d-1}} + \frac{r_0}{a^d}\right)$$

$$+ \left(\frac{r_{d-1}}{a \cdot a^d} + \frac{r_{d-2}}{a^2 \cdot a^d} + \cdots + \frac{r_1}{a^{d-1} \cdot a^d} + \frac{r_0}{a^d \cdot a^d}\right)$$

$$+ \left(\frac{r_{d-1}}{a \cdot a^{2d}} + \cdots\right) + \cdots$$

$$= \left(.\overline{r_{d-1} r_{d-2} \cdots r_1 r_0}\right)_a.$$

Thus u/t is strictly repeating every d digits.

Now suppose u/t in base a is strictly repeating every e digits for some e. That is, suppose

$$\frac{u}{t} = \left(\overline{.r_{e-1}r_{e-2} \cdots r_1 r_0} \right)_a.$$

Then, setting $w = r_{e-1}a^{e-1} + r_{e-2}a^{e-2} + \cdots + r_1 a + r_0$,

$$\frac{u}{t} = w \left[\frac{1}{a^e} + \frac{1}{a^{2e}} + \cdots \right] = \frac{w}{a^e - 1}.$$

Since $(u, t) = 1$ and $tw = u(a^e - 1)$, t divides $a^e - 1$, so $a^e \equiv 1 \pmod{t}$.

It follows that the set of numbers $d \geqslant 1$ such that $a^d \equiv 1 \pmod{t}$ is the same as the set of numbers $e \geqslant 1$ such that u/t repeats in base a every e digits. So the least d, namely the order of $a \bmod t$, is the least e, namely the period of u/t in base a.

Finally, since $a^{\phi(t)} \equiv 1 \pmod{t}$, the order of $a \bmod t$ divides $\phi(t)$ by Proposition 1, Chapter I-11E, so the period of u/t in base a divides $\phi(t)$. That completes the proof.   □

We will complete our discussion on the possible values of d in Chapter III-2.

E3. Show that in base a, $1/5$ has a period of length 1, 2, or 4.

E4. Expand $1/9$ in base 8.

E5. If $14/77 = (.\overline{b_1 \ldots b_m})_8$, what is m?

E6. What is the length of $13/80$ in base 10? in base 8?

E7. In base a, $3/11$ can have a period of length 1, 2, 5, or 10. Find bases such that each of these possibilities actually occurs.

E8. What are the possibilities for the length of the period of $1/23$ in base a? What is the length of $1/23$ in base 2?

†***E9.** Find a fraction whose decimal expansion has a period of at least 60.

Error-correcting Codes, I ## 13

Error-correcting codes is an application of $\mathbb{Z}_2$ and other finite fields which was discovered only around 1948. Our exposition will assume some acquaintance with matrices and vectors. For a review of notation see Chapter 9A.

The problem is the following. Suppose a message consisting of blocks of digits, or *words*, is to be transmitted through a channel to a receiver. If the channel is "noisy" and tends to introduce random errors into what was sent, i.e., change digits, how can the receiver determine what was sent?

The basic idea for the solution is to send messages with redundant data, that is, messages with digits which are repeated, partially repeated, or presented in a certain special format. The receiver can detect or even correct errors in the digits of the message received, by seeing how what was received varies from the format in which the message was known to be originally sent.

An example. Suppose the message consists of 4-digit decimal numbers. Here is a simple way of checking for an error in one digit, analogous to "casting out 9's." Instead of transmitting "*abcd*," send out "*abcde*," where 9 divides the base 10 number *abcde*. Thus, instead of sending 3856, send 38565. If the receiver receives *ABCDE*, where 9 does not divide the base 10 number *ABCDE*, then the receiver knows that at least one of the digits *A*, *B*, *C*, *D*, *E* is not a digit that was transmitted. On the other hand, if the receiver receives *ABCDE* and 9 divides *ABCDE*, then either *ABCDE* = *abcde*, or a digit 0 was changed to a 9 or vice versa, or there are errors in at least two digits. If the probability is low that an error occurs in any given digit, and if *ABCDE* is divisible by 9, then the probability is high that no errors occurred, and the receiver would have confidence that no errors occurred. (For example, suppose that the probability of an error in any digit is .1. If k, $0 \leqslant k \leqslant 5$, is the number of digits of $ABCDE$ which equal 0 or 9, then the probability of no errors if 9 divides $ABCDE$ turns out to be approximately $(97.5 - k) \cdot 10^{-2}$.)

Here is another example. Suppose we are sending out 1-digits numbers. Suppose we wish to send a. Code as follows. Send out (for example) the 5 digit word $aaaaa$. The receiver would think that the digit A was sent if he receives a word with at least three A's in it, and he would be misled only if at least three of the a's sent had been erroneously changed to A's, where $A \neq a$.

This last code is one in which the receiver not only detects errors, but also is able to determine what was sent, that is, to *correct* the received word, despite the presence of up to two errors. A correcting capability is very desirable for codes in certain situations. One situation is where the receiver is receiving data from a measuring device (such as a space probe), which is transmitting measurements as it moves through the thing being studied (such as Jupiter's magnetic field) and cannot retransmit data which the receiver knows is erroneous. (The channel in this situation would be the space through which the radio waves pass, and the noise would be radio noise, or static.) Another situation would be where the transmitting consists of the placing of data into the memory of a computer—the channel here is the memory, which may contain imperfectly manufactured components, and the receiving is the retrieval of the data.

In the two situations just described, the information sent is numerical, and might naturally be in numbers expressed in base 2. Since all of the mathematics tends to be easiest in base 2 also, we shall henceforth assume that we are sending words written in base 2.

In base 2 here are two examples of codes analogous to those we just looked at in base 10:

EXAMPLE 1. The parity check code. Given n information digits $abcd \ldots e$, let $f \equiv a + b + c + d + \cdots + e \pmod 2$ and send $abcd \ldots ef$. The receiver receives $ABCD \ldots EF$. If $A + B + C + D + \cdots + E \not\equiv F$ (mod 2), then there is an odd number of errors, while if $\equiv$, then there is an even number, none, or two, or $\ldots$. If more than one error is extremely unlikely, the receiver would have confidence that if $A + B + C + D + \cdots + E \equiv F \pmod 2$, then no errors occurred, while if $\not\equiv$, the receiver would be able to detect the presence of an error. The receiver would not be able to tell where the error is. Nonetheless, this is a very efficient code, since $n/(n + 1)$ of each code word is information, and only one digit in each word is redundant.

EXAMPLE 2. The repetition code. Given one information digit a, send the word of length n, $aaa \ldots a$. The receiver receives $ABCD \ldots E$. If the number of 1's among $AB \ldots E$ exceeds the number of 0's, the receiver decides that $a = 1$; otherwise $a = 0$ (unless n is even and the number of 1's = the number of 0's). The receiver will decode incorrectly only if there are at least $n/2$ errors in $AB \ldots E$. This code then corrects up to $n/2$

errors in the sense that if there are less than $n/2$ errors the receiver can determine correctly the transmitted word. This code is, however, quite inefficient, for only $1/n$ of each code word is information and $n - 1$ digits in each word are redundant.

The development of codes has tended to proceed from the assumption that it is desirable to have codes which are both efficient (that is, the ratio of information digits per word to word length is "large") and capable of correcting errors in a small proportion of the digits of each word.

In the rest of this chapter we describe two examples of efficient codes constructed using matrices with entries in $\mathbb{Z}_2$. These codes are examples of codes described by R. W. Hamming of Bell Telephone Laboratories and published in 1950. In Chapter III-9 we shall describe other codes.

Code I. Here is an example of a code which corrects one error in words of length 7, where each word has 4 information bits.

We work with elements of $\mathbb{Z}_2$. We write $[0]_2 = 0$, $[1]_2 = 1$, so $\mathbb{Z}_2 = \{0, 1\}$. Our words are 7-tuples with entries in $\mathbb{Z}_2$.

Let

$$\mathbf{H} = \begin{bmatrix} 1 & 0 & 1 & 0 & 1 & 0 & 1 \\ 0 & 1 & 1 & 0 & 0 & 1 & 1 \\ 0 & 0 & 0 & 1 & 1 & 1 & 1 \end{bmatrix}.$$

(This matrix $\mathbf{H}$ has the property that for r, s, t in $\mathbb{Z}_2$ not all zero,

$$\begin{pmatrix} r \\ s \\ t \end{pmatrix}$$

is the $(tsr)_2$-th column of $\mathbf{H}$. In particular, all columns of $\mathbf{H}$ are different, an important fact.)

Let (a, b, c, d) be a typical word which we wish to transmit, where a, b, c, d are in $\mathbb{Z}_2$. We call (a, b, c, d) the information word. Form the vector

$$\mathbf{C} = \begin{bmatrix} x \\ y \\ a \\ z \\ b \\ c \\ d \end{bmatrix}$$

with x, y, z chosen so that $\mathbf{HC} = \mathbf{0}$, that is, so that (in $\mathbb{Z}_2$)

$$\begin{aligned} x + a + b + d &= 0 \\ y + a + c + d &= 0 \\ z + b + c + d &= 0. \end{aligned} \tag{1}$$

Then x, y, z are uniquely and quickly determined from a, b, c, d using $\mathbf{H}$. Here (x, y, z) is the redundant part of the word, and $\mathbf{C}$, a vector made up of the information word (a, b, c, d) and the redundancy (x, y, z), is the coded word. The word $\mathbf{C}$ is what is transmitted. Suppose the receiver receives

$$\mathbf{R} = \begin{bmatrix} b_1 \\ b_2 \\ b_3 \\ \vdots \\ b_7 \end{bmatrix}.$$

Case 0. Suppose $\mathbf{R} = \mathbf{C}$. Then $\mathbf{HR} = \mathbf{0}$, because $\mathbf{HC} = \mathbf{0}$.

Case 1. Suppose $\mathbf{R}$ differs from $\mathbf{C}$ in at most one entry. Then $\mathbf{R} - \mathbf{C} = \mathbf{E}$ has 1 in some entry and 0 in all other entries. Then $\mathbf{HE}$ is the column of $\mathbf{H}$ corresponding to the location of the 1 in $\mathbf{E}$; when the receiver computes $\mathbf{HR}$, she gets:

$$\mathbf{HR} = \mathbf{HC} + \mathbf{HE} = \mathbf{0} + \mathbf{HE} = \text{(the column of } \mathbf{H} \text{ corresponding to}$$
$$\text{where the 1 is in } \mathbf{E}).$$

Since all columns of $\mathbf{H}$ are distinct, if there is one error, the receiver can determine where the error is by examining $\mathbf{HR}$; once she knows $\mathbf{E}$, she knows $\mathbf{R} - \mathbf{E} = \mathbf{C}$, the word which was transmitted.

Case 2. If $\mathbf{R}$ differs from $\mathbf{C}$ in two or more entries, then $\mathbf{HR} = \mathbf{HC} + \mathbf{HE}$ $= \mathbf{0} + \text{(sum of two or more columns of } \mathbf{H})$. Since the sum of two or more columns of $\mathbf{H}$ is either $\mathbf{0}$ or a column of $\mathbf{H}$, the receiver will decode inaccurately if she assumes no errors or one error occurred.

Thus this code is capable of correcting exactly one error. That is, if the receiver can confidently assume that at most one error occurred in the transmission of a word, then the receiver will be able to confidently determine what was sent. She will be misled iff more than one error occurs in a given word. (If p, the probability of an error in any given digit, is $p = .1$, then the probability of at most one error in a word is $e = (1 - .1)^7 + 7(1 - .1)^6(.1) = .85$; if $p = .01$, $e = .998$.)

To illustrate the decoding of code I, suppose

$$\mathbf{R} = \begin{bmatrix} 1 \\ 0 \\ 1 \\ 1 \\ 0 \\ 1 \\ 0 \end{bmatrix}.$$

Then

$$\mathbf{HR} = \begin{bmatrix} 1 & 0 & 1 & 0 & 1 & 0 & 1 \\ 0 & 1 & 1 & 0 & 0 & 1 & 1 \\ 0 & 0 & 0 & 1 & 1 & 1 & 1 \end{bmatrix} \begin{bmatrix} 1 \\ 0 \\ 1 \\ 1 \\ 0 \\ 1 \\ 1 \end{bmatrix} = \begin{bmatrix} 1 \\ 1 \\ 1 \end{bmatrix},$$

so assuming one error, it must be the last digit:

$$\mathbf{C} = \begin{bmatrix} 1 \\ 0 \\ 1 \\ 1 \\ 0 \\ 1 \\ 0 \end{bmatrix}.$$

$$\text{If } \mathbf{R} = \begin{bmatrix} 1 \\ 1 \\ 1 \\ 0 \\ 0 \\ 1 \\ 0 \end{bmatrix}, \quad \mathbf{HR} = \begin{bmatrix} 0 \\ 1 \\ 1 \end{bmatrix}, \quad \text{so} \quad \mathbf{C} = \begin{bmatrix} 1 \\ 1 \\ 1 \\ 0 \\ 0 \\ 0 \\ 0 \end{bmatrix}.$$

$$\text{If } \mathbf{R} = \begin{bmatrix} 0 \\ 0 \\ 0 \\ 1 \\ 0 \\ 0 \\ 1 \end{bmatrix}, \quad \mathbf{HR} = \begin{bmatrix} 1 \\ 1 \\ 0 \end{bmatrix}, \quad \text{so} \quad \mathbf{C} = \begin{bmatrix} 0 \\ 0 \\ 1 \\ 1 \\ 0 \\ 0 \\ 1 \end{bmatrix}.$$

$$\text{If } \mathbf{R} = \begin{bmatrix} 0 \\ 1 \\ 0 \\ 1 \\ 0 \\ 1 \\ 0 \end{bmatrix}, \quad \mathbf{HR} = \begin{bmatrix} 0 \\ 0 \\ 0 \end{bmatrix}, \quad \text{so} \quad \mathbf{C} = \mathbf{R}.$$

Code II. A modification of Code I will enable the receiver to detect the presence of two errors, as well as to correct one error.

Let

$$\mathbf{H} = \begin{bmatrix} 1 & 1 & 1 & 1 & 1 & 1 & 1 & 1 \\ 0 & 1 & 0 & 1 & 0 & 1 & 0 & 1 \\ 0 & 0 & 1 & 1 & 0 & 0 & 1 & 1 \\ 0 & 0 & 0 & 0 & 1 & 1 & 1 & 1 \end{bmatrix},$$

essentially the matrix of Code I with an additional row of 1's on the top. If (a, b, c, d) is the information word, let

$$\mathbf{C} = \begin{bmatrix} w \\ x \\ y \\ a \\ z \\ b \\ c \\ d \end{bmatrix} \quad \text{with } \mathbf{HC} = \mathbf{0}.$$

Then x, y, z satisfy equations (1) of Code I, and w satisfies

$$0 = w + x + y + z + a + b + c + d.$$

Adding this equation to equations (1) of Code I yields the simpler equation defining w:

$$w + a + b + c = 0.$$

Transmit the vector **C**.

Suppose the receiver receives **R**.

Case 0. **R** = **C**. No errors. Then **HR** = **0**.

Case 1. **R** − **C** = **E** has one nonzero entry, one error. Then **HR** = **HE** since **HC** = **0**, and **HE** is the column of **H** corresponding to where the error occurred. Since all columns of **H** are distinct, the location of the error can be found, and the error can be corrected.

Case 2. **R** − **C** = **E** has two nonzero entries, two errors. Then **HR** = **HE** is the sum of two columns of **H**. It cannot be determined which two columns of **H** make up the sum. For example, here are two sums of columns of **H**:

$$\begin{pmatrix} 1 \\ 1 \\ 1 \\ 0 \end{pmatrix} + \begin{pmatrix} 1 \\ 1 \\ 0 \\ 1 \end{pmatrix} = \begin{pmatrix} 0 \\ 0 \\ 1 \\ 1 \end{pmatrix} = \begin{pmatrix} 1 \\ 0 \\ 1 \\ 0 \end{pmatrix} + \begin{pmatrix} 1 \\ 0 \\ 0 \\ 1 \end{pmatrix}.$$

But what is certain is that **HE** is not a column of **H**, since the sum of two columns of **H** always has top entry = 0, and every column of **H** has top entry = 1.

This, then, is a code which corrects one error and detects two errors in words of length 8 with 4 information digits. The receiver will be misled only if there are 3, 5, or 7 errors.

E1. Here is a collection of received words which were transmitted after being encoded with code II. For each word assume there are 0, 1, or 2 errors. Decode each word.

$$\begin{pmatrix} 1 \\ 0 \\ 1 \\ 1 \\ 0 \\ 1 \\ 1 \\ 1 \end{pmatrix}, \begin{pmatrix} 1 \\ 1 \\ 1 \\ 1 \\ 0 \\ 0 \\ 0 \\ 1 \end{pmatrix}, \begin{pmatrix} 1 \\ 0 \\ 1 \\ 1 \\ 1 \\ 0 \\ 0 \\ 1 \end{pmatrix}, \begin{pmatrix} 0 \\ 1 \\ 0 \\ 1 \\ 1 \\ 0 \\ 1 \\ 1 \end{pmatrix}$$

*E2. What is the maximum allowable probability of error in a typical digit in order that Code II can be used with 99.9% probability that the receiver will not be misled (i.e., 3 or more errors occur) in a single word?

*E3. Define a code, analogous to code II, which uses a 5×16 matrix **H**, and sends out binary words of length 16 (of which 11 are information digits) such that the receiver can correct one error and detect two errors.

E4. In code II, do there exist received words which the receiver can determine with certainty have at least 3 errors?

E5. If C_1 and C_2 are two 8-tuples, let $d(C_1 - C_2)$, the distance between C_1 and C_2, be the number of 1's in $C_1 - C_2$. Can you determine the minimum distance between two coded words in Code II? Do you see any relationship between the minimum distance and the error correcting ability of the code?

14 The Chinese Remainder Theorem

A. The Theorem

This is a famous theorem of number theory, so called because it was known to the ancient Chinese. Its proof is another application of the fact that the greatest common divisor of two numbers can be written as a linear combination of the two numbers.

Chinese remainder theorem. *If m_1, m_2, ..., m_n are pairwise relatively prime natural numbers > 1, and a_1, a_2, ..., a_n are any integers, then there is a solution of the simultaneous congruences*

$$x \equiv a_1 \pmod{m_1},$$
$$x \equiv a_2 \pmod{m_2},$$
$$\vdots$$
$$x \equiv a_n \pmod{m_n}. \tag{1}$$

If x and x' are two solutions, then $x \equiv x' \pmod{M}$, where $M = m_1 m_2 \ldots m_n$.

PROOF. We first solve the special case of (1) where i is some fixed subscript, $a_i = 1$, and $a_1 = a_2 = \cdots = a_{i-1} = a_{i+1} = \cdots = a_n = 0$.

Let $k_i = m_1 m_2 \cdots m_{i-1} m_{i+1} \cdots m_n$. Then k_i and m_i are relatively prime (why?) so we can find integers r and s such that $rk_i + sm_i = 1$. This gives the congruences $rk_i \equiv 0 \pmod{k_i}$, $rk_i \equiv 1 \pmod{m_i}$. Since

$m_1, m_2, \ldots, m_{i-1}, m_{i+1}, \ldots, m_n$ all divide k_i, it follows that $x_i = rk_i$ satisfies

$$x_i \equiv 0 \pmod{m_1},$$
$$x_i \equiv 0 \pmod{m_2},$$
$$\vdots$$
$$x_i \equiv 0 \pmod{m_{i-1}},$$
$$x_i \equiv 1 \pmod{m_i},$$
$$x_i \equiv 0 \pmod{m_{i+1}},$$
$$\vdots$$
$$x_i \equiv 0 \pmod{m_n}.$$

For each subscript i, $1 \leqslant i \leqslant n$, we find such an x_i. To solve the system of congruences (1), set $x = a_1 x_1 + a_2 x_2 + \cdots + a_n x_n$. Then $x \equiv a_i x_i \equiv a_i$ $\pmod{m_i}$ for each i, $1 \leqslant i \leqslant n$, so x is a solution to the congruences (1).

As to uniqueness, suppose also $x' \equiv a_i \pmod{m_i}$ for each i. Then $x - x' \equiv 0 \pmod{m_i}$ for each i. So m_i divides $x - x'$ for each i; hence the least common multiple of all the m_i's divides $x - x'$. But since the m_i are pairwise relatively prime, this least common multiple is the product M. So $x \equiv x' \pmod{M}$. □

E1. Why are k_i and m_i relatively prime?

Another way to solve these congruences is to do them two at a time.

The congruence $x \equiv a_1 \pmod{m_1}$ is solved by any x of the form $x = a_1 + m_1 u_1$, where u_1 is arbitrary.

If also x is to satisfy $x \equiv a_2 \pmod{m_2}$, then x must satisfy $x = a_1 + m_1 u_1 = a_2 + m_2 t_2$ for some u_1, t_2. We solve this, if we can, for u_1, t_2 by Bezout's identity, and let $x_2 = a_1 + m_1 u_1 = a_2 + m_2 t_2$. Then $x = x_2$ solves

$$x \equiv a_1 \pmod{m_1},$$
$$x \equiv a_2 \pmod{m_2}.$$

The general solution to these congruences is then $x = x_2 + [m_1, m_2] u_2$, where u_2 is arbitrary and $[m_1, m_2]$ is the least common multiple of m_1 and m_2.

If x is required also to satisfy $x \equiv a_3 \pmod{m_3}$, then x must satisfy $x = x_2 + [m_1, m_2] u_2 = a_3 + m_3 t_3$ for some u_2, t_3. We solve $x_2 = [m_1, m_2] u_2 = a_3 + m_3 t_3$, if we can, for u_2, t_3. Let $x_3 = a_3 + m_3 t_3 = x_2 + [m_1, m_2] u_2$. Then the general solution to the first three congruences is $x = x_3 + [[m_1, m_2], m_3] u_3$, where u_3 can be any integer.

If also x is required to satisfy $x \equiv a_4$ (mod m_4), then we solve $x_3 +$ $[[m_1, m_2], m_3]u_3 = a_4 + m_4 t_4$ for u_3, t_4, if we can; we set $x_4 = a_4 + m_4 t_4$ and get the general solution $x = x_4 + [[[m_1, m_2], m_3], m_4]u_4$, etc.

This approach has the advantage of giving solutions, if there are any, even if the moduli $m_1, \ldots, m_r$ are not relatively prime. See Exercise E5.

EXAMPLE. We solve
$$x \equiv 3 \pmod{11},$$
$$x \equiv 6 \pmod{8},$$
$$x \equiv -1 \pmod{15}.$$

By the method in the proof we find x_1, solving
$$x_1 \equiv 1 \pmod{11},$$
$$x_1 \equiv 0 \pmod{8},$$
$$x_1 \equiv 0 \pmod{15}$$

by solving $r(120) + s(11) = 1$: $r = -1$, $s = 11$, therefore $x_1 = 120r = -120$.
Then we solve find x_2, solving
$$x_2 \equiv 0 \pmod{11},$$
$$x_2 \equiv 1 \pmod{8},$$
$$x_2 \equiv 0 \pmod{15}$$

by solving $r(165) + s(8) = 1$: $r = -3$, $s = 62$, therefore $x_2 = 165r = -495$.
Then we find x_3, solving
$$x_3 \equiv 0 \pmod{11},$$
$$x_3 \equiv 0 \pmod{8},$$
$$x_3 \equiv 1 \pmod{15}$$

by solving $r(88) + s(15) = 1$: $r = 7$, $s = -41$, therefore $x_3 = 88r = 616$.
Then $x = 3(-120) + 6(-495) + (-1)(616) = -3946$ is a solution to the original congruences. So is $x = -3946 + u(1320)$ for any u, so the smallest positive solution is with $u = 3$, $x = 14$.

By the alternate method, we solve $x \equiv 3$ (mod 11): $x_1 = 3 + 11u_1$ is the general solution. Then we solve also $x \equiv 6$ (mod 8): $x_2 = 6 + 8t_2 = 3 + 11u_1$. Solving for t_2, u_1 in $3 = 11u_1 - 8t_2$, we get $u_1 = t_2 = 1$. Then $x_2 = 14 + 88u_2$ is the general solution of the first two congruences.

Then we solve also $x \equiv -1$ (mod 15): $x_3 = 14 + 88u_2 = -1 + 15t_3$. Solving $15 = 15t_3 - 88u_2$, we get $t_3 = 1$, $u_2 = 0$. So $x_3 = 14 + 1320u_3$ is the general solution to all three congruences.

E2. Here is an old Chinese problem.

   Three rice farmers raised their rice collectively, and divided it equally at harvest time. One year each of them went to a different market to sell his share of the rice. Each of the three markets only bought rice in multiples of a certain base weight, which differed at each of the three markets. The first farmer sold his rice at a market where the base weight was 87 lbs. He sold all he could, and returned with 18 lbs of rice. The second farmer sold all the rice he could at a market where the base weight was 170 lbs, and returned with 58 lbs. The third farmer sold all the rice he could at a market where the base weight was 143 lbs and returned (at the same time as the other two) with 40 lbs. How much rice did they raise together?

E3. Prove that for any integer a, a^5 and a have the same final digit, in base 10.

E4. Use the Chinese remainder theorem to prove that $\phi(mn) = \phi(m)\phi(n)$ if m and n are relatively prime.

E5. Suppose m_1, m_2 are not pairwise relatively prime. Then there are numbers a_1 and a_2 for which there is no solution to the congruences

$$x \equiv a_1 \pmod{m_1},$$
$$x \equiv a_2 \pmod{m_2}. \tag{2}$$

Give an example. For which (a_1, a_2) can (2) be solved? For many pairs (a_1, a_2) with $0 \leqslant a_1 < m_1$, $0 \leqslant a_2 < m_2$, is there a solution to (2)? Generalize to the case of more than two congruences.

E6. Solve

$$\begin{cases} x \equiv 32 \pmod{63}, \\ x \equiv 33 \pmod{64}, \\ x \equiv 34 \pmod{65}. \end{cases}$$

E7. A crate of eggs is given. It is known that if the eggs are removed 2 at a time there remains 1 egg; if 3 at a time there remain 2; if 4 at a time there remain 3; if 5 at a time there remain 4; if 6 at a time there remain 5; and if 7 at a time none are left. How many eggs are in the crate?

E8. Find t, u, v so that $33t + 15 = 20u + 5 = 29v + 13$.

E9. Solve

$$\begin{cases} x \equiv 18 \pmod{11}, \\ x \equiv 3 \pmod{18}, \\ x \equiv 7 \pmod{25}, \\ x \equiv 11 \pmod{32}. \end{cases}$$

E10. Here is a problem of Yih-hing, 717 A.D. Solve

$$\begin{cases} x \equiv 1 \pmod{2}, \\ x \equiv 2 \pmod{5}, \\ x \equiv 5 \pmod{6}, \\ x \equiv 5 \pmod{12}. \end{cases}$$

E11. The English troops faced the Scottish troops in a recreation of the battle of Dumfries. The two sets of troops lined up 50 metres apart. Each line in unison fired a salvo of blanks from its muskets, then reloaded, all with great style and ritual. Each side started off with the same total number x of blanks. The English line contained 100 musketeers, the Scottish 67 men. After firing as many salvos as possible, the English side was left with 13 blanks; after firing as many salvos as they could, the Scottish side had 32 left. Assuming that at each salvo all troops on a side fired, what was (the smallest possible) x?

†E12. (a) Let R, S be two commutative rings. Consider the set $R \times S$ of all ordered pairs (a, b) where a is in R and b is in S. Make $R \times S$ into a commutative ring via the following definitions:

$(a_1, b_1) + (a_2, b_2) = (a_1 + a_2, b_1 + b_2)$ (addition);
$(a_1, b_1) \cdot (a_2, b_2) = (a_1 \cdot a_2, b_1 \cdot b_2)$ (multiplication);
$(0, 0)$ is the zero element;
$(1, 1)$ is the identity element;
$(-a, -b)$ is the negative of (a, b).

Show that $R \times S$ is a commutative ring.
(b) Write down addition and multiplication tables for $\mathbb{Z}_2 \times \mathbb{Z}_3$.
(c) Define a function f from $\mathbb{Z}_6$ to $\mathbb{Z}_2 \times \mathbb{Z}_3$ by setting $f([a]_6) = ([a]_2, [a]_3)$. Write down the value of f on each element of $\mathbb{Z}_6$. Observe that f maps onto $\mathbb{Z}_2 \times \mathbb{Z}_3$.
(d) Now let m_1, m_2 be any two relatively prime integers $\geqslant 2$. Define a function f from $\mathbb{Z}_{m_1 m_2}$ to $\mathbb{Z}_{m_1} \times \mathbb{Z}_{m_2}$ as in part (c). Use the Chinese remainder theorem to show that f is onto.

B. A Generalization of Fermat's Theorem

The next result answers a question raised in Chapter 10, Exercise E7, on secret codes.

A natural number is *squarefree* if it is a product of distinct primes, that is, it is not divisible by any square > 1.

Proposition. *Given a natural number m, there is some $q > 1$ (depending on m) such that for all integers a, $a^q \equiv a$ (mod m) if and only if m is squarefree. In fact, if $m = p_1 p_2 \ldots p_r$, a product of distinct primes, and*

$$q = k[\, p_1 - 1, p_2 - 1, \ldots, p_r - 1\,] + 1$$

for any natural number k, then $a^q \equiv a$ (mod m) for all a.

PROOF. Suppose $m = p_1 p_2 \ldots p_r$ with $p_1, p_2, \ldots, p_r$ all distinct primes. Then for any a in $\mathbb{Z}$ and any $q > 1$,

$$a^q \equiv a \ (\mathrm{mod}\ m)$$

iff

$$a^q \equiv a \ (\mathrm{mod}\ p_i) \quad \text{for each } i, 1 \leqslant i \leqslant r.$$

This is by the uniqueness part of the Chinese remainder theorem.

Now let $q = q_0 + 1$, where q_0 is any common multiple of $p_1 - 1, p_2 - 1, \ldots, p_r - 1$. Then for each i between 1 and r, $q_0 = (p_i - 1)q_i$ for some q_i,

so

$$a^q = a^{q_0+1} = a^{(p_i-1)q_i+1} = a^{(p-1)q_i} \cdot a.$$

If a and p_i are relatively prime, then $a^{p_i-1} \equiv 1 \pmod{p_i}$ (by Fermat); otherwise $a \equiv 0 \pmod{p_i}$. In either case we get that

$$a^q = (a^{p_i-1})^{q_i} a \equiv a \bmod p_i)$$

for each i between 1 and r. Thus $a^q \equiv a \pmod m$.

On the other hand, suppose $m = p_1^{e_1} p_2^{e_2} \cdots p_r^{e_r}$ with some $e_i > 1$. Suppose, for example, that $e_1 > 1$. Then, using the Chinese remainder theorem, let $x = a$ be a solution of

$$x \equiv p_1^{e_1-1} \pmod{p_1^{e_1}},$$

$$x \equiv 0 \pmod{p_2^{e_2}},$$

$$\vdots$$

$$x \equiv 0 \pmod{p_r^{e_r}}.$$

Then m does not divide a because $p_1^{e_1}$ does not divide a, but m does divide a^q for any $q > 1$. So $a \not\equiv 0 \pmod m$, but $a^q \equiv 0 \pmod m$ for any $q > 1$; hence $a^q \not\equiv a \pmod m$ for any $q > 1$. $\qquad\square$

EXAMPLE. Since $26 = 13 \cdot 2$, the corollary applies to $\mathbb{Z}_{26}$, and we get that for any integer a, $a^{13} \equiv a \pmod{26}$.

E13. Find q and verify that $a^q \equiv a \pmod m$ for all a where

   (i) $m = 6$,
   (ii) $m = 10$,
   (iii) $m = 30$.

†*E14. Use the proposition to prove that if m is squarefree, then for any a and any natural number k,

$$a^{k\phi(m)+1} \equiv a \pmod m.$$

*E15. Use E14 to prove Euler's theorem for m any squarefree integer.

E16. Show that for any m, squarefree or not, there exists a natural number r such that for any integer a and any k,

$$a^{r+k\phi(m)} \equiv a^r \pmod m.$$

Show that if $(a, m) = 1$, then r can be chosen to be 0 (Euler's theorem).

15 Secret Codes, II

This chapter may be read independently of Chapter 10, and may be omitted without loss of continuity.

Euler's theorem may be used as the basis of an effective cipher, or secret code. This cipher was suggested by R. L. Rivest, A. Shamir and L. Adleman of M.I.T. in 1977 based on ideas of Diffee and Hellman (1976).

We first show how the cipher works, and then try to explain why it is effective.

Assuming the messages are in English, associate space and letters to base 100 digits, or 2-digit base 10 numbers, in any convenient way, such as the following.

A	B	C	...	O	...	T	...	Y	Z	(*space*)
01	02	03		15		20		25	26	00

We thereby replace the English message by a sequence of base 100 digits. Then we divide up the sequence of digits into "words," or base 100 numbers, each containing a fixed number $r \geqslant 1$ of base 100 digits.

EXAMPLE. The message RETREAT becomes the sequence of base 100 digits 18 05 20 18 05 01 20. If we want r-digit words with, say, $r = 3$, we collect the digits into 3-digit base 100 numbers (or 6-digit ordinary numbers): 180520 180501 200000 (where the zeros at the end denote spaces).

Now choose some number q so that no prime dividing q is smaller than the largest possible word. This requirement on q guarantees that all the words of our message are relatively prime to q.

Encode the message as follows. Pick some number s which is relatively prime to $\phi(q)$ ($=$ Euler's ϕ function). Then there is some number t such that $st \equiv 1 \pmod{\phi(q)}$; hence for any w relatively prime to q, $(w^s)^t \equiv w \pmod{q}$.

E1. Prove this.

Encode the message by taking a word w and finding $z < q$ so that $w^s \equiv z$ (mod q). Then z is the corresponding word in the coded message.

Given a coded word z, the receiver decodes it by finding w' with $w' < q$ and $z^t \equiv w'$ (mod q).

If the encoder started with $w < q$ and the receiver ends up with $w' < q$, then $w' \equiv (w^s)^t \equiv w$ (mod q); since both w and w' are $< q$, they must be equal.

To illustrate with a code easy to do by hand computations, let $r = 1$ and let $q = 101$, a prime. Then $\phi(q) = 100$. Let $s = 3$. Then for $t = 67$, $st = 201 \equiv 1$ (mod 100). Suppose the message is HELLO. The words are 08, 05, 12, 12, 15. The sender encodes using the pair (101, 3). He takes each word and raises it to the third power mod 101,

$$8^3 \equiv 7 \quad (\text{mod } 101),$$

$$5^3 = 24 \ (\text{mod } 101),$$

$$12^3 \equiv 11 \ (\text{mod } 101),$$

$$15^3 \equiv 42 \ (\text{mod } 101),$$

to get the coded message, 07, 24, 11, 11, 42. The receiver decodes by using the pair (101, 67): she takes each of the coded numbers and raises it to the 67th power mod 101. One can verify by hand that $7^{67} \equiv 8$ (mod 101), etc.

Or for a more complicated example which is still easy to do with a hand calculator, let $r = 2$ and $q = 2803$, a prime. Then $\phi(2803) = 2802 = 2 \cdot 3 \cdot 467$. Let $s = 113$. Then we solve $113t + 2802x = 1$ as in Chapter 3, to get $t = -1339$, $x = 54$. Since $-1339 \equiv 1463$ (mod 2802), we find that $113 \cdot 1463 \equiv 1$ (mod 2802).

We encode the message GO, or 0715. To do that we find the least nonnegative residue of 715^{113} (mod 2803) by the technique of Chapter 12: we write $113 = (1110001)_2$, and obtain 715^{113} by the sequence of operations $XSXSXSSSSX$ where X means "multiply by 715 and reduce mod 2803," and S means "square the result and reduce mod 2803," working from left to right beginning with 1:

Operation	Power	Power mod 2803
X	$715^1 \equiv 715 \equiv 715$	
S	$715^2 \equiv 511225 \equiv 1079$	
X	$715^3 \equiv 771485 \equiv 660$	
S	$715^6 \equiv 435600 \equiv 1135$	
X	$715^7 \equiv 811525 \equiv 1458$	
S	$715^{14} \equiv 2125764 \equiv 1090$	
S	$715^{28} \equiv 1188100 \equiv 2431$	
S	$715^{56} \equiv 5909761 \equiv 1037$	
S	$715^{112} \equiv 1075369 \equiv 1820$	
X	$715^{113} \equiv 1301300 \equiv 708.$	

So the encoded word is 0708, or, in English, *GH*. (In general the coded numerical word will not translate into English.)

We decode by raising 708 to the 1463 power mod 2803. We omit the decoding calculations, which should yield 0715.

E2. Encode the message NO using this code.

In practice, coding is done using a computer. Choose r to be large, and q to be the product of two primes, p_1 and p_2, each of at least $2r + 1$ decimal digits. Choose s to be relatively prime to $\phi(q) = (p_1 - 1)(p_2 - 1)$. To encode, divide the message into words with r letters, which translate into numbers with r base 100 digits, and raise each to the sth power mod q. The receiver, if she knows $\phi(q)$, can find t with $st \equiv 1 \pmod{\phi(q)}$, and can decode by taking the received numerical words and raising each to the tth power (mod q) to get the original numerical message.

Why is the code effective?

Suppose someone wished to read an intercepted message, encoded using q and s, and suppose even that the interceptor knew q and s. All he would have to do is factor q, find $\phi(q)$, and solve $st \equiv 1 \pmod{\phi(q)}$. But the first task turns out to be a formidable problem, and that is the key to the secrecy of the code.

For example, suppose $r = 31$ and q were the product of two unknown 63-digit prime numbers. (For finding such large primes, see Part III, Chapter 3.) According to the designers of this code, if the factorization of q is unknown, it will take many centuries of computer time on the fastest known computer to factor q. (On the other hand, if the factors of q are known, it is a matter of a few seconds of computer time to find t and decode the message.) Thus unless the factorization of q is revealed, the code is in practice unbreakable.

Thus one could make q and s (but not the prime factorization of q) public knowledge. For example, if agent A wished to send a message to agent B, A could take the message, encode it using q and s, which are public, and send the message. Only B would know t, so only B could decode the message.

If A and B wished to exchange messages and be sure that no third party were either intercepting messages or sending spurious messages, they could each be assigned a pair of numbers: to A, q_A and s_A; to B, q_B and s_B. Both would know (q_A, s_A) and (q_B, s_B), but only A would know t_A and only B would know t_B. To send a message to B, A would encode it twice: first by using the pair (q_A, t_A), then by using (q_B, s_B). B would decode first using (q_B, t_B), then using (q_A, s_A). Since A encoded using the secret exponent t_A, which B was able to decode, B would know the message was sent by A. Since only B knows the secret exponent t_B, A would know that only B could decode the message. Thus communication between A and B would be secure.

For more discussion of this cipher, see Gardner (1977). For an indication of how useful ciphers such as this may be, see Shapley and Kolata (1977).

E3. Let $r = 1$, $q = 29$, $s = 4$. Encode and decode the message RETREAT.

*E4. Fermat's theorem says that if q is prime, for any a with $(a, q) = 1$, $a^{q-1} \equiv 1$ (mod q). For testing to see whether a number q is prime it would be useful if the converse were true, namely, that if for all a with $(a, q) = 1$, $a^{q-1} \equiv 1$ (mod q), the q is prime. Show that the converse just described is false by showing that $q = 561$ is a composite number, and for all a with $(a, q) = 1$, $a^{q-1} \equiv 1$ (mod q).

E5. Suppose q is squarefree. Use Exercise E14, Chapter 14, to show that if s is any integer relatively prime to $\phi(q)$, there is some integer t such that $(w^s)^t \equiv w$ (mod q) for *all* w, not just for w relatively prime to q.

E6. (a) Suppose given q, s, t satisfying (as in E5) the property that for all w, $(w^s)^t \equiv w$ (mod q). Show that one can use (q, s) and (q, t) to encode and decode numerical words of any size $< q$, not just words of size $<$ any prime divisor of q, as specified in the text.

(b) Set up such a code for $q = 7 \cdot 11 \cdot 37 = 2849$. Use it to encode and decode the message FOOL, making it form two coded words.

II. POLYNOMIALS

This part of the book is about polynomials. Many of the results we obtained for integers in Part I have analogues for polynomials. For example, every polynomial factors uniquely into a product of irreducible polynomials. We show this in Chapter 2. Much of the rest of Part II is devoted to trying to determine irreducible polynomials and find factorizations of polynomials with coefficients in various fields. Ideas from Part I such as congruences and the Chinese remainder theorem reappear for polynomials as useful tools in studying factorizations. The analogue for polynomials of congruence classes will appear in Part III.

Polynomials

1

This part of the book is about polynomials, with coefficients in a field F. You are already familiar with polynomial functions with real coefficients from calculus. They are the simplest functions arising in calculus: examples are $p(x) = 2 - 3x + x^2$, $p(x) = x - \frac{1}{2}x^3$, $p(x) = 2$, etc. A general example is

$$p(x) = a_0 + a_1x + a_2x^2 + \cdots + a_nx^n,$$

where $a_0, \ldots, a_n$ are real numbers, and n is some integer $\geqslant 0$.

In writing a polynomial function in this way, we are using the convention customarily used in calculus of describing a function by what it does to a "typical" or "variable" or "indeterminate" real number x. In this notation we know how to add two polynomial functions, multiply a polynomial by a constant scalar, or multiply two polynomials—it is the same as for any function: x is thought of as being a typical real number, and the sum (product) of two functions, applied to x, is defined to be the sum (product) of the values of the functions at x. Thus if

$$p(x) = a_0 + a_1x + \cdots + a_nx^n,$$

and

$$q(x) = b_0 + b_1x + \cdots + b_mx^m,$$

then

$$p(x) + q(x) = (a_0 + a_1x + \cdots + a_nx^n) + (b_0 + b_1x + \cdots + b_mx^m);$$

if, say, $m > n$, we can collect terms and get

$$p(x) + q(x) = (a_0 + b_0) + (a_1 + b_1)x$$
$$+ \cdots + (a_n + b_n)x^n + b_{n+1}x^{n+1} + \cdots + b_mx^m. \quad (1)$$

Similarly,

$$p(x) \cdot q(x) = (a_0 + \cdots + a_n x^n)(b_0 + \cdots + b_m x^m)$$

$$= a_0 b_0 + (a_0 b_1 + a_1 b_0)x + \cdots + \left(\sum_{i+j=k} a_i b_j \right) x^k \cdots + a_n b_m x^{n+m}.$$

$$(2)$$

We abstract from the "calculus" way of describing functions to get a general definition of a polynomial.

A *polynomial* (*not* a polynomial function) with coefficients in a field F is defined to be an expression of the form

$$p(x) = a_0 + a_1 x + \cdots + a_n x^n,$$

where $a_0, a_1, \ldots, a_n$ are in F and x is an indeterminant—not an indeterminant element of F, but simply a formal symbol. Thus two polynomials

$$p(x) = a_0 + a_1 x + \cdots + a_n x^n$$

and

$$q(x) = b_0 + b_1 x + \cdots + b_m x^m$$

are equal precisely when $a_0 = b_0, a_1 = b_1, \ldots, a_i = b_i, \ldots$ for all i. (In particular, if $m < n, a_{m+1} = a_{m+2} = \cdots = a_n = 0$.)

Equality of polynomials, as we have just defined it, is not the same as equality for functions. It is true that any polynomial $p(x)$ in $F[x]$ defines a function on F by substitution: the function defined by $p(x)$ sends a in F to $p(a)$. However, two functions on F are defined to be equal precisely when the two functions have equal values on every element of the set F. It is possible for two polynomials to define equal functions on F without being equal as polynomials. Here is an example. Let $F = \mathbb{Z}_2$, let $p(x) = x - 1$, let $q = x^3 - 1$. Then $p(x) \neq q(x)$ as polynomials, but $p(0) = q(0) = 1, p(1) = q(1) = 0$ so $p(x)$ and $q(x)$ are equal as functions on $\mathbb{Z}_2$. Later we will prove that if F has infinitely many elements, like $F = \mathbb{R}$, then two polynomials which define the same function on F must be equal as polynomials.

E1. Using Fermat's theorem, find for each prime number p two different polynomials with coefficients in $\mathbb{Z}_p$ which agree as functions on $\mathbb{Z}_p$.

The set of all polynomials with coefficients in F is denoted $F[x]$.

We say $p(x)$ has degree n if $p(x) = a_0 + a_1 x + \cdots + a_n x^n$ and $a_n \neq 0$. The polynomial $0 = 0 + 0x + 0x^2 + \cdots$, by convention, has degree -1. If $p(x) \neq 0$, its degree, denoted $\deg(p(x))$, is ≥ 0.

The field F is contained in $F[x]$: an element a of F can be thought of as a polynomial of degree 0, namely, $a + 0x + 0x^2 + \cdots$.

If we define addition and multiplication of polynomials by formulas (1) and (2), it is easy to see that $F[x]$ is a commutative ring, and has no zero divisors. To see the latter, suppose $p(x)$, $q(x)$ are two nonzero polynomials in $F[x]$, where $p(x) = a_m x^m + a_{m-1} x^{m-1} + \cdots + a_1 x + a_0$ and $q(x) = b_n x^n + b_{n-1} x^{n-1} + \cdots + b_1 x + b_0$. We show $p(x) \cdot q(x) \neq 0$. By 0 is meant the polynomial $0 + 0x + 0x^2 + \cdots$, and $p(x)q(x) = 0$ means that each coefficient of $p(x)q(x)$ is equal to zero.

Assume $p(x)$ has degree m and $q(x)$ has degree n, so $a_m \neq 0$, $b_n \neq 0$. It follows that $a_m b_n \neq 0$ since a_m, b_n are in F, a field. But $a_m b_n$ is the coefficient of x^{m+n} in $p(x)q(x)$. Therefore, $p(x)q(x) \neq 0$, and $F[x]$ has no zero divisors.

Although we defined polynomials in terms of the indeterminant, or formal symbol x, it is possible, and in some cases more convenient, to define a polynomial without using x. We may define a polynomial with coefficients in F as an infinite sequence

$$(a_0, a_1, \ldots, a_n, \ldots)$$

of elements of F satisfying the condition that all a_n from some point on are equal to 0:

$$(a_0, a_1, \ldots, a_n, \ldots) = (a_0, a_1, \ldots, a_d, 0, 0, \ldots)$$

for some d, the degree of the polynomial.

We add two polynomials in this notation as follows:

$$(a_0, a_1, \ldots, a_n, \ldots) + (b_0, b_1, \ldots, b_n, \ldots)$$
$$= (a_0 + b_0, a_1 + b_1, \ldots, a_n + b_n, \ldots);$$

$$(1')$$

and we multiply two polynomials as follows:

$$(a_0, a_1, \ldots, a_n, \ldots)(b_0, b_1, \ldots, b_n, \ldots)$$
$$= \left(a_0 b_0, a_0 b_1 + a_1 b_0, \ldots, \sum_{i+j=k} a_i b_j, \ldots \right).$$

$$(2')$$

We can identify these sequences with the polynomials described in terms of the indeterminate x, by identifying $(a_0, a_1, \ldots, a_d, 0, 0, \ldots)$ with $a_0 + a_1 x + \cdots + a_d x^d$. Thus to x^2 is associated $(0, 0, 1, 0, 0, \ldots)$; to $4 + x + 3x^4 - x^6$ is associated $(4, 1, 0, 0, 3, 0, -1, 0, 0, \ldots)$. Addition and multiplication are the same in either version. (This identification is the kind of thing which happens in linear algebra when one writes coordinates of a vector with respect to a basis. Here the vector space is $F[x]$, the scalars are F, the basis is $1, x, x^2, x^3, \ldots$, and $(a_0, a_1, \ldots, a_d, 0, \ldots)$ is the tuple of coordinates of the vector $a_0 + a_1 x + \cdots + a_d x^d$ with respect to the basis $1, x, x^2, \ldots$.)

We shall ordinarily use an indeterminant in describing polynomials, but there will be times, especially when multiplying or dividing, when it is useful to recall that a polynomial may be thought of as a tuple of coefficients, and "detach the coefficients."

It is reasonable to consider polynomials with coefficients from any commutative ring, such as not only $\mathbb{R}$, $\mathbb{C}$, $\mathbb{Q}$, or $\mathbb{Z}_p$ (which are fields), but also $\mathbb{Z}$ or $\mathbb{Z}_n$, or, for that matter, $F[x]$. (In Chapter I-10 we considered matrices with entries in $\mathbb{Z}_{26}[t]$.) When we consider polynomials with coefficients in $F[x]$, if we call the indeterminant by y, we get $F[x][y]$, which consists of elements of the form

$$a_0(x) + a_1(x)y + a_2(x)y^2 + \cdots + a_d(x)y^d,$$

that is, polynomials in two variables. Examples: $x^2 + y^2 - 1$, $y - x^2$, $3x^4 - 5xy^2 + \frac{1}{2}$, etc. Similarly, polynomials in three variables may be thought of as polynomials in one variable with coefficients in $F[x][y]$, etc.

E2. For which, if any, primes p do $x^6 + 2x^2 + x$ and $x^9 + 8x^3 + x$ agree as functions on $\mathbb{Z}_p$?

†**E3.** Prove that for $p(x)$ in $F[x]$, there is some $q(x)$ with $p(x)q(x) = 1$ iff $p(x)$ has degree 0.

E4. If we consider infinite sequences $(a_0, a_1, \ldots, a_n, \ldots)$ of elements of a field F and do not assume that from some point on $a_n = 0$, we get something called a formal power series. For example, the sequence

$$\left(1, 1, \frac{1}{2}, \frac{1}{3!}, \frac{1}{4!}, \cdots\right)$$

is a formal power series. Using powers of x as with polynomials this looks like

$$1 + x + \frac{x^2}{2} + \frac{x^3}{3!} + \cdots + \frac{x^n}{n!} + \cdots$$

(the Taylor's series for e^x). We can treat formal power series without worrying about convergence: indeed, convergence when the field F is not a subfield of $\mathbb{C}$ may make no sense. We add and multiply using formulas (1') and (2'). If we denote the set of all formal power series with coefficients in F by $F[[X]]$, then $F[[X]]$ is a commutative ring. Note: $F[X] \subset F[[X]]$.

(a) Prove that $F[[X]]$ has no zero divisors. (*Hint*: Consider lowest degree nonzero terms, not highest.)

*(b) Suppose $(a_0, a_1, \ldots) = \sum_{i=0}^{\infty} a_i x^i$ is a formal power series. Show that there is another formal power series $\sum_{i=0}^{\infty} b_i x^i$ with $(\sum_{i=0}^{\infty} a_i x^i)(\sum_{i=0}^{\infty} b_i x^i) = 1$, iff $a_0 \neq 0$. ($\sum b_i x^i$ is called the inverse of $\sum a_i x^i$.)

(c) Find the inverse of $1 - x$.

Unique Factorization

2

A. Division Theorem

Recall that if the polynomial $p(x) = a_0 x + \cdots + a_d x^d$, where $a_d \neq 0$, then $d = \deg(p(x))$ is the degree of $p(x)$. The coefficient a_d is called the leading coefficient of $p(x)$; if $a_d = 1$, then $p(x)$ is called a *monic* polynomial.

We shall sometimes call a polynomial of degree $\leqslant 0$ a constant, or a scalar, since it can be considered as an element of F, the field of coefficients.

The simple fact that we can assign to a polynomial an integer $\geqslant -1$, its degree, is extremely useful, because it permits for polynomials proofs by induction of the same kind that we did for $\mathbb{Z}$. In fact, the whole sequence of arguments in Chapter I-3, which showed that $\mathbb{Z}$ has unique factorization, is valid for $F[x]$ if F is a field. We give in this chapter a concise account of how that goes. You may want to refer back to the arguments for $\mathbb{Z}$ at this point to see how we proceeded there.

We will often let f, g, p, q, r, etc., denote polynomials, omitting the "(x)" in "$f(x)$."

The first step, as with the theory for $\mathbb{Z}$, is the

Division Theorem for $F[x]$. *Let F be a field. Let f, g be two polynomials in $F[x]$, with $f \neq 0$. There are polynomials q, r, with $\deg r < \deg f$, such that $g = fq + r$. If also $g = fq_1 + r_1$, with $\deg r_1 < \deg f$, then $q = q_1, r = r_1$.*

PROOF. We use induction (2) on the degree of the dividend g. Fix f, a polynomial of degree $\geqslant 0$. If $\deg g < \deg f$, then $g = f \cdot 0 + g$ satisfies the division theorem. Suppose $\deg g \geqslant \deg f$. Let $f = a_d x^d + \cdots + a_0$, $g = b_{d+s} x^{d+s} + \cdots + b_0$, where $s \geqslant 0$. Let $g' = g - (b_{d+s}/a_d)x^s f$. Then $\deg(g') < \deg(g)$, since we have canceled off the highest power of x. By induction, $g' = fq_0 + r$ for some q_0, r with $\deg r < \deg f$. Then $g = f(q_0 + (b_{d+s}/a_d)x^s) + r$, proving the existence of q, r for f and g.

For uniqueness, suppose $g = fq + r = fq' + r'$, $\deg r < \deg f$, $\deg r' < \deg f$. Then $0 = f(q - q') + (r - r')$, so $f(q' - q) = r - r'$. The left side is either 0 or a polynomial of degree at least that of f; the right side has degree $< \deg f$. Thus equality can only occur if both sides $= 0$, so $r = r'$, $q = q'$.   □

The argument in getting g from g' in the proof is the first step in the process of long division of polynomials, which actually obtains q and r. We illustrate this familiar procedure by an example.

$$
\begin{array}{r}
x^2 - 3x + 7 \\
x^3 + 3x^2 + x + 5 \overline{)\, x^5 \qquad - x^3 + \ x^2 \qquad\quad + 7} \\
x^5 + 3x^4 + \ x^3 + \ 5x^2 \\
\hline
-3x^4 - 2x^3 - \ 4x^2 \qquad\quad + 7 \\
-3x^4 - 9x^3 - \ 3x^2 - 15x \\
\hline
7x^3 - \ x^2 + 15x + \ 7 \\
7x^3 + 21x^2 + \ 7x + 35 \\
\hline
-22x^2 + \ 8x - 28
\end{array}
$$

Much writing can be saved by omitting powers of x, or detaching coefficients, so that the division looks like:

$$
\begin{array}{r}
1 - 3 \quad 7 \\
1\ 3\ 1\ 5 \overline{)\, 1 \quad 0 - 1 \quad 1 \quad 0 \quad 7} \\
1 \quad 3 \quad 1 \quad 5 \\
\hline
-3 - 2 \ -4 \quad 0 \\
-3 - 9 \ -3 - 15 \\
\hline
7 \ -1 \quad 15 \quad 7 \\
7 \quad 21 \quad 7 \quad 35 \\
\hline
-22 \quad 8 - 28
\end{array}
$$

Thus

$$x^5 - x^3 + x^2 + 7 = (x^3 + 3x^2 + x + 5)(x^2 - 3x + 7) - 22x^2 + 8x - 28.$$

(Note that in detaching the coefficients for dividing, the numbers correspond from left to right with descending powers of x, which is opposite to the way we did it in Chapter 1.)

We shall say that a polynomial f *divides* a polynomial g, notation, $f\,|\,g$, if $g = fq$ for some polynomial q. For example, $x^2 - 1$ divides $x^4 - 1$, $x^2 + x + 1$ divides $x^6 - 1$, while $x - 1$ does not divide $x^3 - 2$.

Here is a criterion for finding factors of a polynomial. In it the number $f(a)$ is obtained by thinking of $f(x)$ as a function and evaluating it at a.

Root Theorem. *If $f(x)$ is a polynomial with coefficients in a field F, and if a is in F with $f(a) = 0$, then $x - a$ divides $f(x)$.*

PROOF. Write $f(x) = (x - a)q(x) + r(x)$, using the division theorem. Since $\deg(r(x)) < \deg(x - a) = 1$, $r(x)$ must be a scalar, call it r, in F. So

$$f(x) = (x - a)q(x) + r.$$

View both sides as functions on F and evaluate at $x = a$. The equation becomes $0 = r$. So $x - a$ divides $f(x)$. $\qquad\square$

Corollary 1. *A nonzero polynomial $f(x)$ of degree n in $F[x]$, F a field, has at most n distinct roots in F.*

PROOF. Induction on n. If $n = 0$, $f(x)$ is a nonzero constant polynomial, and so has no roots in F. Now suppose $f(x)$ is a polynomial of degree n, $n > 0$, and suppose it has distinct roots $a_1, \ldots, a_r$ in F. Then $f(a_r) = 0$, so by the root theorem $f(x) = (x - a_r)g(x)$ where $g(x)$ has degree $n - 1$. Since $f(a_i) = 0 = (a_i - a_r)g(a_i)$ for $1 \leqslant i \leqslant r - 1$, and since $a_i - a_r \neq 0$, therefore $g(a_i) = 0$, and so $g(x)$ has roots $a_1, \ldots, a_{r-1}$. By induction $r - 1 \leqslant n - 1 = \deg(g(x))$. So $r \leqslant n$. $\qquad\square$

Corollary 2. *If F is a field with infinitely many elements and $f(x)$ and $g(x)$ are any two polynomials with coefficients in F, then $f(x) = g(x)$ as polynomials if and only if $f(x) = g(x)$ as functions on F.*

This says that no confusion will arise if one thinks of polynomials in $\mathbb{R}[x]$ as functions.

The condition that F have infinitely many elements is necessary, as we have observed in Chapter II-1.

PROOF OF COROLLARY 2. If $f(x) = g(x)$ as polynomials, then for a any element of F, $f(a) = g(a)$, so $f(x)$ and $g(x)$ are equal as functions on F.

Conversely, if $f(x)$ and $g(x)$ are two polynomials, and $f(a) = g(a)$ for every a in F, then $f(x) - g(x) = h(x)$ is a polynomial with the property that $h(a) = 0$ for every a in F. Now $h(x)$ has degree n for some finite number n; if $h(a) = 0$ for every element a in F, and F has infinitely many elements, then $h(x)$ has more than n roots in F. So $h(x)$ must be the zero polynomial, and thus $f(x) = g(x)$ as polynomials, completing the proof. $\qquad\square$

E1. If $f(x)$ is in $\mathbb{R}[x]$ and a_1, a_2 are in $\mathbb{R}$, $a_1 \neq a_2$, show that there is a polynomial $q(x)$ such that

$$f(x) = [(x - a_1)(x - a_2)]q(x) - \frac{f(a_2)(x - a_1) - f(a_1)(x - a_2)}{a_1 - a_2}.$$

E2. Does $x - 3$ divide $x^4 + x^3 + x + 4$ in $\mathbb{Q}[x]$? $\mathbb{Z}[x]$? $\mathbb{Z}_2[x]$? $\mathbb{Z}_3[x]$? $\mathbb{Z}_5[x]$? $\mathbb{Z}_7[x]$? $\mathbb{Z}_4[x]$?

E3. Find all m such that $x^3 + 3$ divides $x^5 + x^3 + x^2 - 9$ in $\mathbb{Z}_m[x]$.

B. Greatest Common Divisors

We continue with results worked out in Chapter I-3 for natural numbers.

Let f, g be in $F[x]$, F a field. A polynomial p in $F[x]$ is a *greatest common divisor* (g.c.d.) of f and g if p divides f and p divides g, and any q in $F[x]$ which also divides both f and g must satisfy $\deg q \leqslant \deg p$.

We can find a greatest common divisor of two polynomials by using the division theorem many times. The process, called *Euclid's algorithm for polynomials*, goes back at least to Simon Stevin, 1585, and works as follows:

Given two polynomials f, g in $F[x]$ with $f \neq 0$, perform the sequence of divisions:

$$g = fq_0 + r_0,$$
$$f = r_0 q_1 + r_1,$$
$$r_0 = r_1 q_2 + r_2,$$
$$\vdots$$
$$r_{n-2} = r_{n-1} q_n + r_n,$$
$$r_{n-1} = r_n q_n + 0.$$

Since $\deg r_0 < \deg f$, $\deg r_1 < \deg r_0$, $\deg r_2 < \deg r_1$, etc., the sequence of divisions ends after at most $\deg f$ steps.

Theorem. *In the Euclidean algorithm, the last nonzero remainder r_n is a greatest common divisor of f and g.*

E4. Prove this.

Note that we carefully said "*a* greatest common divisor," rather than "*the* greatest common divisor." Two polynomials may have many greatest common divisors. If we were to have defined "greatest" for integers as "greatest in absolute value," then the greatest common divisor of two integers wouldn't be unique either; for example, -8 and 6 have two greatest common divisors in the absolute sense: 2 and -2. Similarly for polynomials: $x^2 - 1$ and $5x^2 + 10x + 5$ have many greatest common divisors: $2x + 2$, $\pi x + \pi$, $x + 1$, $x/17 + 1/17$, etc. But all the greatest common divisors divide each other. This is in general a consequence of

Theorem. *If d is the greatest common divisor of f and g, obtained by Euclid's algorithm, and e is any other common divisor of f and g, then e divides d.*

E5. Prove this.

From this it follows that two greatest common divisors e, d of f and g can differ only by a scalar multiple: $d = ae$ for some a in F. For let d be the greatest common divisor obtained by Euclid's algorithm, and suppose it has degree r. If e is a common divisor of f and g, then e divides d. If e is a greatest common divisor of f and g, then e also has degree r. Since e divides d, $d = ae$ where a is some polynomial, and $\deg d = \deg a + \deg e$. Therefore a has degree 0, so must be a nonzero scalar. Then $d/a = e$, so d and e divide each other.

We shall say that two polynomials d, e, such that each is a nonzero scalar multiple of the other, are *associates*.

EXAMPLE. $x^2 + 3x + 1$ and $5x^2 + 15x + 5$ are associates.

Any polynomial is an associate of exactly one monic polynomial.

If we lapse into sloppy terminology and talk about *the* greatest common divisor of two polynomials, we will usually have in mind the unique monic polynomial which is a greatest common divisor of the two polynomials. Generally it won't matter. Any greatest common divisor will serve the role of *the* greatest common divisor.

Continuing the development as with $\mathbb{Z}$, we have

Bezout's lemma. *Any greatest common divisor d of f and g can be written as $d = af + bg$ for some polynomials a, b in $F[x]$.*

E6. Prove this.

Say f and g are *relatively prime* if any greatest common divisor is a constant. In that case 1 is a greatest common divisor, so we can write $1 = af + bg$ for some polynomials a, b.

E7. Write the greatest common divisor of $f(x) = x^4 + 2x^3 - 6x - 9$ and $g(x) = 3x^4 + 8x^3 + 14x^2 + 8x + 3$ as a linear combination of f and g.

Definition. A polynomial p is a *unit* if there is another polynomial q with $pq = 1$.

The only polynomials in $F[x]$, F a field, which are units, are polynomials of degree 0, that is, nonzero constant polynomials or scalars, polynomials which look just like elements of F (Exercise E3, Chapter II-1).

A polynomial p is *irreducible* if p is not a unit, and if $p = fg$, then f or g must be a unit.

EXAMPLES.

$x + a$ is irreducible in $F[x]$ for any field F;
$x^2 + 1$ is irreducible in $\mathbb{R}[x]$;
$x^3 + 2$ is irreducible in $\mathbb{Q}[x]$;
$x^2 + 1$ is irreducible in $\mathbb{Z}_3[x]$;
$x^2 + x + 1$ is irreducible in $\mathbb{Z}_2[x]$.

We shall study the question of which polynomials are irreducible in much of this part of the book.

E8. Show that for f, g, h in $F[x]$, if $(f, g) = 1$ and $h | f$, then $(h, g) = 1$.

E9. For f, g, h in $F[x]$, show that if $(f, g) = 1$, then $(fh, g) = (h, g)$.

E10. Find the g.c.d. of
(a) $x^2 + 1$ and $x^5 + 1$ in $\mathbb{Z}_2[x]$,
(b) $x^2 - x + 4$ and $x^3 + 2x^2 + 3x + 2$ in $\mathbb{Z}_3[x]$.

E11. In $\mathbb{Q}[x]$ solve:
(a) $p(x)(x^2 - 3x + 2) + q(x)(x^2 + x + 1) = 1$;
(b) $p(x)(2x^3 - 7x^2 + 7x - 2) + q(x)(2x^3 + x^2 + x - 1) = 2x - 1$.

E12. Find the g.c.d. in $\mathbb{Q}[x]$ of
(a) $x^6 - 1$ and $x^7 - 2x^3 + 3x^2 - 2x + 1$,
(b) $x^9 - 1$ and $x^{11} - 1$,
(c) $x^8 + 6x^6 - 8x^4 + 1$ and $x^{12} + 7x^{10} - 3x^8 - 3x^2 - 2$. (See E14.)

E13. Let f, g be in $F[x]$. Prove that if d_1 is the greatest common divisor obtained from f and g by Euclid's algorithm and d_2 is the greatest common divisor obtained from rf and sg where r, s are constants, then d_1 is an associate of d_2.

E14. Using Exercise 13, do Exercise 12(c) so as to avoid fractions.

*E15. Prove that the g.c.d. in $\mathbb{Q}[x]$ of $1 + x + \cdots + x^n$ and $1 + x + \cdots + x^m$ is $1 + x + \cdots + x^d$, where $d + 1 = $ g.c.d. $(m + 1, n + 1)$.

†E16. Show that if f, g are in $F[x]$, $d = (f, g)$, and $d = fa + gb$ for some a, b in $F[x]$, then a and b may be chosen so that deg $a < $ deg g, deg $b < $ deg f.

C. Factorization into Irreducible Polynomials

Irreducible polynomials play the same role as prime numbers do in the factorization theory of integers.

Theorem. *Any polynomial of degree $\geqslant 1$ in $F[x]$ is irreducible or factors into a product of irreducible polynomials.*

E17. Give a proof, using induction on the degree of a polynomial.

Theorem. *If p is an irreducible polynomial and p divides fg, then p divides f or p divides g.*

E18. Prove this.

Theorem. *If $f = p_1 p_2 \cdots p_s = q_1 q_2 \cdots q_t$ are two factorizations of the polynomial f into a product of irreducible polynomials in $F[x]$, then $s = t$ and* *there is a* 1–1 *correspondence between* $p_1, \ldots, p_s$ *and* $q_1, \ldots, q_t$, *where, if* p_i *corresponds with* q_j, *then* p_i *and* q_j *are associates.*

E19. Prove this by induction on s, just as with $\mathbb{Z}$.

This is the unique factorization theorem for polynomials with coefficients in a field F. If we require that f and all p_i and q_j be monic polynomials, then we can state the theorem replacing the last word "associates" by the word "equal."

We have left most of the proofs above as exercises, because the theorems and the proofs are so close to those in Chapter I-3 on unique factorization of integers that you will learn more by trying to work them out yourself than you would were they written out for you.

Just as with integers, we can write the factorization of a polynomial f in $F[x]$ as

$$f = p_1^{e_1} p_2^{e_2} \cdots p_r^{e_r}.$$

If any e_i is bigger than 1 we shall say that f has a *multiple factor*: thus $f(x) = (x^2 + 2)^2(x + 1)$ in $\mathbb{R}[x]$ has a multiple factor, while $f(x) = (x^2 + 2)(x + 1)$ does not. If $f(x)$ has a multiple linear factor, e.g., $f(x) = (x^2 + 2)(x + 1)^2$, we shall say $f(x)$ has a *multiple root* in F.

E20. In $F[x]$ show that if p, q are irreducible and monic and $p|q$, then $p = q$.

E21. Show that in $\mathbb{R}[x]$, no polynomial of odd degree is irreducible and that in $\mathbb{C}[x]$ no polynomial of degree 2 is irreducible.

E22. Factor $x^5 - x$ into a product of irreducible polynomials in $\mathbb{Z}_5[x]$.

E23. In $\mathbb{Z}_2[x]$ factor into products of irreducible polynomials:
    (a) $x^8 + x^7 + x^6 + x^4 + 1$;
    (b) $x^6 + x^4 + x^3 + x^2 + 1$;
    (c) $x^{16} - x$;
    (d) $x^7 + x^6 + x^4 + 1$.

E24. Find the g.c.d. of $(t - 2)^3(t - 3)^4(t - i)$ and $(t - 1)(t - 2)(t - 3)^3$ in $\mathbb{C}[x]$.

E25. If $f(x)$ in $\mathbb{Q}[x]$ has in $\mathbb{C}[x]$ a factor of the form $(x - \alpha)^2$, show that $f(x)$ and $f'(x) = (d/dx)f(x)$ are not relatively prime.

E26. Prove again Corollary 1, Section A, using uniqueness of factorization.

E27. Is there a division algorithm for polynomials in $\mathbb{Z}_{15}[x]$? If not, give a counterexample.

†**E28.** In $\mathbb{Z}_{15}[x]$, $x^2 - 1 = (x - 1)(x - 14) = (x - 4)(x - 11)$. Does this contradict uniqueness of factorization?

3 The Fundamental Theorem of Algebra

We have seen that if F is a field, every nonconstant polynomial in $F[x]$ factors in an essentially unique way into the product of irreducible polynomials. Thus irreducible polynomials relate to all polynomials in the same way that primes do to natural numbers. We naturally ask: Which polynomials are irreducible?

The question clearly depends on the field F.

For example, consider the polynomial $x^3 - 2$. This is a polynomial with coefficients in $\mathbb{Q}$, and $\mathbb{Q} \subset \mathbb{R} \subset \mathbb{C}$, so we can ask about its factorizations in $\mathbb{Q}[x]$, $\mathbb{R}[x]$, and $\mathbb{C}[x]$. Here they are:

In $\mathbb{Q}[x]$, $x^3 - 2$ is irreducible.

In $\mathbb{R}[x]$, $x^3 - 2 = (x - \sqrt[3]{2})(x^2 + \sqrt[3]{2}\,x + \sqrt[3]{4})$.

In $\mathbb{C}[x]$, $x^3 - 2 = (x - \sqrt[3]{2})(x - \omega\sqrt[3]{2})(x - \omega^2\sqrt[3]{2})$, where $\omega = e^{2i\pi/3} = -(1/2) + (i\sqrt{3}/2)$ satisfies $\omega^3 = 1$.

In this book we shall look at the question of which polynomials are irreducible when $F = \mathbb{C}$, $\mathbb{R}$, $\mathbb{Q}$, and $\mathbb{Z}_p$ for p prime. In this chapter we shall consider $F = \mathbb{C}$.

A. Irreducible Polynomials in $\mathbb{C}[x]$

The field $\mathbb{C}$ was invented to contain roots of irreducible real polynomials; for example, $i = \sqrt{-1}$ is a root of $x^2 + 1$. It is not hard to see that each polynomial $g(x) = x^2 + ax + b$ with a, b in $\mathbb{C}$ has a root in $\mathbb{C}$. In fact, write

$(a^2/4) - b = re^{i\theta}$ (recall Chapter I-1) and let $d = \sqrt{r}\, e^{i\theta/2}$. Then $d^2 = (a^2/4) - b$, and

$$g(x) = x^2 + 2\left(\frac{a}{2}\right)x + \left(\frac{a}{2}\right)^2 - \left(\frac{a^2}{4} - b\right)$$

$$= \left(x + \frac{a}{2}\right)^2 - d^2$$

$$= \left(x + \frac{a}{2} + d\right)\left(x + \frac{a}{2} - d\right).$$

(This is the quadratic formula from elementary algebra.) So every polynomial in $\mathbb{C}[x]$ of degree 2 factors into a product of two linear factors.

It is possible to give a similar explicit factorization of any polynomial of degree 3 or 4 in $\mathbb{C}[x]$, based on the fact that we can always find a root in $\mathbb{C}$ of $x^n - c$ for any c in $\mathbb{C}$ (write $c = re^{i\theta}$, set $d = \sqrt[n]{r}\, e^{i\theta/n}$, then $d^n = c$). Here is how to find roots of any polynomial of degree 3.

Let $f(z) = z^3 + a_2 z^2 + a_1 z + a_0$. Set $x = z + a_2/3$; then $f(x - a_2/3) = p(x)$ has the form $p(x) = x^3 + qx + r$ for some q, r. Let $\omega = e^{2\pi i/3}$, a cube root of 1. As checked by multiplying out and noting that $1 + \omega + \omega^2 = 0$, the following identity holds:

$$x^3 + x(-3yz) + (y^3 + z^3) = (x + y + z)(x + \omega y + \omega^2 z)(x + \omega^2 y + \omega z).$$

Therefore, if we can solve for y and z in the two equations

$$q = -3yz, \qquad r = y^3 + z^3,$$

it follows by the identity that $x^3 + qx + r$ will have roots $-y - z$, $-\omega y - \omega^2 z$, $-\omega^2 y - \omega z$.

Now if $r = y^3 + z^3$, $-q^3/27 = y^3 z^3$, then the polynomial $(t - y^3)(t - z^3) = t^2 - rt - q^3/27$. Thus, by the quadratic formula,

$$y^3 = \frac{r}{2} + \sqrt{\frac{r^2}{4} + \frac{q^3}{27}}, \qquad z^3 = \frac{r}{2} - \sqrt{\frac{r^2}{4} + \frac{q^3}{27}},$$

and so the cube roots of

$$\frac{r}{2} + \sqrt{\frac{r^2}{4} + \frac{q^3}{27}} \quad \text{and} \quad \frac{r}{2} - \sqrt{\frac{r^2}{4} + \frac{q^3}{27}}$$

are the desired y and z.

E1. Find all roots of
  (a) $x^3 + 3x + 5$,
  (b) $x^3 + 2x^2 + 4x + 2$,
  (c) $x^3 + 3x + 1$.

E2. Find all complex fourth roots of -16; of $2i$; of $e^{2\pi i/3}$.

E3. Freely adapted from Cardano (1545, Book I). You lend $1000 (the principal) to a (current) friend for 3 years. The bargain you make with him is as follows: the interest is compounded annually. At the end of 3 years, your friend owes you and is to pay back (in principal and interest) half the principal, plus half of what was owed (in principal and interest) after the first year, plus half of what was owed after the second year. What is the interest rate? How much profit do you make?

It is interesting that while there is a procedure for finding roots of any polynomial of degree 4 analogous to that for degree 3, there is no corresponding way to find the roots of a polynomial of degree $\geqslant 5$: that is, there is a polynomial $p(x)$ of degree 5 whose roots cannot be described by taking the coefficients of $p(x)$ and manipulating them by the usual algebraic operations together with the operations of taking nth roots (forming radicals), in the way we did for polynomials of degree 2 or 3. This famous theorem is due to Abel, the Norwegian mathematician for whom the term abelian ("abelian group") was named.

Despite Abel's theorem there is an even more famous theorem, first proved by Gauss, which tells us what we want all at once.

Fundamental Theorem of Algebra. *Every polynomial $p(x)$ in $\mathbb{C}[x]$ of degree $\geqslant 1$ has a root in $\mathbb{C}$.*

Thus the only irreducible polynomials in $\mathbb{C}[x]$ are of degree 1.

B. Proof of the Fundamental Theorem

The rest of this chapter is devoted to one of the half-dozen or more distinctly different proofs of the fundamental theorem of algebra. The proof we present is essentially a proof of Argand, 1814. It involves a minimal acquaintance with functions of two (real) variables, and may be omitted without loss of continuity.

In the proof we shall assume that $p(z)$ is monic, that is, has leading coefficient $= 1$.

Before beginning the proof we describe some facts we need which go into the proof. Let $z = x + iy$. We think of the complex z-plane as the same as the real xy-plane. A polynomial $p(z)$ in $\mathbb{C}[z]$ may be written as

$$p(z) = p(x + iy) = p_1(x, y) + ip_2(x, y),$$

where $p_1(x, y)$ and $p_2(x, y)$ are real polynomials in the real variables x, y. Then $|p(z)|$ may be written as

$$|p(z)| = \sqrt{p_1(x, y)^2 + p_2(x, y)^2}.$$

Since $p_1(x, y)$ and $p_2(x, y)$ are real polynomials in x, y, $p_1(x, y)^2 + p_2(x, y)^2$ is a nonnegative real-valued continuous function of x, y. Since $\sqrt{t}$ is a continuous function of t for $t \geqslant 0$, $|p(z)| = \sqrt{p_1(x, y)^2 + p_2(x, y)^2}$ is continuous as a function of x, y.

A basic fact from calculus is that a function continuous on a closed disc $D = \{(x, y) | x^2 + y^2 \leqslant R\}$ in the xy-plane has a minimum value in D.

Our proof that $p(z)$ has a root in $\mathbb{C}$ has two parts.

(1) There is a point z_0 in the complex plane such that $|p(z_0)| \leqslant |p(z)|$ for all z in $\mathbb{C}$ (not just in some disc).

(2) If z_0 is the point found in part 1, where $|p(z_0)|$ is a minimum, then $p(z_0) = 0$.

We recall the triangle inequality:

$$|a + b| \leqslant |a| + |b|. \tag{Δ}$$

From the triangle inequality we get

$$|a| = |a + b - b| \leqslant |a + b| + |b|$$

so

$$|a + b| \geqslant |a| - |b|. \tag{*}$$

Part 1.

Lemma. *Let $f(z)$ in $\mathbb{C}[z]$ have degree $\geqslant 1$. Given any positive real number M there is some real number $R > 0$ such that for all z with $|z| > R$, $|f(z)| \geqslant M$.*

PROOF. We proceed by induction on the degree d of $f(z)$.

First, $d = 1$, $f(z) = a + bz$, $b \neq 0$. Then

$$|f(z)| = |a + bz| \geqslant |bz| - |a| \quad \text{(by (*))}$$
$$= |b| |z| - |a|.$$

Given M, choose $R = (M + |a|)/|b|$. Then if $|z| > R$, $|f(z)| > M$.

Assume the lemma is true for polynomials of degree $d - 1$. Suppose $f(z)$ has degree d. Then $f(z) = a + zf_1(z)$, where $f_1(z)$ has degree $d - 1$.

Given M, choose $R \geqslant 1$ so that for $|z| > R$, $|f_1(z)| > M + |a|$, by induction. Then for $|z| > R$,

$$|f(z)| = |a + zf_1(z)|$$
$$\geqslant |z| |f_1(z)| - |a| \quad \text{(by (*))}$$
$$\geqslant |f_1(z)| - |a| \quad \text{(since } |z| > R \geqslant 1)$$
$$\geqslant M + |a| - |a| = M,$$

proving the lemma by induction. $\qquad \square$

To prove the first part of the fundamental theorem, let $p(z) = z^n + a_{n-1}z^{n-1} + \cdots + a_1 z + a_0$. Choose R so that for $|z| > R$, $|p(z)| \geqslant 1 + |a_0|$. Let $D = \{z | |z| \leqslant R\}$. From calculus it is known that there is some z_0 in D such that $|p(z_0)| \leqslant |p(z)|$ for all z in D. Now, by the way that we have chosen D, $|p(z_0)| \leqslant |p(z)|$ for all z. For if z is not in D, $|z| > R$, so $|p(z)| \geqslant 1 + |a_0| > |p(0)|$. Since 0 is in D, $|p(0)| \geqslant |p(z_0)|$. Thus $|p(z_0)| \leqslant |p(z)|$ for all z, in D or not. That completes the first part of the proof.

Part 2. Let z_0 be the point found in part 1 such that $|p(z_0)| \leqslant |p(z)|$ for all z. We are going to make two changes of variables to put z_0 at the origin and make our polynomial look nice.

First make a change of variables $w = z - z_0$. Then $p(z) = p(w + z_0) = q_1(w)$ is a polynomial in w and $|q_1(0)| = |p(z_0)| \leqslant |p(z)| = |q_1(w)|$ for all w : $|q_1(w)|$ has its minimum at $w = 0$.

We want to show that $q_1(0) = p(z_0) = 0$. If that is the case, we are done. So for the rest of the proof we assume that $q_1(0) = a \neq 0$; from that assumption we shall reach a contradiction.

Assuming $a \neq 0$, let $q_2(w) = (1/a)q_1(w)$. Then $|q_2(w)|$ has a minimum at $w = 0$ iff $|q_1(w)|$ does. Now $q_2(w)$ has the form

$$q_2(w) = 1 + bw^m + b_1 w^{m+1} + \cdots + b_k w^{m+k},$$

where $m + k = n =$ the degree of $q_2(w) =$ the degree of $p(z)$.

Let r be an mth root of $-1/b$. Then $r^m b = -1$. Let $w = ru$, and set $q(u) = q_2(ru) = q_2(w)$. Then $|q(u)|$ has a minimum at $u = 0$ iff $|q_2(w)|$ has a minimum at $w = 0$. Now $q(u)$ has the form

$$q(u) = 1 + b(ru)^m + b_1(ru)^{m+1} + \cdots + b_k(ru)^{m+k}$$

$$= 1 - u^m + u^{m+1}Q(u) \qquad (\text{since } r^m b = -1),$$

where

$$Q(u) = c_1 + c_2 u + \cdots + c_k u^{k-1}$$

is a polynomial in u, with $c_j = b_j r^{m+j}$ for each j, $1 \leqslant j \leqslant k$.

Note that $q(0) = 1$, so 1 is the minimum value of $|q(u)|$.

Let t be a real number > 0. Setting $u = t$,

$$|Q(t)| = |c_1 + c_2 t + \cdots + c_k t^{k-1}|$$

$$\leqslant |c_1| + |c_2|t + \cdots + |c_k|t^{k-1}$$

by the triangle inequality. We call this last polynomial $Q_0(t)$. It is a polynomial with real coefficients, and is > 0 when t is real and > 0.

As $t \to 0$, $tQ_0(t) \to 0$. Choose t with $0 < t < 1$ so that $tQ_0(t) < 1$.

We show that for this choice of t, setting $u = t$ gives $|q(t)| < 1 = |q(0)|$, contradicting the assumption that $|q(u)|$ had its minimum at $u = 0$. Here is why $|q(t)| < 1$:

$$|q(t)| = |1 - t^m + t^{m+1}Q(t)|$$

$$\leqslant |1 - t^m| + |t^{m+1}Q(t)| \qquad (\text{by the triangle inequality})$$

$$= (1 - t^m) + t^m t |Q(t)| \qquad (\text{since } 0 < t < 1)$$

$$\leqslant (1 - t^m) + t^m(tQ_0(t)).$$

Since t is chosen so that $tQ_0(t) < 1$, this last number is

$$< (1 - t^m) + t^m = 1 = |q(0)|.$$

Since $t \neq 0$, $|q(u)|$ does not have its minimum at $u = 0$. We have reached a contradiction, and the proof is complete. $\square$

The above proof is given in articles by C. Fefferman (1967) and F. Terkelson (1976), and is essentially in Chrystal (1898-1900, Chapter XII).

For a rather different proof, which gives a computational procedure for locating a root of $f(z)$ in $\mathbb{C}$, see Kuhn (1974).

†**E4.** (a) Show that if $p(z) = z^n + a_{n-1}z^{n-1} + \cdots + a_2z^2 + a_1z + a_0$ and $|z| > 1 + |a_{n-1}| + \cdots + |a_1| + |a_0|$, then $|p(z)| > 0$.

    (b) Give an example where $|z| > |a_{n-1}| + \cdots + |a_1| + |a_0|$ but $p(z) = 0$.

E5. Let $f(z) = z^4 + 2z^3 - 6z^2 - 22z + 65$. Prove that $f(z)$ does not have a minimum at $z = 2i$ by letting $z_0 = 2i$ in the second part of the proof and following through the argument of that proof to find some z with $|f(z)| < |f(2i)|$.

E6. Prove that if F is an infinite field and has some irreducible polynomial $p(x)$ of degree d, then it has infinitely many. (*Hint*: Try $p(x - a)$.)

4

Irreducible Polynomials in $\mathbb{R}[x]$

Given the validity of the fundamental theorem of algebra, determining the irreducible polynomials in $\mathbb{R}[x]$ is a simple application of the division theorem.

Theorem. *If $f(x)$ is an irreducible polynomial with real coefficients, then either*

$f(x)$ *has degree* 1, *or*
$f(x) = ax^2 + bx + c$ *with* $b^2 - 4ac < 0$.

PROOF. Clearly any polynomial of degree 1 is irreducible.
Suppose $f(x) = ax^2 + bx + c$. Then $f(x)$ factors in $\mathbb{C}[x]$ into

$$f(x) = a\left(x + \frac{b}{2a} + \frac{1}{2a}\sqrt{b^2 - 4ac}\right)\left(x + \frac{b}{2a} - \frac{1}{2a}\sqrt{b^2 - 4ac}\right).$$

If $b^2 - 4ac < 0$ then these roots of $f(x)$ are not real, so $f(x)$ has no real roots (why?) and is therefore irreducible. Thus the two kinds of real polynomials we claimed were irreducible in $\mathbb{R}[x]$ are in fact irreducible.

For the converse, suppose $f(x)$ is an irreducible polynomial of degree > 1. Then it has no real roots. Let α be a nonreal complex root of $f(x) : f(\alpha) = 0$. Say $\alpha = r + is$, r, s real, $s \neq 0$. Let $\bar{\alpha} = r - is$. Then

$$g(x) - (x - \alpha)(x - \bar{\alpha}) = (x - (r + is))(x - (r - is))$$
$$= x^2 - 2r + (r^2 + s^2)$$

is in $\mathbb{R}[x]$. Apply the division theorem with f and g:

$$f(x) = g(x)q(x) + r(x), \quad \text{where } q(x), r(x) \quad \text{are in } \mathbb{R}[x].$$

Then $r(x)$ is a real polynomial of degree $\leqslant 1$. Think of this equation as an equality of functions on $\mathbb{C}$, and set $x = \alpha$. Then $0 = f(\alpha) = g(\alpha)q(\alpha) +$

$r(\alpha)$; $g(\alpha) = 0$ by the way we constructed $g(x)$, so $r(\alpha) = 0$. Now $r(x) = ax + b$ with a, b real, so if $r(\alpha) = 0$, then $a\alpha + b = 0$. Since α is not real this equation can hold only when $a = b = 0$. Thus $r(x) = 0$, and we get

$$f(x) = g(x)q(x).$$

This means that $f(x)$ is not irreducible, unless $q(x)$ is a constant. If $q(x)$ is a constant, $f(x)$ has degree 2 and has no real roots, so has already been accounted for. This completes the proof. □

Of course, we used in the proof the fundamental theorem of algebra. Thus while we know which polynomials are irreducible in $\mathbb{R}[x]$ (or $\mathbb{C}[x]$) it is substantially harder to see how to factor or to find the roots of a polynomial we know is not irreducible. In general the roots must be obtained by approximation. Such problems form an important part of the subject of numerical analysis. We shall consider a special case of this problem in Chapter 7.

E1. Show that a polynomial of odd degree in $\mathbb{R}[x]$ with no multiple roots must have an odd number of real roots.

E2. Show that if $f(x)$ is a polynomial with real coefficients and $z = r + is$ in $\mathbb{C}$ is a root of $f(x)$, then so is $\bar{z} = r - is$, the complex conjugate of z.

E3. Given that $f(x) = x^4 - 4x^3 + 3x^2 + 14x + 26$ has roots $3 + 2i$ and $-1 - i$, factor f into a product of irreducible polynomials in $\mathbb{R}[x]$.

E4. Suppose we look for the roots of $f(x) = x^3 - 8x + 2$. Since $f(-3) = -1$, $f(0) = 2$, $f(1) = -5$, $f(3) = 5$, a sketch of the graph of $y = f(x)$, or the intermediate value theorem from calculus, shows that f has three real roots. But the technique described in Section 3A gives roots expressed as sums of complex numbers. What are the three real roots of $f(x)$?

5 Partial Fractions

A. Rational Functions

In the same way that the field of rational numbers is formed from the ring of integers, the field of rational functions may be constructed from the set of polynomials with coefficients in a field. A *rational function* with coefficients in the field F is an expression of the form $f(x)/g(x)$ where $f(x)$ and $g(x)$ are in $F[x]$ and $g(x) \neq 0$; two rational functions are equal,

$$\frac{f(x)}{g(x)} = \frac{h(x)}{k(x)},$$

if $k(x)f(x) = g(x)h(x)$ in $F[x]$. Call the set of rational functions with coefficients in F by $F(x)$ (as opposed to $F[x]$, which denotes the set of polynomials with coefficients in F).

Addition and multiplication of rational functions is defined by the usual rules for fractions (we drop "(x)"):

$$f/g + h/k = (fk + hg)/gk; \qquad (f/g)(h/k) = fh/gk.$$

It is very easy to verify that $F(x)$ is a field. A polynomial f may be viewed as a rational function by thinking of it as $f/1$.

The terminology "rational function" is somewhat misleading. The elements of $F(x)$ are not functions on the field F, but formal expressions in the same sense as polynomials are. One can evaluate a rational function $f(x)/g(x)$ at any element a of F at which $g(a) \neq 0$, but two rational functions may agree when evaluated at every element of F and yet be different elements of $F(x)$, such as $x^3/(x^2 + x + 1)$ and x in $\mathbb{Z}_2[x]$; and there may exist rational functions in $F(x)$ which cannot be defined as functions on F at all, such as $x^3 - x + 1/(x^3 - x)$ in $\mathbb{Z}_3(x)$, whose denominator gives zero when evaluated at any element of $\mathbb{Z}_3$. However, as with polynomials, it can be proved that if F is an infinite field, then two rational functions which have the same values when evaluated on infinitely many elements of F must be equal.

E1. Prove this last assertion.

B. Partial Fractions

The method of partial fractions is a way of decomposing a rational function f/g into a sum of terms with denominators of degrees smaller than g when a factorization of g is known. In case f/g is a rational function with real coefficients, viewed as a real-valued function, then partial fractions becomes an important technique of integration. In this section we shall describe the general method. In the two following sections we apply the method to integration and to counting the number of solutions to certain equations.

We assume $f(x)$ and $g(x)$ are in $F[x]$, where F is an arbitrary field.

Given f/g, we first use the division theorem, if necessary, to write $f = gq + r$, with $\deg r < \deg g$. Then $f/g = q + r/g$. The basic problem, to write f/g as a sum of terms with "nice" denominators, remains for r/g. So we shall assume that we started out with f/g, where $\deg f < \deg g$.

Here is the general theorem on partial fractions.

Theorem. *Let $g = p_1^{e_1} p_2^{e_2} \cdots p_r^{e_r}$ be a factorization of g into a product of powers of relatively prime polynomials p_i, and suppose that $\deg f < \deg g$.*

Then there are unique polynomials h_i, $i = 1, \ldots, r$ with $\deg h_i < \deg p_i^{e_i}$, such that

$$\frac{f}{g} = \frac{h_i}{p_1^{e_1}} + \frac{h_2}{p_2^{e_2}} + \cdots + \frac{h_r}{p_r^{e_r}}.$$

PROOF. Induction on r, $r = 1$ being trivial.

In order to pass from $r - 1$ to r, and thus prove the theorem, we let $a = p_1^{e_1} \cdots p_{r-1}^{e_{r-1}}$, $b = p_r^{e_r}$ and prove the following, which is the induction step.

Lemma. *Let $g = ab$ where a and b are relatively prime, and suppose $\deg f < \deg g$. Then there are unique polynomials r, s with $\deg r < \deg a$, $\deg s < \deg b$, so that*

$$\frac{f}{g} = \frac{r}{a} + \frac{s}{b}.$$

The theorem follows. For using the lemma, we may write

$$\frac{f}{g} = \frac{r}{p_1^{e_1} \cdots p_{r-1}^{e_{r-1}}} + \frac{s}{p_r^{e_r}};$$

use induction to write

$$\frac{r}{p_1^{e_1} \cdots p_{r-1}^{e_{r-1}}} = \frac{h_1}{p_1^{e_1}} + \cdots + \frac{h_{r-1}}{p_{r-1}^{e_{r-1}}},$$

and set $h_r = s$. $\qquad\qquad\square$

To prove the lemma, we use Bezout's lemma: since a and b are relatively prime, there are polynomials s, r such that

$$as + br = 1,$$

where we can choose $\deg s < \deg b$, $\deg r < \deg a$ by Exercise E16, Chapter II-2. Now we divide by $g = ab$:

$$\frac{f}{g} = \frac{r}{a} + \frac{s}{b}.$$

To show r, s are unique is easy. $\qquad\qquad\square$

Once we have a rational function written as a sum of terms of the form f/p^e, we can further decompose f/g by representing the numerator in base p.

To write f in base p means to write $f = r_0 + r_1 p + r_2 p^2 + \cdots + r_k p^k$, where p is any polynomial, and $\deg r_i < \deg p$ for all i. In case $p(x) = x$, writing the polynomial $f(x)$ in base x is the way we usually write $f(x)$. If we could write f in base p, then f/p^e would decompose as

$$\frac{f}{p^e} = \frac{r_0 + r_1 p + \cdots + r_k p^k}{p^e}$$
$$= \frac{r_0}{p^e} + \frac{r_1}{p^{e-1}} + \cdots + \frac{r_k}{p^{e-k}}$$

with $\deg r_i < \deg p$ for all i. In case $p(x) = x - r$, this would mean that all of the r_i would be constants.

We write f in base p just as we did with integers, as follows. Divide p into f,

$$f = pq_0 + r_0,$$

then divide p into the quotients, successively:

$$q_0 = pq_1 + r_1,$$
$$q_1 = pq_2 + r_2,$$
$$\vdots$$
$$q_{k-1} = pq_k + r_k,$$

where $\deg q_{k-1} < \deg p$. Then $f = r_k p^k + r_{k-1} p^{k-1} + \cdots + r_1 p + r_0$, as can be seen by substituting each successive equation into the first equation. Since the quotient and remainder in the division algorithm are unique, the representation of f in base p is unique.

Representation in base p for polynomials is thus essentially the same as for integers.

Writing the numerator in base p and reducing to lowest terms completes the decomposition of f/g.

EXAMPLE. Let

$$\frac{f(x)}{g(x)} = \frac{3x^4 + 5}{(x^2 + 1)^2 x}$$

in $\mathbb{R}(x)$. We can follow the method of partial fractions as just described, but it is easier simply to use the result: by the general theory, we know that the decomposition should be

$$\frac{3x^4 + 5}{(x^2 + 1)^2 x} = \frac{ax + b}{x^2 + 1} + \frac{cx + d}{(x^2 + 1)^2} + \frac{e}{x}$$

for some real numbers a, b, c, d, e. The most direct method is simply to put the right side under a common denominator, collect coefficients, equate them to the coefficients on the left side, and solve:

$$\frac{3x^4 + 5}{(x^2 + 1)^2 x} = \frac{(a + e)x^4 + bx^3 + (a + c + 2e)x^2 + (b + d)x + e}{(x^2 + 1)^2 x}$$

so

$$
\begin{aligned}
a + e &= 3 \\
b &= 0 \\
a + c + 2e &= 0 \\
b + d &= 0 \\
e &= 5
\end{aligned}
$$

with solution $a = -2$, $b = d = 0$, $c = -8$, $e = 5$.

E2. Write x^5 in base $x + 1$.

E3. Write $(x^2 + 3x + 1)^4$ in base $x + 2$; in base $x^2 + x + 1$.

E4. Decompose into partial fractions:

$$\frac{t + 1}{(t - 1)(t + 2)}; \qquad \frac{1}{(t + 1)(t^2 + 2)}; \qquad \frac{x^2 + 4}{(x + 1)^2(x - 2)(x + 3)}$$

*E5. What is the analogue of partial fractions in $\mathbb{Z}$? Illustrate it with $17/180$.

E6. Where in the proof of the fundamental theorem of algebra did we write a polynomial in a new base?

E7. Prove that there are infinitely many irreducible polynomials in $K[x]$ for K any field, as follows: Suppose $p_1, \ldots, p_n$ are all the irreducible polynomials in $K[x]$. Form

$$\frac{1}{p_1} + \frac{1}{p_2} + \cdots + \frac{1}{p_n} = \frac{a}{b}$$

where a, b are polynomials with $(a, b) = 1$. Argue that the degree of a is at least $= 1$ and none of the irreducible polynomials $p_1, \ldots, p_n$ can divide a, so that there must be some other irreducible polynomial.

E8. Let p be a polynomial, f/g a rational function. By analogy with infinite decimals, obtain the expansion of f/g in base p:

$$\frac{f}{g} = \sum_{i=-n}^{\infty} \frac{a_i}{p^i}.$$

Illustrate with

$$\frac{f}{g} = \frac{x^2}{x^3 + 3x + 1}, \qquad p(x) = x,$$

to "5 places."

*E9. Let f/g be a rational function. By Exercise E8 we can write

$$\frac{f(1/x)}{g(1/x)} = \frac{f_0(x)}{g_0(x)} = \sum_{k=-n}^{\infty} \frac{a_k}{x^k}.$$

Replacing x by $1/x$ we get

$$\frac{f(x)}{g(x)} = \sum_{k=-n}^{\infty} a_k x^k.$$

(a) If $f(0) \neq 0$ and $g(0) \neq 0$ show that $n = 0$, so that $f(k)/g(k) = \sum_{k=0}^{\infty} a_k x^k$.
(b) What connection, if any, is there between the expression $f(x)/g(x) = \sum_{k=0}^{\infty} a_k x^k$ and the Taylor series for $f(x)/g(x)$ about $x = 0$ ($F = \mathbb{R}$)?

C. Integrating

The standard use of partial fractions in calculus is in finding the indefinite integral of a rational function $f(x)/g(x)$, that is, finding a function $H(x)$ such that $H'(x) = f(x)/g(x)$.

It is assumed that the factorization of $g(x)$ in $\mathbb{R}[x]$ is known. Then by partial fractions, $f(x)/g(x)$ can be written as a polynomial $q(x)$ plus a sum of terms of the forms

$$\frac{a}{(x - d)^r}$$

and

$$\frac{bx + c}{(x^2 + e^2)^s} = \frac{bx}{(x^2 + e^2)^s} + \frac{e}{(x^2 + e^2)^s}.$$

Thus to integrate $f(x)/g(x)$ involves finding $\int q(x)\,dx$, where $q(x)$ is a polynomial (very easy), and integrals of the form

$$\int \frac{a\,dx}{(x-d)^r},\tag{1}$$

$$\int \frac{bx\,dx}{(x^2+e^2)^s},\tag{2}$$

and

$$\int \frac{c\,dx}{(x^2+e^2)^s}.\tag{3}$$

We show how to do these integrals.

The first is not hard: if $r\neq 1$ the integral (1) equals

$$\frac{a}{(1-r)(x-d)^{r-1}}+C;$$

if $r=1$ it equals $a\log(x-d)+C$.

If we let $x^2=u$ in integral (2), it becomes

$$\int \frac{bx\,dx}{(x^2+e^2)^s}=\frac{b}{2}\int \frac{du}{(u+e^2)^s},$$

which is of the form (1).

The remaining integral, integral (3), is more difficult. In calculus books an integral of type (3) is usually done by setting $x=e\tan t$ to transform it into

$$\int \frac{ce\sec^2 t\,dt}{(e^2\sec^2 t)^s}=\frac{c}{e^{2s-1}}\int \cos^{2s-2}t\,dt,$$

which is then done by a recurrence formula. But it can also be done using partial fractions in $\mathbb{C}[x]$. For convenience we first substitute $x=eu$ in (3) so that (3) becomes

$$\int \frac{ce\,du}{(e^2(1+u^2))^s}=\frac{c}{e^{2s-1}}\int \frac{du}{(1+u^2)^s}.$$

We then want to solve

$$\int \frac{du}{(1+u^2)^s}.\tag{4}$$

In $\mathbb{C}[x]$, $1+u^2=(1+iu)(1-iu)$. So by partial fractions, there exist complex numbers $a_1,\ldots,a_s,b_1,\ldots,b_s$ in $\mathbb{C}$ such that

$$\frac{1}{(1+u^2)^s}=\frac{a_1}{1+iu}+\frac{a_2}{(1+iu)^2}+\cdots+\frac{a_s}{(1+iu)^s}$$

$$+\frac{b_1}{1-iu}+\frac{b_2}{(1-iu)^2}+\cdots+\frac{b_s}{(1-iu)^s}.\tag{5}$$

So to find (4) we would solve for $a_1, \ldots, a_s, b_1, \ldots, b_s$ and then find

$$\int \frac{a_k \, du}{(1 + iu)^k} \quad \text{and} \quad \int \frac{b_k \, du}{(1 - iu)^k} \quad \text{for } k = 1, \ldots, s.$$

The formulas for these are essentially the same as in the real case, except when $k = 1$. To avoid difficulties with the complex logarithm function, we first do the case $s = 1$.

For $s = 1$ an easy computation shows that

$$\frac{1}{1 + u^2} = \frac{1}{2(1 + iu)} + \frac{1}{2(1 - iu)}.$$

Hence

$$\int \frac{du}{1 + u^2} = \frac{1}{2}\left[\int \frac{du}{1 + iu} + \int \frac{du}{1 - iu} \right].$$

On the other hand, the substitution $u = \tan t$ solves $\int du/(1 + u^2)$ without passing to the complex numbers. So we get

$$\frac{1}{2}\left[\int \frac{du}{1 + iu} + \int \frac{du}{1 - iu} \right] = \text{arc tan } u + C.$$

Now for any s, $s \geqslant 1$, we have that in formula (5) $a_1 = b_1$. For when we put the right side of (5) under the common denominator $(1 + u^2)^s$, the equation for the coefficient of u^{2s-1} is

$$0 = a_1 i^{s-1}(-i)^s + b_1(-i)^{s-1}i^s,$$

or, after canceling,

$$0 = a_1 - b_1.$$

In fact, we have $a_k = b_k$ for all $k = 1, \ldots, s$ (E11). Hence,

$$\int \frac{du}{(1 + u^2)^s} = a_1\left[\int \frac{du}{1 + iu} + \int \frac{du}{1 - iu} \right]$$

$$+ \sum_{k=2}^{s} a_k\left[\int \frac{du}{(1 + iu)^k} + \int \frac{du}{(1 - iu)^k} \right]. \qquad (6)$$

Integrating, we get

$$2a_1 \text{ arc tan } u + \sum_{k=2}^{s} a_k\left(\frac{1}{1 - k}\right)\left[\frac{1}{i(1 + iu)^{k-1}} - \frac{1}{i(1 - iu)^{k-1}} \right] + C.$$

For u real, each term in brackets is fixed under complex conjugation, so is a real function of u.

In particular, for $s = 2$, formula (5) becomes

$$\frac{1}{(1 + u^2)^2} = \frac{a_1}{1 + iu} + \frac{a_2}{(1 + iu)^2} + \frac{b_1}{(1 - iu)} + \frac{b_2}{(1 - iu)^2}.$$

Solving, we get $a_1 = a_2 = b_1 = b_2 = 1/4$. So formula (6) is

$$\int \frac{du}{(1+u^2)^2} = \frac{1}{2}\arctan u + \frac{1}{4}\left[\frac{-1}{i(1+u)} + \frac{1}{i(1-iu)}\right] + C.$$

The term in brackets is easily seen to be equal to $2u/(1+u^2)$. So

$$\int \frac{du}{(1+u^2)^2} = \frac{1}{2}\arctan u + \frac{u}{2(1+u^2)} + C.$$

This is an example of how complex numbers may be used to solve real problems in a nice way. We saw this also in solving real cubic equations in Chapters 3 and 4, and we shall see it again in the next section.

E10. Using complex numbers, find
    (a) $\int du/(1+u^2)^3$,
    (b) $\int dx/(1+x^4)$,
    (c) $\int dx/(1-x^4)$.

E11. Show that in solving integral (4) by partial fractions, $a_1, \ldots, a_s, b_1, \ldots, b_s$ are real, and $a_1 = b_1, \ldots, a_s = b_s$.

D. A Partitioning Formula

In this section we give an application of partial fractions which does not involve integration.

A problem which has classically been of interest in number theory is that of counting the number of ways a natural number can be partitioned into sums of natural numbers. For example, here are all the ways of partitioning the number 5:

$$\begin{aligned}
5 &= 1 + 1 + 1 + 1 + 1 \\
&= 1 + 1 + 1 + 2 \\
&= 1 + 2 + 2 \\
&= 1 + 1 + 3 \\
&= 2 + 3 \\
&= 1 + 4 \\
&= 5.
\end{aligned}$$

There is also the problem of determining the number of ways a natural number can be partitioned, subject to restrictions on the numbers making up the sum. For example, in how many ways can the number n be partitioned into sums involving only the numbers 1, 2, and 3?

Combinatorial problems such as these are related to problems of practical present-day interest. Here is an example. A manufacturer of paper makes standard rolls of paper of length N, and sells paper in rolls of smaller lengths $n_1, \ldots, n_r$ which one obtains by cutting up the standard rolls:

$$\overline{\quad n_1 \quad n_1 \quad n_1 \quad n_2 \quad n_2 \quad n_3 \cdots N \quad}$$

How can one cut up the standard rolls in such a way that there is (little or) no waste? If all the numbers are integers, and the cutting can be done with no waste, then each solution is a (physical) partition of N into a sum of integers chosen from among $n_1, \ldots, n_r$. (This problem was actually posed in the 1950s by a paper manufacturer, with the extra complication of involving a demand for various quantities of rolls of each length n_i. The problem was solved by the invention of a new kind of algorithmic technique, related to linear programming, called integer programming.)

In this section we illustrate the use of partial fractions in the study of partitions, by finding an explicit formula for the number of ways a number n can be partitioned into sums of 1's, 2's, and 3's. This number of ways is the same as the number of solutions (r, s, t) of the equation $r \cdot 1 + s \cdot 2 + t \cdot 3 = n$ with $r, s, t \geqslant 0$.

Our starting point is a trick of Euler:

If $a(n)$ is the number of solutions of $r + 2s + 3t = n$, $r, s, t \geqslant 0$, then

$$(1 + x + x^2 + \cdots)(1 + x^2 + x^4 + \cdots)(1 + x^3 + x^6 + \cdots)$$

$$= 1 + \sum_{n=1}^{\infty} a(n) x^n.$$

For multiplying out the left side, and defining $a(0) = 1$,

$$\left(\sum_{r=0}^{\infty} x^r \right) \left(\sum_{s=0}^{\infty} x^{2s} \right) \left(\sum_{t=0}^{\infty} x^{3t} \right) = \sum_{r, s, t=0}^{\infty} x^{r+2s+3t}$$

$$= \sum_{n=0}^{\infty} \left(\sum x^{r+2s+3t} \right) = \sum_{n=0}^{\infty} a(n) x^n.$$

Here the inside sum in the middle term runs through all (r, s, t), $r, s, t > 0$, with $r + 2s + 3t = n$. For such (r, s, t), $r + 2s + 3t = n$, of course, and there are $a(n)$ solutions. So the inside sum is $a(n) x^n$. What we have done is rearrange the series, which is perfectly permissible if $x > 0$.

Now the series we multiplied together are geometric series in x, x^2, and x^3 respectively, and we have for $|x| < 1$,

$$1 + x + x^2 + \cdots = \sum_{r=0}^{\infty} x^r = \frac{1}{1-x},$$

$$1 + x^2 + x^4 + \cdots = \sum_{s=0}^{\infty} x^{2s} = \frac{1}{1-x^2},$$

$$1 + x^3 + x^6 + \cdots = \sum_{t=0}^{\infty} x^{3t} = \frac{1}{1-x^3}.$$

So we get the formula

$$\sum_{n=0}^{\infty} a(n)x^n = \frac{1}{(1-x)(1-x^2)(1-x^3)}.$$ (1)

We use partial fractions to implement the idea that while it is hard to find the coefficient of x^n in a *product* of power series, it is easy in a *sum* of power series. So we shall use partial fractions on the right side of (1) to get a nice expression for $a(n)$.

The right side of (1) factors over the complex numbers

$$\frac{1}{(1-x)(1-x^2)(1-x^3)} = \frac{1}{(1-x)^3(1+x)(1+x+x^2)}$$

$$= \frac{1}{(1-x)^3(1+x)(1-\omega x)(1-\omega^2 x)}$$

where $\omega = e^{2\pi i/3}$ is a complex cube root of 1, satisfying $\omega + \omega^2 = -1$.
By partial fractions in $\mathbb{C}[x]$, the last term equals

$$\frac{A}{1-x} + \frac{B}{(1-x)^2} + \frac{C}{(1-x)^3} + \frac{D}{1+x} + \frac{E}{1-\omega x} + \frac{F}{1-\omega^2 x}$$ (2)

for some A, B, C, D, E, F in $\mathbb{C}$. Putting (2) under a common denominator and equating coefficients of the numerators, we get the system of linear equations:

$$1 = A + B + C + D + E + F,$$
$$0 = B + 2C - 2D - (2+\omega^2)E - (2+\omega)F,$$
$$0 = -A + 2C + D + 2\omega^2 E + 2\omega F,$$ (3)
$$0 = -A - B + C - D + 2E + 2F,$$
$$0 = -B + 2D - (2\omega^2 + 1)E - (2\omega + 1)F,$$
$$0 = A - D + \omega^2 E + \omega F,$$

which eventually yields solutions (see E14)

$$A = \frac{17}{72}, \quad B = \frac{1}{4}, \quad C = \frac{1}{6},$$
$$D = \frac{1}{8}, \quad E = \frac{1}{9}, \quad F = \frac{1}{9}.$$

So

$$\sum_{n=0}^{\infty} a(n)x^n = \frac{(17/72)}{1-x} + \frac{(1/4)}{(1-x)^2} + \frac{(1/6)}{(1-x)^3} + \frac{(1/8)}{1+x} + \frac{(1/9)}{1-\omega x} + \frac{(1/9)}{1-\omega^2 x}.$$

We now expand each of these terms as an infinite series. This is easy to do.

$$\frac{1}{1-x} = 1 + x + x^2 + \cdots = \sum_{n=0}^{\infty} x^n,$$

$$\frac{1}{1-\omega x} = 1 + \omega x + \omega^2 x^2 + \cdots = \sum_{n=0}^{\infty} (\omega x)^n, \qquad \frac{1}{1-\omega^2 x} = \sum_{n=0}^{\infty} (\omega^2 x)^n,$$

$$\frac{1}{1+x} = 1 - x + x^2 - x^3 \cdots = \sum_{n=0}^{\infty} (-1)^n x^n.$$

To find $1/(1-x)^2$ and $1/(1-x)^3$ observe that

$$\frac{1}{(1-x)^2} = \frac{d}{dx}\left(\frac{1}{1-x}\right) = \frac{d}{dx}\left(\sum_{n=0}^{\infty} x^n\right) = \sum_{n=0}^{\infty} (n+1)x^n$$

and

$$\frac{1}{(1-x)^3} = \frac{1}{2}\frac{d}{dx}\left(\frac{1}{(1-x)^2}\right)$$

$$= \frac{1}{2}\frac{d}{dx}\left(\sum_{n=0}^{\infty} (n+1)x^n\right) = \frac{1}{2}\sum_{n=0}^{\infty} (n+1)(n+2)x^n$$

since an absolutely convergent power series (which these are for $|x| < 1$) can be differentiated term by term.

Thus

$$\sum_{n=0}^{\infty} a(n)x^n = \frac{17}{72}\left(\sum_{n=0}^{\infty} x^n\right) + \frac{1}{4}\left(\sum_{n=0}^{\infty} (n+1)x^n\right)$$

$$+ \frac{1}{6}\left(\sum_{n=0}^{\infty} \frac{(n+1)(n+2)}{2} x^n\right)$$

$$+ \frac{1}{8}\left(\sum_{n=0}^{\infty} (-1)^n x^n\right) + \frac{1}{9}\sum_{n=0}^{\infty} (\omega^n + \omega^{2n})x^n,$$

where

$$\omega^n + \omega^{2n} = \varepsilon_n = \left\{\begin{matrix} 2 \\ -1 \end{matrix}\right\} \quad \text{if} \quad n\left\{\begin{matrix} \equiv \\ \not\equiv \end{matrix}\right\} 0 \ (\text{mod } 3).$$

Collecting terms, and setting $a(0) = 1$, we get an explicit formula for $a(n)$:

$$a(n) = \frac{17}{72} + \frac{(n+1)}{4} + \frac{(n+1)(n+2)}{12} + \frac{(-1)^n}{8} + \frac{1}{9}\varepsilon_n.$$

In particular, for $n = 0, \ldots, 5$ we have $a(0) = 1$, $a(1) = 1$, $a(2) = 2$, $a(3) = 3$, $a(4) = 4$, $a(5) = 5$, which are correct.

We can simplify the expression for $a(n)$ by comparing it with $(n+3)^2/12$. We have

$$\frac{(n+3)^2}{12} - a(n)$$

$$= \frac{(n+3)^2}{12} - \left[\frac{17}{72} + \frac{(n+1)}{4} + \frac{(n+1)(n+2)}{12} + \frac{(-1)^n}{8} + \frac{1}{9}\varepsilon_n \right]$$

$$= \frac{n^2+6n+9}{12} - \left[\frac{3(n+1)+(n+1)(n+2)}{12} + \frac{17}{72} + \frac{(-1)^n}{8} + \frac{1}{9}\varepsilon_n \right].$$

Since

$$\frac{3(n+1)+(n+1)(n+2)}{12} = \frac{(n+1)(n+5)}{12} = \frac{n^2+6n+5}{12},$$

we get

$$\frac{(n+3)^2}{12} - a(n) = \frac{n^2+6n+9}{12} - \frac{n^2+6n+5}{12} - \left[\frac{17+(-1)^n 9 + 8\varepsilon_n}{72} \right]$$

$$= \frac{1}{3} - \left[\frac{17+(-1)^n 9 + 8\varepsilon_n}{72} \right]$$

$$= \frac{7-(-1)^n 9 - 8\varepsilon_n}{72}.$$

This last expression is $\leqslant (7+9+8)/72 = 1/3$ and $\geqslant (7-9-16)/72 = -1/4$. So

$$\frac{-1}{4} \leqslant \frac{(n+3)^2}{12} - a(n) \leqslant \frac{1}{3}.$$

Since $a(n)$ is an integer, we can obtain $a(n)$ easily by letting $a(n) =$ the integer nearest to $(n+3)^2/12$. After this effort we get quite a simple formula for $a(n)$!

E12. Count the number of solutions of $r + 2s = n$ with $r, s > 0$.

E13. Do the same for $r + 3s = n$.

E14. An easy way to solve (2) is to put (2) under a common denominator and equate numerators:

$$1 = A(1-x)^2(1+x)(1+x+x^2)$$
$$+ B(1-x)(1+x)(1+x+x^2)$$
$$+ C(1+x)(1+x+x^2)$$
$$+ D(1-x)^3(1+x+x^2)$$
$$+ E(1-x)^3(1+x)(1-\omega^2 x)$$
$$+ F(1-x)^3(1+x)(1-\omega x).$$

Then set $x = 1$, $x = -1$, $x = \omega$, $x = \omega^2$. You get the values of C, D, F, and E instantly, and the values of A, B from the last two equations of (3). Verify by this method that the values of A, B, C, D, E, F in the text are correct.

The Derivative of a Polynomial

6

In the last chapter we described some algebra which arises in integrating. In this chapter we observe that differentiating can be done purely algebraically, and can give a partial criterion for deciding whether a polynomial has a repeated factor. Since we can differentiate without using limits, we can assume that the polynomials have coefficients in any field, not necessarily the field of real numbers.

Let F be a field, $f(x)$ a polynomial with coefficients in F. Define $D(f) = f'$, another polynomial in $F[x]$, by the two rules:

(1) *For a in F, $n \geqslant 0$, $D(ax^n) = a(x^{n-1} + \cdots + x^{n-1})$ (n times) = nax^{n-1}.*

This is a slightly misleading formula, for the exponent "n" in ax^n is a nonnegative integer, whereas the coefficient "n" in nax^{n-1} denotes $1 + 1 + \cdots + 1$ (n times) in F. Thus if $F = \mathbb{Z}_3$, $D(x^3) = 3x^2 = 0$ since $3 = 0$ in $\mathbb{Z}_3$.

(2) $D(f + g) = Df + Dg$.

These rules define the derivative f' of f for any polynomial f in $F[x]$.

Proposition. $D(f \cdot g) = fD(g) + D(f)g$.
This is the familiar product rule.

SKETCH OF PROOF. By rule (2), if $f = f_1 + f_2$, then $D(fg) = D((f_1 + f_2)g) = D(f_1 g + f_2 g) = D(f_1 g) + D(f_2 g)$. A similar result holds if g is a sum. Since any polynomial is a sum of terms of the form $a_n x^n$ for $n \geqslant 0$, it suffices to prove the theorem when $f(x) = a_n x^n$, $g(x) = b_m x^m$. But using rule (1) this is easy.   □

Here is an algebraic reason for wanting the derivative of a polynomial.

Theorem. *Let $f(x)$ be in $F[x]$, where F is a field contained in the complex numbers. Then $f(x)$ has no multiple factors iff the greatest common divisor of f and f' is 1.*

(Recall our comments about greatest common divisors of polynomials. The last phrase of the theorem means: "The only common divisors of f and f' are constants.")

PROOF. Let $f(x) = p(x)^e q(x)$ with $e > 1$. Then

$$f'(x) = p(x)^e q'(x) + ep(x)^{e-1} q(x) = p(x)^{e-1}(p(x)q'(x) + eq(x)).$$

Thus $p(x)^{e-1}$ is a nontrivial common divisor and $f(x)$ and $f'(x)$ are not relatively prime.

Conversely, if $d(x)$ is the greatest common divisor of $f(x)$ and $f'(x)$ and has degree $\geqslant 1$, let $p(x)$ be an irreducible factor of $d(x)$. Then $f(x) = p(x)g(x)$. So $f'(x) = p'(x)g(x) + p(x)g'(x)$. Since $p(x)$ is irreducible, $p(x)$ divides $p'(x)$ or $g(x)$.

Let $p(x) = a_n x^n + \cdots + a_1 x + a_0$. Then $p'(x) = na_n x^{n-1} + \cdots + a_1$. Since $n > 0$ (which it must be if $p(x)$ is irreducible) and $F \subseteq \mathbb{C}$, $na_n \neq 0$. Thus $p'(x)$ is a nonzero polynomial of degree less than that of $p(x)$. Therefore $p(x)$ cannot divide $p'(x)$, and so $p(x)$ divides $g(x)$.

Since $p(x)$ divides $g(x)$, $g(x) = p(x)h(x)$ for some $h(x)$, and so $f(x) = p(x)^2 h(x)$, and $f(x)$ has a multiple factor. That completes the proof.   □

EXAMPLE. Let $F = \mathbb{R}$. If $f(x) = x^4 - 2x^3 + 3x^2 - 2x + 1$, then $f'(x) = 4x^3 - 6x^2 + 6x - 2$, and $(f(x), f'(x)) = x^2 - x + 1$. Since $x^2 - x + 1$ is irreducible, it is a multiple factor of $f(x)$, so necessarily $f(x) = (x^2 - x + 1)^2$.

E1. For each prime p, find a nonconstant polynomial with coefficients in $\mathbb{Z}_p$ whose derivative $= 0$.

E2. Test for multiple factors in $\mathbb{Q}[x]$:

$$x^4 + 3x^3 + 6x^2 + 7x + 3,$$
$$x^4 + 2x^3 + 8x^2 + 6x + 9,$$
$$x^4 + 5x^2 + 9.$$

E3. If $f(x)$ is in $\mathbb{Q}[x]$, $f(x)$ can also be thought of as having complex coefficients. Show that $f(x)$ has a multiple factor in $\mathbb{Q}[x]$ iff $f(x)$ has a multiple factor in $\mathbb{C}[x]$.

†**E4.** Prove: If F is any field (not necessarily a subset of the complex numbers) and $(f(x), f'(x)) = 1$, then $f(x)$ has no multiple factors.

E5. Let $f(x)$ be a polynomial with complex coefficients, with derivative $f'(x)$. Let

$$D_0 = \text{the greatest common divisor of } f \text{ and } f'$$
$$D_1 = \text{the greatest common divisor of } D_0 \text{ and } D_0',$$
$$D_2 = \text{the greatest common divisor of } D_1 \text{ and } D_1',$$

etc. Suppose D_{m-1} is constant, so $D_m = 0$.

Set $f_1 = f/D$, $f_2 = D/D_1$, $f_3 = D_1/D_2$, $\ldots$, $f_m = D_{m-2}/D_{m-1}$.

Set $g_1 = f_1/f_2$, $g_2 = f_2/f_3$, $g_4 = f_3/f_4$, $\ldots$, $g_m = f_m$.

Show:

(a) each of $g_1, \ldots, g_m$ has no multiple roots;

(b) $g_1, \ldots, g_m$ are pairwise relatively prime;

(c) $f = g_1 g_2^2 g_3^3 \cdots g_m^m$.

Thus $g_r(x) = 0$ has as roots precisely those roots of $f(x)$ which have multiplicity r.

(d) Write $f(x) = x^6 - 3x^5 + 6x^3 - 3x^2 - 3x + 2$ in the form of part (c).

†*E6. Prove that $f^n + g^n = h^n$ has no solutions in $\mathbb{R}[x]$ with $n > 2$, f, g, h of degree ≥ 1, and $(f, g, h) = 1$.

7 Sturm's Algorithm

We discussed in Chapter 5 the technique of computing the integral of a rational function $a(x)/b(x)$ where $a(x)$, $b(x)$ are in $\mathbb{R}[x]$ whenever the factorization of $b(x)$ can be found. Unfortunately there is no algebraic procedure for computing the factorization of a polynomial $f(x)$ in $\mathbb{R}[x]$ in general, and so the factorization of $f(x)$ must be approximated.

Finding the factorization of $f(x)$ in $\mathbb{R}[x]$ is equivalent to finding the roots of $f(x)$ in $\mathbb{C}$. To go very much into the techniques available for approximating the roots of a complex polynomial is beyond the scope of this book. You may wish to consult a book on numerical analysis, or Henrici (1974) or Dejon and Henrici (1969).

We shall, however, look at a technique for finding the number of real roots of $f(x)$ in $\mathbb{R}[x]$ between any two given numbers. The result, Sturm's theorem, is a clever application of Euclid's algorithm.

To make the argument easier, we first replace $f(x)$ by $g(x) = f(x)/d(x)$, where $d(x)$ is the greatest common divisor of $f(x)$ and $f'(x)$. Then $g(x)$ has exactly the same roots as $f(x)$, but each occurs in $g(x)$ with multiplicity 1. We show this as follows. Suppose a is a root of $f(x)$, $f(x) = (x - a)^e k(x)$ with $k(a) \neq 0$. Then $f'(x) = (x - a)^{e-1} h(x)$ with $h(a) \neq 0$, so $d(x) = (x - a)^{e-1} j(x)$ for some $j(x)$ dividing $h(x)$; hence $j(a) \neq 0$. Since $j(a) \neq 0$, $j(x)$ and $x - a$ are relatively prime, so $j(x)$ divides $k(x)$, and $f(x)/d(x) = (x - a)k(x)/j(x)$. This means that if a is a root of $f(x)$, then a is a root of $g(x) = f(x)/d(x)$ of multiplicity 1. Conversely, if a is a root of $g(x)$, then, since $g(x)$ divides $f(x)$, a is a root of $f(x)$, and thus is a root of $g(x)$ of multiplicity 1.

If we want to locate the roots of $f(x)$, then, it is enough to locate the roots of $g(x)$. Since $g(x)$ has no multiple roots, $g(x)$ and $g'(x)$ are relatively prime by the theorem of Chapter 6. So when we apply Euclid's algorithm to $g(x)$ and $g'(x)$ we end up with a constant polynomial as a final remainder.

By replacing $f(x)$ by $g(x)$ if necessary, we shall assume through the rest of this chapter that $f(x)$ has no multiple roots.

Sturm's algorithm works as follows. Write down the Euclidean algorithm in the following slightly modified form:

$$f_0(x) = f(x),$$
$$f_1(x) = f'(x),$$
$$f_0(x) = q_1(x)f_1(x) - f_2(x),$$
$$f_1(x) = q_2(x)f_2(x) - f_3(x),$$
$$\vdots$$
$$f_{r-1}(x) = q_r(x)f_r(x) - 0.$$

Since $(f(x), f'(x)) = 1$, $f_r(x)$ is a constant.

Consider the sequence of functions

$$\mathscr{P}(x) = \{f_0(x), f_1(x), \ldots, f_r(x)\}.$$

For a not a root of $f(x) = f_0(x)$, let $w(a)$ be the number of changes of sign (omitting zeros) in the sequence $\mathscr{P}(a) = \{f_0(a), f_1(a), \ldots, f_r(a)\}$.

Sturm's Theorem. *If $b < c$ and $f(b) \neq 0$, $f(c) \neq 0$, then the number of distinct roots of $p(x)$ between b and c is $w(b) - w(c)$.*

Application of Sturm's theorem is aided by the following exercise.

†**E1.** Let $f(x) = a_0 + a_1x + \cdots + a_{n-1}x^{n-1} + x^n$ be a monic polynomial with real coefficients and let M be either 1 or $|a_0| + |a_1| + \cdots + |a_{n-1}|$, whichever is larger. Show that all real roots of $f(x)$ lie between $-M$ and $+M$.

EXAMPLE. Let $f(x) = x^4 + 2x^3 + 3x^2 + 1$—all roots are between $-M$ and M where, by Exercise E1, $M = 6$. The modified Euclidean algorithm with $f(x)$ and $f'(x)$ gives the sequence of functions

$$\mathscr{P}(x) = \begin{bmatrix} x^4 + 2x^3 + 3x^2 + 1 \\ 4x^3 + 6x^2 + 6x \\ -\dfrac{3}{4}x^2 + \dfrac{3}{4}x - 1 \\ -\dfrac{32}{3}x + \dfrac{40}{3} \\ 158/128 \end{bmatrix}.$$

Evaluated at -6, these functions have the following signs:

$$[+, -, -, +, +].$$

Thus $w(-6) = 2$. Evaluated at 0, these functions have the following signs:

$$[+, 0, -, +, +].$$

Thus $w(0) = 2$. Evaluated at 6, these functions have the following signs:

$$[+, +, -, -, +].$$

Thus $w(6) = 2$. It follows that $w(-6) - w(6) = 0 =$ the number of roots of $f(x)$ between -6 and 6. Since all roots must be between -6 and 6, $f(x)$ has no roots at all.

(This can be checked by computing, by usual calculus techniques, that $f(x)$ has a minimum at $x = 0$, and $f(0) = 1$, so $f(x) > 0$ for all x.)

EXAMPLE. Let $f(x) = x^3 - 5x^2 + 8x - 8$. Then the sequence of functions $\mathcal{P}(x)$ is

$$\mathcal{P}(x) = \begin{bmatrix} x^3 - 5x^2 + 8x - 8 \\ 3x^2 - 10x + 9 \\ x + 16 \\ -1 \end{bmatrix}.$$

Here $M = 21$. Evaluated at $x = -21$, the sequence of numbers $\mathcal{P}(-21)$ has the following signs: $[-, +, -, -]$. At $x = 0$, the sequence of numbers $\mathcal{P}(0)$ has the signs $[-, +, +, -]$. At $x = 21$ the sequence of numbers $\mathcal{P}(21)$ has the signs: $[+, +, +, -]$. Thus $w(-21) = 2$, $w(0) = 2$, $w(21) = 1$. This means that there is only one real root of $f(x)$, and it lies between 0 and 21. We can then forget the sequence $\mathcal{P}(x)$ and just observe that $f(0) < 0$, $f(21) > 0$, and somewhere in between it crosses the x-axis. We try some values of x: $f(12)$ is $+$, $f(6)$ is $+$, $f(3)$ is $-$, $f(4)$ is $+$, so the root is between 3 and 4. To pin it down more precisely at this point one would use another technique, like Newton's method (which can be found in any calculus text).

PROOF OF STURM'S THEOREM. Given a polynomial $f(x)$ with no multiple roots, let $\mathcal{P}(x) = [f_0(x), f_1(x), \ldots, f_r(x)]$ be the sequence of functions obtained by the modification of Euclid's algorithm. Then $f_r(x)$, the greatest common divisor of $f(x)$ and $f'(x)$, is a nonzero constant.

In the interval between b and c indicate all the points $d_1, \ldots, d_s$ which are roots of one or more of $f_0(x), f_1(x), \ldots, f_{r-1}(x), f_r(x)$. Note that $f_r(x)$ is a constant $\neq 0$ so has no roots.

$$\vdash\!\!\!-\!\!\!-\!\!\!-\!\!\!-\!\!\!-\!\!\!-\!\!\!-\!\!\!-\!\!\!-\!\!\!-\!\!\!-\!\!\!-\!\!\!-\!\!\!-\!\!\!\dashv$$
$$b \quad d_1 \quad d_2 \quad \cdots \quad d_{i-1} \quad d_i \quad d_{i+1} \cdots d_s \quad c$$

It may be that b or c is a root of some one or more of $f_1(x), \ldots, f_r(x)$, but we assume that b and c are not roots of $f_0(x)$.

Claim: d_i cannot be a root of both $f_j(x)$ and $f_{j+1}(x)$ for some j. For if it were, then, since $f_j(x) = f_{j+1}(x)q_{j+1}(x) - f_{j+2}(x)$, $f_{j+2}(d_i) = 0$; since $f_{j+1}(x) = f_{j+2}(x)q_{j+2}(x) - f_{j+3}(x)$, $f_{j+3}(d_i) = 0$; continuing in this way, we get $f_r(d_i) = 0$. But $f_r(x)$ is a nonzero constant. Thus the claim is true.

We shall consider how the function $w(x)$, the number of changes in sign in the sequence of numbers $\mathcal{P}(x)$, changes as we move along the x-axis from left to right from b to c.

First of all, if we let x vary within an interval between two consecutive d_i's then $w(x)$ is constant. For suppose x' and x'' are two numbers between, say, d_{i-1} and d_i. Then $f_j(x')$ and $f_j(x'')$ must have the same sign, for each j. For otherwise they have opposite signs, and since $f_j(x)$ is a continuous function (being a polynomial), there must be some d between x' and x'' where $f_j(d) = 0$. But then $d = d_k$ for some k. Thus, if x', x'' are both within an interval between two consecutive d_i's, then the signs of the numbers in the sequence

$$\mathcal{P}(x') = [f_0(x'), f_1(x'), \ldots, f_r(x')]$$

are the same as the signs of the corresponding numbers in the sequence

$$\mathcal{P}(x'') = [f_0(x''), f_1(x''), \ldots, f_r(x'')],$$

and hence the number of changes in sign in the two sequences $\mathcal{P}(x')$ and $\mathcal{P}(x'')$ are equal, that is, $w(x') = w(x'')$.

So as we move along the x-axis between two consecutive d_i's, $w(x)$ remains constant.

Now we see what happens to $w(x)$ when we move past some d_i. So we look at x', x'' as illustrated.

First suppose that d_i is a root of $f_0(x)$. Then d_i is not a root of $f_1(x) = f_0'(x)$, by the argument above, in the third paragraph of the proof. Thus the following cases can arise.

(i) $f_1(d_i)$ is $+$. Then $f_1(x)$ is $+$ for all x between d_{i-1} and d_{i+1} (at least) so $f_0(x)$ is increasing for x between d_{i-1} and d_{i+1}. Since $f_0(d_i) = 0$, $f_0(x')$ is $-$, while $f_0(x'')$ is $+$, and the sequences of signs of $\mathcal{P}(x)$ start

$$\mathcal{P}(x') = [-, +, \ldots], \qquad \mathcal{P}(x'') = [+, +, \ldots].$$

(ii) $f_1(d_i)$ is $-$. Then $f_1(x)$ is $-$ for all x between d_{i-1} and d_{i+1}, so $f_0(x)$ is decreasing for all x between d_{i-1} and d_{i+1}. Since $f_0(d_i) = 0$, $f_0(x')$ is $+$, while $f_0(x'')$ is $-$, and the sequences of signs of $\mathcal{P}(x)$ start

$$\mathcal{P}(x') = [+, -, \ldots], \qquad \mathcal{P}(x'') = [-, -, \ldots].$$

In each of the two cases, the effect on $w(x)$ is that $w(x'') = w(x') - 1$: $w(x)$ lowers by 1 in moving from left to right across a root of $f_0(x)$.

Suppose d_i is a root of $f_j(x)$, $j > 0$. Then d_i is not a root of $f_{j-1}(x)$ or of $f_{j+1}(x)$. So the signs of $f_{j-1}(x)$ and $f_{j+1}(x)$ do not vary between d_{i-1} and d_{i+1}. Now

$$f_{j-1}(x) = f_j(x)q_j(x) - f_{j+1}(x)$$

so, in particular, at d_i, where $f_j(d_i) = 0$, $f_{j-1}(d_i)$ and $f_{j+1}(d_i)$ have opposite signs. Thus $f_{j-1}(x)$ and $f_{j+1}(x)$ have opposite signs for all x between d_{i-1} and d_{i+1}.

So $\mathscr{P}(x')$ and $\mathscr{P}(x'')$ both either look like

$$[\ldots +, ?, -, \ldots]$$

or like

$$[\ldots -, ?, +, \ldots],$$

where ? denotes the sign of f_j. Whatever ? is $(+, 0, \text{ or } -)$ will not make any difference in computing $w(x')$ or $w(x'')$.

Thus the fact that d_i is a root of $f_j(x)$, $j > 0$, does not effect any change in $w(x)$ as x moves from left to right across d_i.

The same argument holds in case $d_1 = b$ or $d_s = c$. If, for example, $d_1 = b$ is a root of $f_j(x)$ for some $j > 0$, then for x'' between d_1 and d_2, the signs of $f_{j-1}(x'')$ and of $f_{j+1}(x'')$ will be the same as their signs at $b = d_1$, and they will be opposite, so if

$$\mathscr{P}(b) = [\ldots +, 0, -, \ldots],$$

then

$$\mathscr{P}(x'') = [\ldots, +, ?, -, \ldots],$$

while if

$$\mathscr{P}(b) = [\ldots -, 0, +, \ldots],$$

then

$$\mathscr{P}(x'') = [\ldots, -, ?, +, \ldots]$$

where ? can be either $+$ or $-$, it does not matter for computing $w(x'')$. So in any case, $w(x)$ is not changed in moving to the right away from b. The case $d_s = c$ is similar.

To sum up, as we let x move from b to c, $w(x)$ behaves as follows:

the value of $w(x)$ does not change on any interval containing no root of any $f_j(x)$;

the value of $w(x)$ is not affected in going past a point d_i by the fact that d_i is a root of $f_j(x)$, if $j > 0$;

the value of $w(x)$ decreases by 1 in going past a point d_i which is a root of $f_0(x) = f(x)$;

if b, c are not roots of $f_0(x)$, hence not roots of $f(x)$, then the value of $w(x)$ does not change on leaving b or arriving at c.

Thus $w(b) - w(c) = $ the number of points d_i between b and c which are roots of $f(x)$, hence $= $ the number of distinct roots of $f(x)$ between b and c.

Sturm's theorem is proved.   □

E2. Find the number of roots of each of the following polynomials. Locate each root between a pair of consecutive integers:
(a) $x^5 - 3x + 7$;
(b) $x^5 - x^4 + x^3 + x^2 - x + 1$;
(c) $x^3 - 3x + 1$.

*E3. If you assume that neither b nor c are roots of $f(x)$, then Sturm's algorithm works perfectly well without assuming f has no multiple factors. Prove this.

8 Factoring in $\mathbb{Q}[x]$, I

In Chapters 3 and 4 we determined all the irreducible polynomials with real or complex coefficients, and in Chapter 7 we made a start toward finding the real roots of a general polynomial in $\mathbb{R}[x]$. In this chapter we begin considering the question of which polynomials with coefficients in the rational numbers $\mathbb{Q}$ are irreducible, and how to find the rational roots, or more generally the irreducible factors, of a polynomial in $\mathbb{Q}[x]$.

The situation in $\mathbb{Q}[x]$ is different, as we shall see, from that in $\mathbb{R}[x]$ or $\mathbb{C}[x]$. While in $\mathbb{R}[x]$ or $\mathbb{C}[x]$ we could describe explicitly all irreducible polynomials, we cannot do that in $\mathbb{Q}[x]$, but can only describe certain criteria which imply irreducibility. On the other hand, while we only took a first step (with Sturm's theorem) toward determining the factorization of a general polynomial in $\mathbb{R}[x]$, in $\mathbb{Q}[x]$ we shall eventually describe an explicit procedure (in fact, two different explicit procedures) for determining the factorization of any given polynomial in a finite number of steps.

The starting point for all the results on $\mathbb{Q}[x]$ is the fact that factoring in $\mathbb{Q}[x]$ is "the same" as factoring in $\mathbb{Z}[x]$. The first part of this chapter is devoted to showing that fact.

A. Gauss's Lemma

Let $f(x) = a_n x^n + \cdots + a_1 x + a_0$ be a polynomial with rational coefficients $a_n, \ldots, a_1, a_0$. We can multiply $f(x)$ by the least common denominator of the coefficients, call it s, and get a polynomial with integer coefficients, $sf(x) = g(x)$. Since $g(x)$ and $f(x)$ are associates, $g(x)$ will be irreducible in $\mathbb{Q}[x]$ if and only if $f(x)$ is. So in studying polynomials in $\mathbb{Q}[x]$, we can always assume that they have integer coefficients.

We can ask for more. We shall say that a polynomial $f(x)$ with rational coefficients is *primitive* if $f(x)$ has integer coefficients and the greatest common divisor of those coefficients is 1. Then any polynomial with integer coefficients is an associate in $\mathbb{Q}[x]$ of a primitive polynomial. For if the greatest common divisor of the coefficients of $f(x)$ is, say, t, then $(1/t)f(x)$ is still a polynomial with integer coefficients, the greatest common divisor of the coefficients of $(1/t)f(x)$ is one (why?), and $f(x)$ and $(1/t)f(x)$ are associates in $\mathbb{Q}[x]$. So any polynomial in $\mathbb{Q}[x]$ is an associate of a primitive polynomial.

Lemma. *The product of two primitive polynomials is again a primitive polynomial.*

PROOF. The product of any two polynomials with integer coefficients is again a polynomial with integer coefficients. We must check the primitivity. Let $a(x)$ and $b(x)$ be two primitive polynomials. Then for any prime p, p does not divide all the coefficients of $a(x)$, and p does not divide all the coefficients of $b(x)$. So

$$a(x) \not\equiv 0 \pmod{p} \quad \text{and} \quad b(x) \not\equiv 0 \pmod{p}.$$

Then

$$a(x)b(x) \not\equiv 0 \pmod{p},$$

so p does not divide all the coefficients of $a(x)b(x)$. Since this is true for any prime p, no prime divides all the coefficients of $a(x)b(x)$, so the greatest common divisor of those coefficients is 1, and $a(x)b(x)$ is primitive. $\square$

The main result of this section is

Gauss's Lemma. *Let $f(x)$ be a polynomial with integer coefficients. If $f(x) = a(x)b(x)$ in $\mathbb{Q}[x]$, then $f(x) = a_1(x)b_1(x)$ with $a_1(x)$ and $b_1(x)$ polynomials with integer coefficients, and with $a_1(x)$ and $b_1(x)$ associates of $a(x)$ and $b(x)$ respectively.*

Gauss's lemma implies that in order to show $f(x)$ in $\mathbb{Z}[x]$ is irreducible in $\mathbb{Q}[x]$, it is enough to show that $f(x)$ does not factor into the product of two polynomials with integer coefficients.

PROOF OF GAUSS'S LEMMA. If $f(x) = a(x)b(x)$ in $\mathbb{Q}[x]$, then $rf(x) = ra(x)b(x)$ for any rational number r. If the result is true for $rf(x)$ it is true for $f(x)$. So we can assume without loss of generality that $f(x)$ is a primitive polynomial. Now if $f(x) = a(x)b(x)$, where $a(x)$ and $b(x)$ are polynomials with rational coefficients, then there are rational numbers s, t, such that $sa(x)$ and $tb(x)$ are primitive. (We showed above that any polynomial with rational coefficients is an associate of such a polynomial.) By the lemma, $sta(x)b(x) = stf(x)$ is then a primitive polynomial. So is $f(x)$.

E1. If r is a rational number such that $rf(x)$ and $f(x)$ are both primitive polynomials, show that $r = 1$ or -1.

By Exercise E1, $st = \pm 1$, so $f(x) = \pm sa(x) \cdot tb(x)$. We can set $a_1(x) = \pm sa(x)$, $b_1(x) = tb(x)$ to complete the proof of Gauss's lemma. □

We shall say that a polynomial in $\mathbb{Z}[x]$ is *irreducible in $\mathbb{Z}[x]$* if it does not factor into the product of two polynomials with degrees $\geqslant 1$ with integer coefficients. Gauss's lemma says that a polynomial with integer coefficients is irreducible in $\mathbb{Z}[x]$ if and only if it is irreducible as a polynomial in $\mathbb{Q}[x]$.

E2. Find a primitive polynomial which is an associate of
(a) $f(x) = (4/3)x^4 + 6x^3 + (2/9)x^2 + (9/2)x + 18$,
(b) $f(x) = 36x^3 + 180x^2 + 24x + 1/7$.

E3. (a) Show: If $f(x)$ in $\mathbb{Z}[x]$ is monic, it is primitive;
(b) Show that if $f(x)$ in $\mathbb{Z}[x]$ is monic and factors as $f(x) = g(x)h(x)$ with $g(x), h(x)$ in $\mathbb{Q}[x]$, then $g(x), h(x)$ are associates of $g_1(x), h_1(x)$ in $\mathbb{Z}[x]$ which are monic and such that $f(x) = g_1(x)h_1(x)$.

B. Finding Roots

Here is a criterion for deciding whether a polynomial with integer coefficients has a linear factor.

Theorem. *If $f(x) = a_n x^n + \cdots + a_1 x + a_0$ is in $\mathbb{Z}[x]$ and $x = r/s$ is a root, with r, s relatively prime integers, then s divides a_n and r divides a_0.*

Note that if $x = r/s$ is a root of $f(x)$, then $sx - r$ is a factor of $f(x)$.

PROOF. Suppose $0 = f(r/s) = a_n(r^n/s^n) + \cdots + a_1(r/s) + a_0$. Multiply through by s^n. Then s must divide $a_n r^n$, and since r and s are relatively prime, s must therefore divide a_n. Similarly, r must divide $a_0 s^n$; for the same reason, r must divide a_0. That completes the proof. □

For example, the only possible roots of the polynomial $x^4 + 8x^3 + 15x^2 - 6x - 9$ are $x = 1, -1, 3, -3, 9$, or -9, since these are the only divisors of -9.

Since a_n and a_0 each have only a finite number of divisors, the theorem limits the possible roots of a polynomial with integer coefficients to only a finite number. (See Exercise E6.)

E4. Find all rational roots of
(a) $x^3 - x + 1$,
(b) $x^3 - x - 1$,
(c) $x^3 + 2x + 10$,
(d) $x^3 - 2x^2 + x + 15$.

E5. Find all rational roots of
(a) $x^7 - 7$,
(b) $2x^2 - 3x + 4$,
(c) $2x^4 - 4x + 3$.

E6. Let $d(n)$ be the number of positive divisors of $n > 1$ (including 1 and n).
(a) Show that $d(a)d(b) = d(ab)$ if $(a, b) = 1$.
(b) Find $d(p^m)$ for p prime. Find $d(n)$ for any n.
(c) Show that if $f(x)$ is a monic polynomial in $\mathbb{Z}[x]$
and $f(0) = n$, then there are $2d(n)$ potential roots of $f(x)$.

C. Testing for Irreducibility

Finding roots is the same as finding factors of degree 1. A harder problem is to find all the factors of any degree of a polynomial with rational coefficients. The remainder of this chapter is an introduction to this problem.

We begin with a criterion for proving that certain polynomials are irreducible.

Theorem (Eisenstein's irreducibility criterion). *If* $f(x) = a_n x^n + a_{n-1} x^{n-1} + \cdots + a_1 x + a_0$ *is in* $\mathbb{Z}[x]$, *and there exists a prime* p *such that* p *does not divide* a_n, p *does divide* $a_{n-1}, a_{n-2}, \ldots, a_1$ *and* a_0, *but* p^2 *does not divide* a_0, *then* $f(x)$ *is irreducible in* $\mathbb{Z}[x]$ *(and hence in* $\mathbb{Q}[x]$).

This theorem shows that there are irreducible polynomials in $\mathbb{Q}[x]$ of any degree, for example, $x^n - 2$.

PROOF. Suppose $f(x) = g(x)h(x)$ in $\mathbb{Z}[x]$, where $g(x) = b_r x^r + \cdots + b_1 x + b_0$, $h(x) = c_s x^s + \cdots + c_1 x + c_0$. Multiplying together $g(x)$ and $h(x)$ and equating coefficients with those of $f(x)$, we get $n + 1$ equations (assume $r \leqslant s$):

$$a_n = b_r c_s,$$
$$\vdots$$
$$a_s = b_r c_{s-r} + b_{r-1} c_{s-r+1} + \cdots + b_0 c_s,$$
$$\vdots$$
$$a_r = b_r c_0 + b_{r-1} c_1 + \cdots + b_0 c_r,$$
$$\vdots$$
$$a_3 = b_3 c_0 + b_2 c_1 + b_1 c_2 + b_0 c_3,$$
$$a_2 = b_2 c_0 + b_1 c_1 + b_0 c_2,$$
$$a_1 = b_1 c_0 + b_0 c_1,$$
$$a_0 = b_0 c_0.$$

Since p^2 does not divide a_0, but p does, p divides b_0 but not c_0, or vice versa. Suppose p divides b_0, and p does not divide c_0. Then since p divides each of $a_1, a_2, \ldots, a_r$, but not c_0, p must divide b_1, hence b_2, hence $b_3, \ldots$, hence b_r. But then p divides a_n, which is impossible. The "vice versa" is similar. So the assumed factorization of $f(x)$ cannot exist.   □

Eisenstein's criterion is very nice when it works, but there are many irreducible polynomials to which it does not apply. For example, consider $p(x) = x^5 + 4x^4 + 2x^3 + 3x^2 - x + 5$: Eisenstein's criterion is useless here, but $p(x)$ may still be irreducible. How do we decide?

One can try a "brute force" approach.

If $p(x)$ factors, it must factor into $a(x)b(x)$, where $a(x)$, $b(x)$ are polynomials in $\mathbb{Z}[x]$ (by Gauss's lemma), where we can assume $a(x)$ has degree 1 and $b(x)$ has degree 4, or $a(x)$ has degree 2 and $b(x)$ degree 3.

If $a(x)$ has degree 1, then $p(x)$ has a rational root, so we can apply the theorem about such roots. Any root must divide 5, so must be 1, -1, 5, or -5. It is easy to check that none of these are roots of $p(x)$. So $p(x)$ has no linear factor in $\mathbb{Q}[x]$.

To test the case where $a(x)$ has degree 2, we write the potential factorization out, as in the proof of Eisenstein's criterion:

$$x^5 + 4x^4 + 2x^3 + 3x^2 - x + 5 = (ax^2 + bx + c)(dx^3 + ex^2 + fx + g)$$

where a, b, c, d, e, f, g are integers. Multiplying out and equating coefficients, we get

$$ad = 1,$$
$$bd + ae = 4,$$
$$be + cd + af = 2,$$
$$ce + bf + ag = 3,$$
$$cf + bg = -1,$$
$$cg = 5,$$

which we try to solve. Since (by Gauss's lemma) all the variables are integers, there are only a few possible values of a, d, c and g, so there is a chance of finding all solutions of these equations.

E7. Show the above set of equations has no solution. (See E15.)

This "brute force" technique is not so successful for polynomials of higher degree. Even on a polynomial of degree 9 or 10 the "brute force" approach gives a system of equations for the coefficients which is rather complicated and difficult to handle in a systematic way. So we look for other techniques.

One test for irreducibility is to reduce mod n. If $f(x)$ is a polynomial with integer coefficients which is primitive, then we can reduce the coefficients mod n for any n and get a nonzero polynomial $\bar{f}(x)$ in $\mathbb{Z}_n[x]$. If

n is an integer which is relatively prime to the leading coefficient of $f(x)$, then $\bar{f}(x)$ in $\mathbb{Z}_n[x]$ has the same degree as $f(x)$.

So suppose $f(x) = a(x)b(x)$, where $a(x)$, $b(x)$ have integer coefficients and are primitive. Then $\bar{f}(x) = \bar{a}(x)\bar{b}(x)$, in $\mathbb{Z}_n[x]$, and so $\bar{f}(x)$ would factor. Therefore

Corollary. *If $\bar{f}(x)$ in $\mathbb{Z}_n[x]$ is irreducible for some n not dividing the leading coefficient of $f(x)$, then $f(x)$ is irreducible in $\mathbb{Q}[x]$.*

Testing irreducibility of $\bar{f}(x)$ in $\mathbb{Z}_n[x]$ is a finite problem, since there are only a finite number of possible divisors of $\bar{f}(x)$.

For example, let $f(x) = 3x^5 - 4x^4 + 2x^3 + x^2 + 18x + 31$; mod 2, $f(x)$ is congruent to $\bar{f}(x) = x^5 + x^2 + 1$. It is easy to check that $\bar{f}(x)$ has no roots in $\mathbb{Z}_2$; if we look for a possible factorization of the form

$$\bar{f}(x) = (ax^2 + bx + c)(dx^3 + ex^2 + fx + g)$$

with $a, b, \ldots, g$ in $\mathbb{Z}_2$, the equations for the coefficients are easily shown to be insolvable. So $\bar{f}(x)$ is irreducible in $\mathbb{Z}_2[x]$, and hence $f(x)$ is irreducible in $\mathbb{Q}[x]$.

The example of Exercise E7 can be shown irreducible in $\mathbb{Z}_3[x]$—see E15.

In Chapters 11 and 13 we shall describe two techniques for systematically factoring a polynomial $f(x)$ in $\mathbb{Q}[x]$. One uses an analogue of the Chinese remainder theorem for polynomials. The other involves factoring mod m. We shall look for a systematic way to factor polynomials mod m, and in particular look for irreducible polynomials in $\mathbb{Z}_p[x]$.

E8. Suppose p divides the leading coefficient of $f(x)$ in $\mathbb{Z}[x]$. Can it be that $f(x)$ in $\mathbb{Z}_p[x]$ is irreducible but $f(x)$ factors?

E9. Give an example of a monic polynomial $f(x)$ in $\mathbb{Z}[x]$ which is irreducible in $\mathbb{Q}[x]$ but which factors modulo 2, 3 and 5.

E10. Suppose $f(x) = a_n x^n + \cdots + a_1 x + a_0$ is in $\mathbb{Z}[x]$. Let $g(x) = a_n^{n-1} f(x/a_n)$.
   (a) Show that $g(x)$ is a monic polynomial with integer coefficients and factors exactly the same way $f(x)$ does.
   (b) If $f(x) = 3x^5 - 4x^4 + 2x^3 + x^2 + 18x + 31$, what is $g(x)$?

E11. Factor $t^4 + 4$ in $\mathbb{Z}[t]$.

E12. Following the proof of Eisenstein's criterion, prove that $2x^4 - 8x^2 + 1$ is irreducible in $\mathbb{Q}[x]$.

E13. If $f(x)$, $g(x)$ are in $\mathbb{Z}[x]$ and $f(x)$ is monic, show that when you do the division algorithm in $\mathbb{Q}[x]$, namely, $g(x) = f(x)q(x) + r(x)$, then $r(x)$ and $q(x)$ are in $\mathbb{Z}[x]$. What if $f(x)$ is not monic?

E14. Show that each of these is irreducible in $\mathbb{Q}[x]$:
   (a) $x^4 + x + 1$;
   (b) $x^4 + 3x + 5$;
   (c) $3x^4 + 2x^3 + 4x^2 + 5x + 1$;
   (d) $x^5 + 5x^2 + 4x + 7$;
   (e) $15x^5 - 2x^4 + 15x^2 - 2x + 15$.

E15. Show that $p(x) = x^5 + 4x^4 + 2x^3 + 3x^2 - x + 5$ is irreducible mod 3, but not mod 2.

E16. (a) Let $\Phi_p(x) = x^{p-1} + x^{p-2} + \cdots + x + 1$ in $\mathbb{Q}[x]$ where p is prime. Show that every nonreal pth root of 1 is a root of $\Phi_p(x)$.
   (b) Let $\Phi_p(x + 1) = g(x)$, where $g(x) = x^{p-1} + a_{p-2}x^{p-2} + \cdots + a_1x + a_0$. Prove that p divides each a_i, $0 \leqslant i \leqslant p - 2$. Prove that $a_0 = p$.
   (c) Using (b), show that $\Phi_p(x)$ is irreducible in $\mathbb{Q}[x]$.

Congruences Modulo a Polynomial 9

Let F be a field, $p(x)$ be a polynomial in $F[x]$. Say that $f(x)$ *is congruent to* $g(x) \bmod p(x)$, written

$$f(x) \equiv g(x) \pmod{p(x)},$$

if $p(x)$ divides $f(x) - g(x)$, that is, if $f(x)$ and $g(x)$ differ by a multiple of $p(x)$.

Congruence $\bmod p(x)$ has identical properties to congruence $\bmod n$ of integers. If you go back to the chapter on congruences (Chapter 6) in Part I, all the facts about congruences obtained there remain valid for congruences $\bmod$ a polynomial. So we shall say no more, and just provide some exercises.

†**E1.** If $p(x)$ has degree d, show that any $f(x)$ is congruent $\pmod{p(x)}$ to a unique polynomial $r(x)$ of degree $< d$.

E2. If $p(x)$ has degree 0, and $f(x)$ and $g(x)$ are any two polynomials, show that $f(x) \equiv g(x) \pmod{p(x)}$.

E3. What are the theorems on canceling $h(x)$ in the congruence $h(x)f(x) \equiv h(x)g(x) \pmod{p(x)}$?

E4. Show that $f(x) \equiv f(a) \pmod{(x - a)}$ for any $f(x)$.

E5. State a theorem about "casting out $x - 1$'s."

E6. Solve, if possible,
   (a) $(x^3 + x + 1)f(x) \equiv 1 \pmod{(x^4 + x + 1)}$ in $\mathbb{Z}_2[x]$,
   (b) $(2x + 1)f(x) = x^3 \pmod{(x^2 + 1)}$ in $\mathbb{Z}_3[x]$,
   (c) $x^9 f(x) \equiv 1 \pmod{(x^2 + 2)}$ in $\mathbb{Q}[x]$,
   (d) $(x^2 + 1)f(x) \equiv x^2 + x + 1 \pmod{(x^4 + 1)}$ in $\mathbb{Z}_2[x]$.

E7. Find the residue of least degree in $\mathbb{Z}_2[x]$:
   (a) of $x^9 \pmod{x^2 + x + 1}$;
   (b) of $x^{13} \pmod{x^4 + x + 1}$;
   (c) of $x^{26} \pmod{x^4 + x + 1}$.

E8. Find a complete set of residues:
   (a) $\mod(x^2 + x + 1)$ in $\mathbb{Z}_2[x]$;
   (b) $\mod(x^2 + 1)$ in $\mathbb{Z}_3[x]$;
   (c) $\mod(x^3 + x + 1)$ in $\mathbb{Z}_2[x]$.

E9. Let $f(x)$, $g(x)$ be monic polynomials with integer coefficients. Show that $f(x)$ and $g(x)$ agree as functions on $\mathbb{Z}_p$ iff $f(x) \equiv g(x) \pmod{x^p - x}$.

Fermat's Theorem, II   **10**

A. The Characteristic of a Field

Fermat's theorem says that if p is a prime number and a is any integer relatively prime to p, then $a^{p-1} \equiv 1 \pmod{p}$. In Chapter I-11 we gave a proof based on the fact that the set of invertible elements of $\mathbb{Z}_p$ forms an abelian group of order $p - 1$. In this chapter we give another proof based on the binomial theorem. We begin by making a definition which relates to the fact that in $\mathbb{Z}_p$, $[p] = 0$.

Let R be a commutative ring, with zero element 0 and identity element 1. Add 1 to itself many times. Two possible situations can arise. One is that $0 \neq 1 + 1 + \cdots + 1$ (n times) for any $n > 0$. Then R is said to have *characteristic* 0. Examples are $\mathbb{Q}$, $\mathbb{R}$, $\mathbb{C}$, $\mathbb{Z}$, $\mathbb{R}[x]$.

The other situation is that there is some $n > 0$ such that $0 = 1 + 1 + \cdots + 1$ (n times). This happens, for example, in $\mathbb{Z}_2$, for in $\mathbb{Z}_2$, $1 + 1 = [1]_2 + [1]_2 = [1 + 1]_2 = [2]_2 = [0]_2 = 0$. Similarly, for any $m > 0$, in $\mathbb{Z}_m$ $1 + 1 + \cdots + 1$ (m times) $= 0$.

Let n_0 be the smallest natural number n for which $0 = 1 + 1 + \cdots + 1$ (n times) in R.

Proposition. *If R has no zero divisors (in particular, if R is a field), then n_0 is a prime number.*

PROOF. If $n_0 = ab$, with a, b natural numbers less than n_0, then

$$0 = 1 + 1 + \cdots + 1 \quad (n_0 \text{ times})$$
$$= \underbrace{(1 + 1 + \cdots + 1)}_{a \text{ times}} \underbrace{(1 + 1 + \cdots + 1)}_{b \text{ times}}$$

since $ab = n_0$. If n_0 is the smallest natural number n for which

$$0 = 1 + 1 + \cdots + 1 \qquad (n \text{ times}),$$

then

$$0 \neq 1 + 1 + \cdots + 1 \qquad (a \text{ times}),$$

and

$$0 \neq 1 + 1 + \cdots + 1 \qquad (b \text{ times}),$$

so R has zero divisors, which is impossible. So n_0 must be prime. □

To signify in the notation that n_0 is prime, we shall relabel n_0 and call it p. If R has no zero divisors, and if

$$0 = 1 + 1 + \cdots + 1 \qquad (p \text{ times}), p \text{ prime},$$

then R is said to have *characteristic p*.

For example, $\mathbb{Z}_p$ has characteristic p. Thus $\mathbb{Z}_2$ has characteristic 2, $\mathbb{Z}_3$ has characteristic 3, etc. Similarly, $\mathbb{Z}_p[x]$ has characteristic p.

E1. Show that if R has characteristic p and $0 = 1 + 1 + 1 + \cdots + 1$ (n times) for $n > 0$, then $p | n$.

E2. Show that if R has characteristic p and a is in R, then $pa = a + a + \cdots + a$ (p times) $= 0$.

B. Applications of the Binomial Theorem

Recall the binomial theorem, Chapter I-2: if $\binom{n}{r} = n!/r!(n-r)!$ for $0 \leqslant r \leqslant n$, $0! = 1$, then $\binom{n}{r}$ is an integer for each n, r, and for any indeterminates x, y,

$$(x+y)^n = \binom{n}{0}x^n + \binom{n}{1}z^{n-1}y + \binom{n}{2}x^{n-2}y^2 + \cdots +$$
$$\binom{n}{r}x^{n-r}y^r + \cdots + \binom{n}{n-1}xy^{n-1} + \binom{n}{n}y^n.$$

Lemma. *If p is prime, then p divides $\binom{p}{r}$ for all r, $0 < r < p$.*

PROOF. For p prime, $\binom{p}{r} = p!/r!(p-r)!$. Now $r!(p-r)!$ divides $p!$ since $\binom{p}{r}$ is an integer. Since $(p, r!(p-r)!) = 1$ for $1 \leqslant r \leqslant p-1$, therefore $r!(p-r)!$ divides $(p-1)!$ and

$$\binom{p}{r} = p\left(\frac{(p-1)!}{r!(p-r)!}\right),$$

an integer multiple of p. □

Corollary. *If p is prime, $(x+y)^p \equiv x^p + y^p \pmod{p}$.*

PROOF. Expand $(x + y)^p$ by the binomial theorem and use the lemma.   □

We shall use this corollary to reprove Fermat's theorem.

Fermat's Theorem. *If p is prime, then for any integer a, $a^p \equiv a$ (mod p).*

Note. If we show that for any integer a, $a^p \equiv a$ (mod p), then, if a and p are relatively prime, we can cancel a from both sides to get Fermat's theorem as stated in Chapter I-11.

PROOF. If we prove $a^p \equiv a$ (mod p) when a is a natural number, then $a^p \equiv a$ (mod p) for a any integer. For any integer a is congruent mod p to some natural number b; if $b^p \equiv b$ (mod p), then

$$a^p \equiv b^p \equiv b \equiv a \pmod{p}$$

and the theorem is true for the integer a.

So it suffices to prove the theorem when a is a natural number. We proceed by induction on a.

For $a = 1$ it is clear. Suppose a is an integer $\geqslant 1$ for which $a^p \equiv a$ (mod p). Consider $(a + 1)^p$. By the corollary, $(a + 1)^p \equiv a^p + 1^p$ (mod p). By induction, this is congruent to $a + 1$ (mod p). That completes the induction step and proves the theorem.   □

If you review the old proof of Fermat's theorem it will be apparent that the old proof and the one we just gave are very different. The old proof was a special case of a result about finite groups, while the new one involves knowing about the binomial theorem mod p. By the ideas of the old proof we were able to prove Euler's theorem, i.e., that if $(a, m) = 1$ then $a^{\phi(m)} \equiv 1$ (mod m). In Chapter I-14, Exercise E15, we showed how to use Fermat's theorem and the Chinese remainder theorem to deduce a proof of Euler's theorem for m squarefree; but to get Euler's theorem for m a prime power by the ideas of the proof we just gave of Fermat's theorem would be difficult.

On the other hand, the corollary to the binomial theorem which we obtained above has other useful applications to finite fields. We first extend the corollary to elements of arbitrary fields of characteristic p. (The corollary as stated above only applies to integers mod p.)

Theorem 1. *If R is a field of characteristic p and a, b are elements of R, then $(a + b)^p = a^p + b^p$.*

PROOF. By the binomial theorem, for y, z any indeterminates, $(y + z)^p = y^p + \binom{p}{1}y^{p-1} + \cdots + \binom{p}{r}y^{p-r}z^r + \cdots + \binom{p}{p-1}y^{p-1}z + z^p$. Set $y = a, z = b, a, b$ in R. Now if $r \neq 0$ or p, $\binom{p}{r} = pq$ for some q, by the lemma, so $\binom{p}{r}ab = q(p \cdot ab) = 0$ by Exercise E2. Thus in R, $(a + b)^p = a^p + 0 + \cdots + 0 + \cdots + 0 + b^p$, as was to be proved.   □

E3. In $GF(9)$, verify that $(1 + 2i)^3 + (2 + 2i)^3 = (1 + 2i + 2 + 2i)^3$.

The next result is an extension of the theorem of Chapter 6 relating the derivative of a polynomial and multiple factors. We had assumed in that theorem that the polynomials had coefficients in a subfield of the complex numbers.

E4. Show that the proof of that theorem was valid if F had characteristic 0.

Theorem 2. *If $f(x)$ is in $F[x]$, where F is a finite field of characteristic p, then $f(x)$ has no multiple factor if and only if $(f(x), f'(x)) = 1$.*

Before reading the proof of Theorem 2 you may wish to review the proof of the theorem of Chapter 6 to see how it went—the proof here begins the same way.

PROOF. If F is any field and $(f(x), f'(x)) = 1$, then by Chapter 6, Exercise E4, f has no multiple roots.

Conversely, suppose $h(x)$ is an irreducible common factor of $f(x)$ and $f'(x)$. Then $f = hg$, $f' = hg' + h'g$. Since h divides f', h divides hg'. Since h is irreducible, h divides g or h divides h'. If h divides g, h is a multiple factor of f, and the theorem is proved. So we suppose the other possibility, that h divides h', and derive a contradiction.

How can h divide h'? Since $\deg h' < \deg h$, this can happen only when $h' = 0$. Now it is easily checked that if $h' = 0$, then

$$h(x) = a_0 + a_1 x^p + a_2 x^{2p} + \cdots + a_r x^{rp}. \tag{1}$$

We show that if $h(x)$ has the form (1), then $h(x)$ cannot be irreducible. That will be a contradiction.

We begin by showing that each a_i is a pth power. To do this, define a function ϕ_p from F to F by setting $\phi_p(b) = b^p$ for each element b of F. We want to show that every element of F is in the image of ϕ_p. To do this it suffices to show that ϕ_p is 1–1. For then the image of ϕ_p would have just as many elements as F, and since the image of ϕ_p is a subset of F and F is a finite set, the image of ϕ_p must be all of F. To show ϕ_p is 1–1, we suppose that $\phi_p(b_1) = \phi_p(b_2)$ and we must show $b_1 = b_2$. If $\phi_p(b_1) = \phi_p(b_2)$, then $b_1^p = b_2^p$, so $b_1^p - b_2^p = 0$. But then $(b_1 - b_2)^p = 0$ by Theorem 1; since F is a field, it follows that $b_1 - b_2 = 0$, so $b_1 = b_2$ and ϕ_p is 1–1.

Thus each element of F is in the image of ϕ_p. In particular, each coefficient of $h(x)$ can be written as $a_i = c_i^p = \phi_p(c_i)$ for some c_i in F, so $h(x) = c_0^p + c_1^p x^p + \cdots + c_r^p x^{rp}$. By Theorem 1 again, we can write this as

$$h(x) = (c_0 + c_1 x + \cdots + c_r x^r)^p.$$

But then $h(x)$ is not irreducible, which is a contradiction. So $h(x)$ does not have the form (1), $h'(x)$ cannot be 0, and $h(x)$ does not divide $h'(x)$. Thus h divides g, h^2 divides f, and the theorem is proved. $\qquad \square$

E5. Find a function ϕ from $\mathbb{Z}$ to $\mathbb{Z}$ which is 1–1 but not onto.

E6. In $GF(9)$, write each element as a third power.

E7. Factor $x^{15} + 3x^{10} + 2x^5 + 4$ in $\mathbb{Z}_5[x]$.

E8. Verify that $(x + y)^{11} \equiv x^{11} + y^{11}$ (mod 11), by explicitly computing the coefficients in the binomial theorem expansion of $(x + y)^{11}$.

11 Factoring in $\mathbb{Q}[x]$, II: Lagrange Interpolation

A. The Chinese Remainder Theorem

In Chapter 8 we showed that when factoring a polynomial with integer coefficients we can always assume the factors have integer coefficients. In this chapter we use that information to describe a systematic procedure for factoring polynomials in $\mathbb{Z}[x]$. The method is attributed to Kronecker c. 1883, but is apparently due originally to F. v. Schubert, 1793. It is based on the Chinese remainder theorem, so we begin by considering how that theorem works in $K[x]$, K any field.

Chinese Remainder Theorem for Polynomials. *Let K be a field. Let $a_1(x), \ldots, a_n(x)$ be arbitrary polynomials, and $m_1(x), \ldots, m_n(x)$ pairwise relatively prime polynomials in $K[x]$. Then there exists a unique polynomial $f(x)$ in $K[x]$ such that*

$$f(x) \equiv a_1(x) \pmod{m_1(x)},$$
$$\vdots \qquad\qquad\qquad (*)$$
$$f(x) \equiv a_n(x) \pmod{m_n(x)},$$

and the degree of $f(x)$ is $<$ the degree of $m_1(x)m_2(x) \cdots m_n(x)$.

PROOF. The proof of this theorem is virtually identical to the Chinese remainder theorem for integers. It goes as follows. Since $m_i(x)$ is relatively prime to $m_j(x)$ for $j \neq i$, $m_i(x)$ is relatively prime to the product

$$l_i(x) = m_1(x)m_2(x) \cdots m_{i-1}(x)m_{i+1}(x) \cdots m_n(x).$$

Thus we can solve the equation

$$1 = h_i(x)m_i(x) + k_i(x)l_i(x)$$

for h_i and k_i by Bezout's lemma. Then $k_i(x)l_i(x)$ satisfies

$$k_i(x)l_i(x) \equiv 0 \pmod{m_j(x)} \quad \text{for all} \quad j \pm i,$$

$$k_i(x)l_i(x) \equiv 1 \pmod{m_i(x)}.$$

So we solve (*) by setting

$$f(x) = a_1(x)k_1(x)l_1(x) + a_2(x)k_2(x)l_2(x) + \cdots + a_n(x)k_n(x)l_n(x).$$

If $f(x)$ has degree $\geqslant \deg(m_1(x) \cdots m_n(x))$ we can use the division theorem to replace $f(x)$ by $r(x) = f(x) - q(x)[m_1(x) \cdots m_n(x)]$, where the degree of $r(x)$ can be made less than the degree of $m_1(x) \cdots m_n(x)$.   □

The Chinese remainder theorem has a different interpretation if we apply the root theorem (Chapter 2) in the following form.

E1. Let $g(x)$ be a polynomial with coefficients in a field F. Show that $g(x) \equiv b$ (mod$[x - a]$) if and only if $g(a) = b$.

Using Exercise E1, we may let $m_i(x) = x - n_i$ in the Chinese remainder theorem and deduce

Corollary. *If $n_0, \ldots, n_d$ are distinct integers and $s_0, \ldots, s_d$ are arbitrary integers, there exists a unique polynomial $q(x)$ in $\mathbb{Q}[x]$ of degree $\leqslant d$ such that $q(n_i) = s_i$ for $i = 0, \ldots, d$.*

We shall show how to write down such a $q(x)$ explicitly in Exercise E3 below.

B. The Method of Lagrange Interpolation

We describe in this section the systematic method of factoring any polynomial $f(x)$ in $\mathbb{Z}[x]$.

First note that if $f(x)$ in $\mathbb{Z}[x]$ of degree n is not irreducible it has a factor of degree $\leqslant n/2$. So let $d = n/2$ if n is even, $d = (n - 1)/2$ if n is odd.

Let $n_0, \ldots, n_d$ be distinct integers, and let $p(n_0) = r_0, \ldots, p(n_d) = r_d$. Since $p(x)$ is in $\mathbb{Z}[x]$, $r_0, \ldots, r_d$ are integers. For each vector $\mathbf{s} = (s_0, \ldots, s_d)$ of integers which are divisors of $(r_0, \ldots, r_d)$ (that is, $s_i | r_i$ for all i, $i = 0, \ldots, d$), use the corollary to the Chinese remainder theorem to get a unique polynomial $a_\mathbf{s}(x)$ in $\mathbb{Q}[x]$ of degree $\leqslant d$ with $a_\mathbf{s}(n_i) = s_i$ for all i. Since each r_i has only a finite number of (positive or negative) divisors s_i, there are only a finite number of possible vectors $\mathbf{s} = (s_0, \ldots, s_d)$ and hence only a finite number of corresponding polynomials $a_\mathbf{s}(x)$, one for each of the possible vectors $\mathbf{s}$.

It turns out that any divisor of $p(x)$ in $\mathbb{Z}[x]$ of degree $\leqslant d$ is an $a_\mathbf{s}(x)$ for some $\mathbf{s}$. For suppose $a(x)$ is in $\mathbb{Z}[x]$ of degree $\leqslant d$ and $a(x)b(x) = p(x)$ for some $b(x)$ in $\mathbb{Z}[x]$. Then for each n_i, $a(n_i)b(n_i) = p(n_i)$ in $\mathbb{Z}$, so $a(n_i)$ divides $p(n_i) = r_i$. Thus the vector $\mathbf{s} = (a(n_0), \ldots, a(n_d))$ is a vector of divisors of $(r_0, r_1, \ldots, r_d)$. The Chinese remainder theorem gives a unique polynomial $a_s(x)$ of degree $\leq d$ with $a_s(n_0) = a(n_0)$, $a_s(n_1) = a(n_1)$, $\ldots$, $a_s(n_d) = a(n_d)$. Since $a(x)$ and $a_s(x)$ both have degree $\leq d$ and have the same value on $d + 1$ elements of $\mathbb{Q}$, they must be equal (for their difference has degree $\leq d$ and has $d + 1$ roots, so must $= 0$). We conclude that any divisor $a(x)$ of $p(x)$ of degree $\leq d$ must be among the finite number of polynomials $a_s(x)$ constructed corresponding to the finite number of vectors $\mathbf{s}$ of divisors of $(r_0, r_1, \ldots, r_d)$.

To determine, then, whether $p(x)$ is irreducible or not in $\mathbb{Z}[x]$, divide $p(x)$ by each of the polynomials $a_s(x)$ to see whether $a_s(x)$ is a divisor of $p(x)$. If some $a_s(x)$ of degree ≥ 1 divides $p(x)$, then an explicit factorization of $p(x)$ has been found. Otherwise, $p(x)$ must be irreducible.

An easy argument by induction on the degree of $p(x)$ then gives

Theorem. *The complete factorization of any polynomial in $\mathbb{Z}[x]$ can be achieved in a finite number of steps.*

E2. Fill in the details of the induction argument for the proof of the theorem.

We call the polynomial $a_s(x)$ a *Lagrange interpolator*. It is constructed as follows.

With $n_0, \ldots, n_d, s_0, \ldots, s_d$ as above, let

$$g(x) = (x - n_0) \cdots (x - n_d),$$

and let $g'(x)$ be its derivative. Then $g(x)/(x - n_i)$, after canceling, is a polynomial of degree d, so $g(x)/(x - n_i)g'(n_i) = h_i(x)$ is a polynomial of degree d such that

$$h_i(n_i) = 1 \quad \text{and} \quad h_i(n_0) = \cdots = h_i(n_{i-1}) = h_i(n_{i+1}) \cdots = h_i(n_s) = 0.$$

E3. Show that (a) $h_i(n_i) = 1$, (b) $a(x) = h_0(x)s_0 + \cdots + h_d(x)s_d$ satisfies $a(n_i) = s_i$ for $i = 0, \ldots, d$, so that $a(x) = a_s(x)$.

Here is an example to illustrate how the factoring method works. Let $f(x) = x^4 + x + 1$. If $f(x)$ factors it must have a factor of degree ≤ 2. (Of course reducing mod 2 we already know it is irreducible!) Now $f(-1) = 1$, $f(0) = 1$, $f(1) = 3$. For each $\mathbf{s} = (s_1, s_0, s_1)$ dividing $(1, 1, 3)$, the corresponding Lagrange interpolator $a_s(x)$ is by E3,

$$a_s(x) = s_{-1}\frac{x(x-1)}{2} + s_0\frac{(x-1)(x+1)}{-1} + s_1\frac{x(x+1)}{2}$$

$$= \left(\frac{s_1}{2} + \frac{s_{-1}}{2} - s_0\right)x^2 + \left(\frac{s_1}{2} - \frac{s_{-1}}{2}\right)x + s_0.$$

Thus for given $\mathbf{s} = (s_{-1}, s_0, s_1)$, $a_{\mathbf{s}}(x)$ is given according to the following table.

$\mathbf{s} = (s_{-1}, s_0, s_1)$	$a_{\mathbf{s}}(x)$
$(1\ 1\ 3)$	$x^2 + x + 1$
$(1\ 1\ 1)$	1
$(1\ 1\ -3)$	$-2x^2 - 2x + 1$
$(1\ 1\ -1)$	$-x^2 - x + 1$
$(-1\ 1\ 3)$	$2x + 1$
$(-1\ 1\ 1)$	$-x^2 + x + 1$
$(-1\ 1\ -3)$	$3x^2 - x + 1$
$(-1\ 1\ -1)$	$-2x^2 + 1$
$(1\ -1\ 3)$	$3x^2 + x - 1$
$(1\ -1\ 1)$	$2x^2 + 1$
$(1\ -1\ -3)$	$-2x - 1$
$(1\ -1\ -1)$	$x^2 - x - 1$
$(-1\ -1\ 3)$	$2x^2 + 2x - 1$
$(-1\ -1\ 1)$	$x^2 + x - 1$
$(-1\ -1\ -3)$	$-x^2 - x - 1$
$(-1\ -1\ -1)$	-1

If $x^4 + x + 1$ factors we can assume all factors have integer coefficients and are primitive. Since $x^4 + x + 1$ is monic, the leading coefficients of any such factors must be ± 1. Now eight of the $a_{\mathbf{s}}(x)$ are primitive but do not have leading coefficient ± 1, so cannot be factors. Two more $a_{\mathbf{s}}(x)$ are uninteresting since they have degree 0. So only six of the $a_{\mathbf{s}}(x)$ must be checked as possible factors of $x^4 + x + 1$:

$$x^2 + x + 1; \qquad -x^2 - x - 1;$$
$$x^2 + x - 1; \qquad -x^2 - x + 1;$$
$$x^2 - x - 1; \qquad -x^2 + x + 1.$$

Since the three on the right are associates of the three on the left, we need only check the three on the left. Three quick divisions show that none are factors.

Notice that the number of possible factors $a_{\mathbf{s}}(x)$ of $f(x)$ depends on d ($= 1/2 \deg f$) but, more significantly, also depends on the number of divisors of $f(n_i)$. The number of possible factors in the example we just did was kept small by the fact that $f(1) = f(0) = 1$, which has only two factors in $\mathbb{Z}$. In general one is not so fortunate, and the number of $a_{\mathbf{s}}(x)$ can become unpleasantly large (see E7). A more efficient method of factoring has been developed in recent years which is based on factoring mod M for an appropriate number M. We shall examine this newer method in Chapter 13.

E4. Factor $f(x) = 3x^4 + 5x^2 - 1$ using Lagrange interpolators.

E5. Factor $x^6 + x^5 - 4x^4 - 4x^3 + 2x^2 + 4x + 1$ using Lagrange interpolators. (If you can use a computer, do!)

E6. Why is it that of the sixteen polynomials $a_s(x)$ arising as Lagrange interpolators for $x^4 + x + 1$, eight are associates of the other eight?

*E7. If $f(x)$ has degree $2d$, and $n_1, \ldots, n_d$ are distinct integers such that $f(n_i) = p_i$ is prime, how many polynomials $a_s(x)$ arise as possible divisors of $f(x)$ using Lagrange interpolation? Do they all pair off as associates?

Factoring in $\mathbb{Z}_p[x]$ 12

Factoring in $\mathbb{Z}_p[x]$, p prime, is interesting, not only for its own sake, but also because it is useful for factoring in $\mathbb{Z}[x]$. See Chapters 8 and 13. For example, $x^4 + 3x + 7$ can be shown to be irreducible in $\mathbb{Z}[x]$ (and therefore in $\mathbb{Q}[x]$) by showing that it is irreducible mod 2. For if $x^4 + 3x + 7 = a(x)b(x)$ then $a(x)b(x)$ can be chosen in $\mathbb{Z}[x]$; then $x^4 + 3x + 7 \equiv a(x)b(x)$ (mod 2) and $x^4 + 3x + 7$ would factor mod 2. But mod 2, $x^4 + 3x + 7 = x^4 + x + 1$, which is easily shown to be irreducible in $\mathbb{Z}_2[x]$.

Any polynomial of degree d in $\mathbb{Z}_p[x]$ can be factored in a finite number of steps, because there are only finitely many possible polynomials of degree $< d$ in $\mathbb{Z}_p$ (p^d, of them, to be precise), and we can simply check them all, using the division theorem. (In fact we need only look for factors of degree $\leq d/2$.) This trial and error approach is very inefficient, however.

In this chapter we present an efficient algorithm for factoring polynomials in $\mathbb{Z}_p[x]$. It was discovered in 1967 by E. R. Berlekamp (see Berlekamp (1967)).

The strategy of Berlekamp's method of factoring a polynomial $f(x)$ in $\mathbb{Z}_p[x]$ is to translate the problem into that of solving a system of linear equations with coefficients in $\mathbb{Z}_p$, and finding greatest common divisors (g.c.d.s). There are very efficient computational methods for doing both problems (row operations and Euclid's algorithm, respectively), so such a translation is a very desirable one.

The idea behind the algorithm is as follows. Suppose $f(x)$ has degree d, and suppose somehow we can find a polynomial $g(x)$ in $\mathbb{Z}_p(x)$ of degree ≥ 1 and $< d$ such that $f(x)$ divides $g(x)^p - g(x)$. Note that if $\deg(g(x)) = e \geq 1$, $g(x)^p - g(x) \neq 0$, for the coefficient of x^{pe} is nonzero. By Fermat's theorem the polynomial $u^p - u$ has p roots in $\mathbb{Z}_p$, namely $u =$

0, 1, 2, . . . , $p - 1$. Thus $u^p - u$ factors mod p into $u^p - u = u(u - 1)(u - 2) \cdots (u - (p - 1))$. Setting $u = g(x)$, we get that

$$g(x)^p - g(x) = g(x)(g(x) - 1)(g(x) - 2) \cdots (g(x) - (p - 1)) \quad (1)$$

in $\mathbb{Z}_p[x]$. Now observe the following.

E1. If a, b are relatively prime polynomials in $F[x]$, F a field, and f is in $F[x]$, prove that g.c.d.$(f, ab) =$ g.c.d.$(f, a) \cdot$ g.c.d.(f, b).

Since $f(x)$ divides $g(x)^p - g(x)$, $f(x)$ is the greatest common divisor of $f(x)$ and $g(x)^p - g(x)$. Since $g(x) - r$ and $g(x) - s$ are relatively prime if $r \neq s$, we have by (1) and Exercise E1 that

$$f(x) = \text{g.c.d.} \left(f(x), g(x)^p - g(x) \right)$$

$$= \prod_{k=0}^{p-1} \text{g.c.d.} \left(f(x), g(x) - k \right) \quad (2)$$

Each factor on the right side has degree at most that of $g(x)$, which is in turn $< d$, the degree of $f(x)$. Thus there must be at least two nontrivial factors of $f(x)$ on the right-hand side of (2), that is, at least two factors of f which have degree $\geqslant 1$, and (2) is a nontrivial factorization of $f(x)$.

We have proved the first part of Berlekamp's theory:

Theorem 1. *Given $f(x)$ in $\mathbb{Z}_p[x]$ of degree d, let $g(x)$ in $\mathbb{Z}_p[x]$ be a polynomial of degree $\geqslant 1$ and $< d$ such that $f(x)$ divides $g(x)^p - g(x)$. Then*

$$f(x) = \text{g.c.d.} \left(f(x), g(x) \right)$$
$$\cdot \text{g.c.d.} \left(f(x), g(x) - 1 \right) \cdots \text{g.c.d.} \left(f(x), g(x) - (p - 1) \right)$$

is a nontrivial factorization of $f(x)$.

To factor $f(x)$, then, we must find a polynomial $g(x)$ such that $f(x)$ divides $g(x)^p - g(x)$. This is done by solving a set of linear equations for the coefficients of $g(x)$. We now obtain these equations.

Suppose $g(x) = b_0 + b_1 x + \cdots + b_{d-1} x^{d-1}$, where $b_0, \ldots, b_{d-1}$ are coefficients to be determined. To see whether $f(x)$ divides $g(x)^p - g(x)$ we first look at $g(x)^p$. By Theorem 1 of Chapter II-10,

$$g(x)^p = b_0^p + b_1^p x^p + \cdots + b_{d-1}^p x^{(d-1)p}.$$

By Fermat's theorem, applied to each of $b_0, b_1, \ldots, b_{d-1}$, we have $b_i^p = b_i$ in $\mathbb{Z}_p$ for all i, so

$$g(x)^p = b_0 + b_1 x^p + b_2 x^{2p} + \cdots + b_{d-1} x^{(d-1)p}. \quad (3)$$

Now use the division theorem: divide $f(x)$ into x^{ip} for $i = 0, 1, 2, \ldots, d - 1$, to get

$$x^{ip} = f(x)q_i(x) + r_i(x), \quad (4)$$

where

$$r_i(x) = r_{i,0} + r_{i,1}x + r_{i,2}x^2 + \cdots + r_{i,d-1}x^{d-1}. \tag{5}$$

Substituting (4) into (3), we get

$$g(x)^p = b_0 r_0(x) + b_1 r_1(x) + \cdots + b_{d-1}r_{d-1}(x) + (\text{multiple of } f(x)).$$

Thus $f(x)$ divides $g(x)^p - g(x)$ if and only if $f(x)$ divides the polynomial

$$b_0 r_0(x) + b_1 r_1(x) + \cdots + b_{d-1}r_{d-1}(x) - \left(b_0 + b_1 x + \cdots + b_{d-1}x^{d-1}\right).$$

But this polynomial has degree $\leqslant d-1$, hence is divisible by $f(x)$ (of degree d) if and only if it is $= 0$.

Thus $f(x)$ divides $g(x)^p - g(x)$ if and only if the coefficients $b_0, \ldots, b_d$ of $g(x)$ satisfy

$$b_0 r_0(x) + b_1 r_1(x) + \cdots + b_{d-1}r_{d-1}(x)$$
$$- \left(b_0 + b_1 x + \cdots + b_{d-1}x^{d-1}\right) = 0. \tag{6}$$

Collecting coefficients of $1, x, x^2, \ldots, x^{d-1}$ in (6), we get d simultaneous linear equations in the d unknowns $b_0, b_1, \ldots, b_{d-1}$. We solve them for $b_0, \ldots, b_{d-1}$ and get the coefficients of a polynomial $g(x)$ such that $f(x)$ divides $g(x)^p - g(x)$.

At this point it is convenient to use matrix notation, to put equation (6) into matrix form. Let $\mathbf{I}$ denote the $d \times d$ identity matrix,

$$\mathbf{I} = \begin{bmatrix} 1 & 0 & \cdots & 0 \\ 0 & 1 & \cdots & 0 \\ \vdots & & & \vdots \\ 0 & 0 & \cdots & 1 \end{bmatrix},$$

and let

$$\mathbf{Q} = \begin{bmatrix} r_{0,0} & r_{0,1} & \cdots & r_{0,d-1} \\ r_{1,0} & r_{1,1} & & \vdots \\ \vdots & \vdots & & \\ r_{d-1,0} & r_{d-1,1} & \cdots & r_{d-1,d-1} \end{bmatrix}$$

be the matrix whose rows are the coefficients of the remainder polynomials $r_0(x), \ldots, r_{d-1}(x)$, as in (5). Then it is easily verified that $(b_0, b_1, \ldots, b_{d-1})$ is a solution of (6) if and only if

$$(b_0, \ldots, b_{d-1})(\mathbf{Q} - \mathbf{I}) = (0, \ldots, 0). \tag{7}$$

Combining all this with Theorem 1 we get

Theorem 2 (Berlekamp's Factoring Algorithm). *To find a nontrivial factorization of $f(x)$ in $\mathbb{Z}_p[x]$ of degree d, find the matrix $\mathbf{Q}$ and find a solution $\mathbf{b} = (b_0, b_1, \ldots, b_{d-1})$ of $\mathbf{b}(\mathbf{Q} - \mathbf{I}) = 0$. Let $g(x) = b_0 + b_1 x + \cdots +$*

$b_{d-1}x^{d-1}$. *If $g(x)$ has degree $\geqslant 1$, then for some s in $\mathbb{Z}_p$, $g(x) - s$ and $f(x)$ have a common factor of degree $\geqslant 1$.*

Here is an example in $\mathbb{Z}_2[x]$. Let $f(x) = 1 + x + x^2 + x^3 + x^4 + x^5 + x^6$. Then $d = 6$. Divide $f(x)$ into x^{2i}, $i = 0, 1, 2, 3, 4, 5$ as in (4) to get $r_i(x)$:

$$x^0 = f(x) \cdot 0 + 1, \qquad\qquad\qquad\qquad r_0(x) = 1;$$

$$x^2 = f(x) \cdot 0 + x^2, \qquad\qquad\qquad\qquad r_1(x) = x^2;$$

$$x^4 = f(x) \cdot 0 + x^4, \qquad\qquad\qquad\qquad r_2(x) = x^4;$$

$$x^6 = f(x) \cdot 1 + (1 + x + x^2 + x^3 + x^4 + x^5), \quad r_3(x) = 1 + x + x^2$$

$$x^8 = f(x) \cdot (x^2 + x) + x, \qquad\qquad\qquad\qquad\qquad\qquad + x^3 + x^4 + x^5;$$

$$x^{10} = f(x) \cdot (x^4 + x^3) + x^3, \qquad\qquad\qquad r_4(x) = x;$$

$$r_5(x) = x^3.$$

The coefficients of $r_0(x), r_1(x), \ldots, r_5(x)$ form the rows of the matrix $\mathbf{Q}$:

$$\mathbf{Q} = \begin{bmatrix} 100000 \\ 001000 \\ 000010 \\ 111111 \\ 010000 \\ 000100 \end{bmatrix}.$$

So

$$\mathbf{Q} - \mathbf{I} = \begin{bmatrix} 000000 \\ 011000 \\ 001010 \\ 111011 \\ 010010 \\ 000101 \end{bmatrix}.$$

To find $g(x) = b_0 + b_1 x + b_2 x^2 + b_3 x^3 + b_4 x^4 + b_5 x^5$, we solve $\mathbf{b}(\mathbf{Q} - \mathbf{I}) = \mathbf{0}$, or:

$$b_3 = 0;$$
$$b_1 + b_3 + b_4 = 0;$$
$$b_1 + b_2 + b_3 = 0;$$
$$b_5 = 0;$$
$$b_2 + b_3 + b_4 = 0;$$
$$b_3 + b_5 = 0.$$

This reduces quickly to $b_3 = b_5 = 0$, $b_1 = b_2 = b_4$. The only solutions with $\deg(g(x)) \geqslant 1$ are $g(x) = b_0 + x + x^2 + x^4$ with $b_0 = 0$ or 1. For either choice of b_0,

$$g(x)^2 - g(x) = x^8 + x = f(x) \cdot (x^2 + x).$$

Thus

$$f(x) = \text{g.c.d.}\,(f(x), x^4 + x^2 + x) \cdot \text{g.c.d.}\,(f(x), x^4 + x^2 + x + 1).$$

It is easy to check that the left factor of the right side, g.c.d.$(f(x), x^4 + x^2 + x)$, is $x^3 + x + 1$, and the right factor is $x^3 + x^2 + x + 1$, and both are irreducible, so the factorization of $f(x)$ into a product of irreducible polynomials in $\mathbb{Z}_2[x]$ is

$$f(x) = x^6 + x^5 + x^4 + x^3 + x^2 + x + 1 = (x^3 + x + 1)(x^3 + x^2 + 1).$$

Using some ideas of linear algebra (see Chapter I-9) we can actually count the number of distinct irreducible factors of $f(x)$. For the rest of this chapter (except for the exercises) we shall assume knowledge of Chapter I-9.

Let N be the set of vectors $\mathbf{b} = (b_0, b_1, \ldots, b_{d-1})$ with $\mathbf{b}(\mathbf{Q} - \mathbf{I}) = \mathbf{0}$. Then N is called the *null space* of $\mathbf{Q} - \mathbf{I}$. Let $\{\mathbf{v}_1, \mathbf{v}_2, \mathbf{v}_3, \ldots, \mathbf{v}_r\}$ be a set of vectors in N such that every vector $\mathbf{b}$ in N is a linear combination of $\mathbf{v}_1, \ldots, \mathbf{v}_r$: for any $\mathbf{b}$ in N there are $x_1, \ldots, x_r$ in $\mathbb{Z}_p$ with $\mathbf{b} = x_1\mathbf{v}_1 + \cdots + x_r\mathbf{v}_r$. The smallest r for which such a set $\{\mathbf{v}_1, \mathbf{v}_2, \ldots, \mathbf{v}_r\}$ exists is called the *dimension* of N.

We observe that the vector $\mathbf{b} = (a, 0, \ldots, 0)$ is always a solution of $\mathbf{b}(\mathbf{Q} - \mathbf{I}) = \mathbf{0}$, since $g(x)^p - g(x) = 0$ when $g(x)$ is the constant polynomial a. If the only solutions of $\mathbf{b}(\mathbf{Q} - \mathbf{I}) = \mathbf{0}$ are of the form $\mathbf{b} = (a, 0, \ldots, 0)$, then the dimension of N is 1, for every vector in the null space of $\mathbf{Q} - \mathbf{I}$ is a multiple of $\mathbf{v}_1 = (1, 0, \ldots, 0)$. To factor $f(x)$ we need to find some other vector $\mathbf{b} = (b_0, \ldots, b_{d-1})$ with $\mathbf{b}(\mathbf{Q} - \mathbf{I}) = \mathbf{0}$.

We have the following information:

Theorem 3. (a) *The dimension of the null space of* $\mathbf{Q} - \mathbf{I} = $ *the number of distinct irreducible factors of* $f(x)$. (b) $f(x)$ *is irreducible in* $\mathbb{Z}_p[x]$ *iff the null space of* $\mathbf{Q} - \mathbf{I}$ *has dimension 1 and* $f(x)$ *and its derivative* $f'(x)$ *are relatively prime.*

The dimension of the null space can be computed in the following way.

Since $\mathbf{Q} - \mathbf{I}$ is a $d \times d$ matrix, the dimension of the null space of $\mathbf{Q} - \mathbf{I}$ can be computed as d minus the row or column rank of $\mathbf{Q} - \mathbf{I}$. The column rank of $\mathbf{Q} - \mathbf{I}$ is equal to the number of nonzero columns after doing column operations on $\mathbf{Q} - \mathbf{I}$ to get it into echelon form. To illustrate with $\mathbf{Q} - \mathbf{I}$ as in the last example, a series of column operations puts $\mathbf{Q} - \mathbf{I}$ into echelon form as follows:

$$\mathbf{Q} - \mathbf{I} = \begin{pmatrix} 000000 \\ 011000 \\ 001010 \\ 111011 \\ 010010 \\ 000101 \end{pmatrix} \overset{(1)}{\to} \begin{pmatrix} 000000 \\ 100100 \\ 010100 \\ 111101 \\ 110000 \\ 000011 \end{pmatrix}$$

$$\overset{(2)}{\to} \begin{pmatrix} 000000 \\ 100000 \\ 010100 \\ 111001 \\ 110100 \\ 000011 \end{pmatrix} \overset{(3)}{\to} \begin{pmatrix} 000000 \\ 100000 \\ 010000 \\ 111101 \\ 110000 \\ 000011 \end{pmatrix}$$

$$\overset{(4)}{\to} \begin{pmatrix} 000000 \\ 100000 \\ 010000 \\ 111000 \\ 110000 \\ 000010 \end{pmatrix} \overset{(5)}{\to} \begin{pmatrix} 000000 \\ 100000 \\ 010000 \\ 001000 \\ 110000 \\ 000100 \end{pmatrix} = \mathbf{E}.$$

Here

(1) consists of several column interchanges,
(2) is adding the first column to the fourth column,
(3) is adding the second column to the fourth column,
(4) is adding the third column to the fourth and sixth, and adding the fifth column to the sixth column,
(5) is adding the third column to each of the first two and interchanging the fourth and fifth.

Since the resulting matrix has 4 nonzero columns, the null space has dimension $6 - 4 = 2$. In fact, a basis of the null space can be obtained by solving $\mathbf{bE} = \mathbf{0}$. (Doing column operations to $\mathbf{Q} - \mathbf{I}$ corresponds to doing manipulations to the equation $\mathbf{b}(\mathbf{Q} - \mathbf{I}) = \mathbf{0}$, manipulations which do not change the set of solutions of $\mathbf{b}(\mathbf{Q} - \mathbf{I}) = \mathbf{0}$ and lead to the simpler equations $\mathbf{bE} = \mathbf{0}$.) In solving $\mathbf{bE} = \mathbf{0}$, b_0 and b_4 may be chosen arbitrarily, and, once chosen, determine any solution $\mathbf{b}$ uniquely. The solutions to $\mathbf{bE} = \mathbf{0}$ have the form:

$$(b_0, b_1, b_2, b_3, b_4, b_5) = (b_0, b_4, b_4, 0, b_4, 0)$$

$$= b_0(1, 0, 0, 0, 0, 0) + b_4(0, 1, 1, 0, 1, 0).$$

We chose $b_4 = 1$ above.

PROOF OF THEOREM 3(a). Suppose $f(x)$ has k distinct irreducible factors, $f(x) = \prod_{i=1}^{k}(p_i(x))^{r_i}$, where $p_i(x)$ is irreducible, and suppose $f(x)$ divides $g(x)^p - g(x)$ for some $g(x)$. If $f(x)$ divides $g(x)^p - g(x) = \prod_{s=0}^{p-1}(g(x) - s)$, then each irreducible factor $p_i(x)$ divides $g(x) - s$ for some s, and, since $g(x) - s$ and $g(x) - t$ are relatively prime for $s \neq t$, the s for which $p_i(x)$ divides $g(x) - s$ is *uniquely determined* by $p_i(x)$—$p_i(x)$ cannot divide both $g(x) - s$ and $g(x) - t$ for $s \neq t$. Denote the unique s, $0 \leqslant s \leqslant p - 1$, such that $p_i(x)$ divides $g(x) - s$, by s_i.

To $g(x)$, then, corresponds the vector $(s_1, \ldots, s_k)$ of elements of $\mathbb{Z}_p$. This defines a function from the set of polynomials $g(x)$ with $f(x)$ dividing $g(x)^p - g(x)$, to the set of k-tuples $(s_1, \ldots, s_k)$ of elements of $\mathbb{Z}_p$.

Now we define the inverse map.

By the Chinese remainder theorem for polynomials (Chapter 11), given any $s_1, \ldots, s_k$ (not necessarily distinct) in $\mathbb{Z}_p$ there is a unique polynomial $g(x)$ of degree $< d$ with

$$g(x) \equiv s_i \pmod{p_i(x)^{r_i}},$$

that is, $p_i(x)^{r_i}$ divides $g(x) - s_i$. For such a $g(x)$, then, $p_i(x)^{r_i}$, for each i, $i = 1, \ldots, k$, divides $\prod_{s=0}^{p-1}(g(x) - s) = g(x)^p - g(x)$. Therefore $f(x) = \prod_{i=1}^{k} p_i(x)^{r_i}$ divides $g(x)^p - g(x)$. Thus given $(s_1, \ldots, s_k)$ we have obtained $g(x)$ of degree $< d$ such that $f(x)$ divides $g(x)^p - g(x)$. This is the inverse of the map constructed in the first paragraph of the proof.

Thus we have a 1–1 correspondence between polynomials $g(x)$ of degree $< d$ with $f(x)$ dividing $g(x)^p - g(x)$, and k-tuples of elements of $\mathbb{Z}_p$, where k is the number of distinct irreducible factors of $f(x)$. Since there are p^k such k-tuples, there must be p^k possible $g(x)$.

Now there is a 1–1 correspondence between vectors in the null space of $\mathbf{Q} - \mathbf{I}$ and polynomials $g(x)$ of degree $< d$ with $f(x)$ dividing $g(x)^p - g(x)$ given by corresponding the polynomial $g(x) = b_0 + b_1 x + \cdots + b_{d-1} x^{d-1}$ with the row vector $(b_0, b_1, \ldots, b_{d-1})$, as in Theorem 2. Thus there are p^k vectors in the null space of $\mathbf{Q} - \mathbf{I}$, and so by Exercise E2 (below) the null space of $\mathbf{Q} - \mathbf{I}$ has dimension k. Since k was the number of distinct irreducible factors of $f(x)$, the proof of part (a) is complete.

E2. If V is a vector space of dimension k over a field with q elements, prove that V has q^k elements.

PROOF OF THEOREM 3(b). From (a), the null space of $\mathbf{Q} - \mathbf{I}$ is one-dimensional iff $f(x) = p(x)^r$, a power of an irreducible polynomial. Then $r = 1$ and $f(x)$ is irreducible iff $f(x)$ and $f'(x)$ are relatively prime. That completes the proof of Theorem 3. $\square$

EXAMPLE. How many distinct irreducible factors divide $f(x) = x^5 + 2x^4 + x^3 + x^2 + 2$ in $\mathbb{Z}_3[x]$? We compute $\mathbf{Q} - \mathbf{I}$:

$$1 = f(x)0 + 1;$$

$$x^3 = f(x)0 + x^3;$$

$$x^6 = f(x)(x + 1) + 1 + x + 2x^2 + x^3;$$

$$x^9 = f(x)(x^4 + x^3 + x) + x;$$

$$x^{12} = f(x)(x^7 + x^6 + x^4) + x^4.$$

So (note that the rows of $\mathbf{Q}$ are the coefficients of the remainders arranged, from left to right, by increasing powers of x)

$$\mathbf{Q} = \begin{pmatrix} 10000 \\ 00010 \\ 11210 \\ 01000 \\ 00001 \end{pmatrix}, \qquad \mathbf{Q} - \mathbf{I} = \begin{pmatrix} 00000 \\ 02010 \\ 11110 \\ 01020 \\ 00000 \end{pmatrix},$$

which, after column operations, becomes

$$\begin{pmatrix} 00000 \\ 10000 \\ 01000 \\ 20000 \\ 00000 \end{pmatrix}.$$

So the null space of $\mathbf{Q} - \mathbf{I}$ has dimension 3, $f(x)$ has three distinct irreducible factors, and any polynomial g of the form

$$(b_0, b_1, b_2, b_3, b_4) = (b_0, b_3, 0, b_3, b_4)$$

is in the null space, where b_0, b_3, b_4 are arbitrary.

E3. Find the three irreducible factors of $f(x)$ in the example.

E4. Try to factor $x^{10} + x^9 + x^7 + x^3 + x^2 + 1$ in $\mathbb{Z}_2[x]$ into a product of irreducible polynomials using Berlekamp's algorithm.

E5. Try to factor $x^8 + x^7 + x^6 + x^4 + 1$ in $\mathbb{Z}_2[x]$.

E6. Show that $x^5 + x^2 + 1$ is irreducible in $\mathbb{Z}_2[x]$.

E7. Find all irreducible polynomials in $\mathbb{Z}_2[x]$ of degree 4 or less.

E8. Show that $x^7 + x^3 + 1$ is irreducible in $\mathbb{Z}_2[x]$.

E9. Show that $7x^7 + 6x^6 + 4x^4 + 3x^3 + 2x^2 + 2x + 1$ is irreducible in $\mathbb{Q}[x]$.

E10. Find the greatest common divisor in $\mathbb{Q}[x]$ of $x^6 + 6x^5 + 9x^4 + 8x^3 + 4x^2 + x + 1$ and $4x^6 + 3x^5 + 4x^4 - x^3 + x^2 + 1$ without doing the Euclidean algorithm in $\mathbb{Q}[x]$.

†*E11. Use Berlekamp's algorithm to factor $x^2 - q$ in $\mathbb{Z}_p[x]$ ($p \nmid q$) and prove that q is a quadratic residue mod p, that is, $x^2 - q \equiv 0 \pmod{p}$ has a root, iff $q^{\frac{p-1}{2}} \equiv 1 \pmod{p}$. This fact is called Euler's lemma (see III-16B).

Factoring in $\mathbb{Q}[x]$, III: mod M 13

One of the best methods currently available for factoring a polynomial with integer coefficients is to factor it mod M for M a large number and then try to "lift" the factorization back to one in $\mathbb{Z}[x]$.

We can assume $f(x)$ is a monic polynomial with integer coefficients. For if $f(x)$ has leading coefficient a, replacing x by x/a and multiplying the resulting polynomial by a^{n-1} gives a monic polynomial with integer coefficients $g(x)$ whose factorization problem is the same as that of $f(x)$. For example, to factor $f(x) = 6x^2 + 5x + 1$, factor $g(x) = 6(6(x/6)^2 + 5(x/6) + 1) = x^2 + 5x + 6 = (x + 3)(x + 2)$; then $f(x)$ factors into $(1/6)(6x + 3)(6x + 2)$.

Given $f(x)$ monic with integer coefficients, the strategy is to factor $f(x)$ mod M, where M is some number which is larger than twice the absolute values of the coefficients of all possible factors of $f(x)$. The reason is the following.

In general, if $f(x) \equiv g(x)h(x) \pmod{m}$ and m is small, there may be many ways this factorization could lift to one of $f(x)$. For example, consider

$$f(x) = x^5 + 17x^4 - 5x^3 - 277x^2 + 144.$$

Modulo 5, $f(x)$ factors as

$$f(x) \equiv (x^3 + 3x + 2)(x^2 + 2x + 2) \pmod{5}.$$

But there are many polynomials which are congruent to $x^2 + 2x + 2$ (mod 5) which could possibly be factors of $f(x)$, such as $x^2 + 2x - 3$, $x^2 + 7x - 3$, $x^2 - 3x + 2$, $x^2 - 13x - 3$, $x^2 + 17x - 8$, etc., and it would take some effort to discover that $f(x)$ factors in $\mathbb{Z}[x]$ as

$$f(x) = (x^3 - 17x + 12)(x^2 + 17x + 12).$$

However, suppose $f(x) \equiv g(x)h(x)$ (mod M), where $g(x)$ and $h(x)$ are in $\mathbb{Z}[x]$, and suppose we know that the coefficients of any possible factor of $f(x)$ of degree that of $g(x)$ must be in absolute value $< M/2$. Suppose we choose $g(x)$ so that each of its coefficients is $> -M/2$ and $\leqslant M/2$. Suppose $f(x) = G(x)H(x)$ in $\mathbb{Z}[x]$, where $G(x) \equiv g(x)$ (mod M) and $H(x) \equiv h(x)$ (mod M). Then it must be that $G(x) = g(x)$. For if $G(x) \equiv g(x)$ (mod M), then either $G(x) = g(x)$ or there is a coefficient of $G(x)$ which differs from the corresponding coefficient of $g(x)$ by a nonzero multiple of M. But then that coefficient of $G(x)$ must be in absolute value $\geqslant M/2$, and is therefore too big in absolute value to qualify $G(x)$ as a possible factor of $f(x)$.

Thus if $f(x) \equiv g(x)h(x)$ (mod M), then either $g(x)$ divides $f(x)$ in $\mathbb{Q}[x]$ or the factorization $f(x) \equiv g(x)h(x)$ (mod M) corresponds to no factorization in $\mathbb{Z}[x]$ at all.

To factor $f(x)$ in $\mathbb{Z}[x]$, then, we want to find a suitably large M, and find all the factorizations of $f(x)$ modulo that M.

A. Bounding the Coefficients of Factors of a Polynomial

Suppose $f(x)$ in $\mathbb{Z}[x]$ is monic, of degree n. Let $g(x)$ be a factor of $f(x)$, which we can assume is also monic and has integer coefficients.

Proposition 1. *Suppose all complex roots of $f(x)$ are in absolute value less than some positive real number R. If $g(x)$ is a factor of $f(x)$, of degree r, then all the coefficients of $g(x)$ are in absolute value at most* $\max\left\{ \binom{r}{k} R^k \,\middle|\, k = 1, 2, \ldots, r \right\}$.

PROOF. Let $g(x) = x^r + b_{r-1}x^{r-1} + \cdots + b_1 x + b_0$, and in $\mathbb{C}[x]$ let $g(x) = \prod_{i=1}^{r}(x + s_i)$ (where $-s_1, \ldots, -s_r$ are the roots of $g(x)$ in the complex numbers).

Multiplying out the factorization $g(x) = \prod_{i=1}^{r}(x + s_i)$, and equating coefficients, we get

$$b_{r-1} = s_1 + s_2 + \cdots + s_r,$$

$$b_{r-2} = s_1 s_2 + s_1 s_3 + s_2 s_3 + s_1 s_4 + \cdots = \sum_{i<1} s_i s_j,$$

$$b_{r-3} = \sum s_i s_j s_k,$$

$$\vdots$$

$$b_0 = s_1 s_2 \ldots s_r.$$

Now each $-s_i$ is a root of $f(x)$ so for each i, $|-s_i| \leqslant R$. Thus, using the triangle inequality, for each i, $|b_i|$ is $\leqslant$ the sum obtained by replacing all the s_i by R. For example, with $r = 4$,

$$|b_3| = |s_1 + s_2 + s_3 + s_4| \leqslant |s_1| + |s_2| + |s_3| + |s_4|$$
$$\leqslant R + R + R + R = 4R,$$
$$|b_2| = |s_1 s_2 + s_1 s_3 + s_1 s_4 + s_2 s_3 + s_2 s_4 + s_3 s_4|$$
$$\leqslant |s_1 s_2| + |s_1 s_3| + |s_1 s_4| + |s_2 s_3| + |s_2 s_4| + |s_3 s_4| \leqslant 6R^2,$$
$$|b_1| = |s_1 s_2 s_3| + |s_1 s_2 s_4| + |s_1 s_3 s_4| + |s_2 s_3 s_4| \leqslant 4R^3,$$
$$|b_0| = |s_1 s_2 s_3 s_4| \leqslant R^4.$$

Replacing each s_i by R amounts to replacing $g(x)$ by $h(x)$, where

$$h(x) = (x + R)^r$$
$$= \sum_{k=0}^{r} \binom{r}{k} R^k x^{r-k} \qquad \text{(binomial theorem)}.$$

So

$$|b_{r-1}| \leqslant \binom{r}{1} R = rR,$$
$$|b_{r-2}| \leqslant \binom{r}{2} R^2 = \frac{r(r-1)}{2} R^2,$$
$$\vdots$$
$$|b_0| \leqslant \binom{r}{r} R^r = R^r.$$

Thus all coefficients of $g(x)$ are $\leqslant \max\left\{\binom{r}{k} R^k | k = 1, \ldots, r\right\}$, as was to be proved. $\qquad\qquad\square$

We observe that if the bound R on the roots is $\geqslant$ the degree r of $g(x)$, then $\binom{r}{k} R^k \leqslant \binom{r}{k+1} R^{k+1}$ for each k, so $\max\left\{\binom{r}{k} R^k | k = 1, \ldots, r\right\}$ $= R^r$ bounds the coefficients of $g(x)$.

E1. Verify this last statement. What is $\max\left\{\binom{r}{k} R^k | k = 1, \ldots, r\right\}$ if $R < r$?

According to Proposition 1, in order to find a good bound on the coefficients of factors of $f(x)$ we must find a good bound on the roots of $f(x)$. We know from Exercise E4, Chapter 4 that if $f(x) = x^n + a_{n-1} x^{n-1} + \cdots + a_1 x + a_0$, then all roots of $f(x)$ are in absolute value less than $R_0 = 1 + |a_{n-1}| + \cdots + |a_1| + |a_0|$. Here is another bound, attributed to H. Zassenhaus.

Proposition 2. *If $f(x) = x^n + a_{n-1}x^{n-1} + \cdots + a_1 x + a_0$ is in $\mathbb{C}[x]$, then all roots of $f(x)$ are in absolute value $\leqslant R_z$ where*

$$R_z = \frac{1}{2^{1/n} - 1} \max\left\{ \left(|a_{n-k}|/\binom{n}{k}\right)^{1/k} | k = 1, \ldots, n \right\}.$$

PROOF. Suppose Q is some number such that for each k,

$$|a_{n-k}| \leqslant Q^k \binom{n}{k},$$

that is,

$$Q \geqslant \left(|a_{n-k}|/\binom{n}{k} \right)^{1/k}.$$

Suppose b is a root of $f(x)$ in $\mathbb{C}$. Then

$$0 = f(b) = b^n + a_{n+1}b^{n-1} + \cdots + a_1 b + a_0,$$

so solving for b^n and taking absolute values,

$$|b^n| = |a_{n-1}b^{n-1} + \cdots + a_1 b + a_0|,$$

and therefore, by the triangle inequality,

$$|b^n| \leqslant |a_{n-1}| |b|^{n-1} + \cdots + |a_1| |b| + |a_0|$$

$$\leqslant \binom{n}{1}Q|b|^{n-1} + \cdots + \binom{n}{n-1}Q^{n-1}|b| + Q^n. \qquad (1)$$

Now

$$(|b| + Q)^n = |b|^n + \binom{n}{1}Q|b|^{n-1} + \cdots + \binom{n}{n-1}Q^{n-1}|b| + Q^n.$$

So the right side of (1) equals

$$(|b| + Q)^n - |b|^n.$$

Hence

$$2|b|^n \leqslant (|b| + Q)^n,$$

or

$$2^{1/n}|b| \leqslant |b| + Q,$$

or

$$(2^{1/n} - 1)|b| \leqslant Q.$$

Thus if b is a root of $f(x)$,

$$|b| \leqslant Q/(2^{1/n} - 1)$$

provided

$$Q \geqslant \left(|a_{n-k}|/\binom{n}{k} \right)^{1/k} \quad \text{for each } k = 1, \ldots, n.$$

The proposition follows immediately. $\square$

Here is another bound.

Proposition 3. *If* $f(x) = x^n + a_{n-1}x^{n-1} + \cdots + a_1x + a_0$ *is in* $\mathbb{C}[x]$, *then all roots of* $f(x)$ *are in absolute value* $\leqslant R_k = 2S$ *where*

$$S = \max\{|a_{n-k}|^{1/k} \,|\, k = 1, \ldots, n\}.$$

PROOF. If $f(b) = 0$, then, as in the proof of Proposition 2,

$$|b|^n \leqslant |a_{n-1}| \,|b|^{n-1} + \cdots + |a_1| \,|b| + |a_0|.$$

Since $|a_{n-k}| \leqslant S^k$ for $k = 1, \ldots, n$,

$$|b|^n \leqslant S|b|^{n-1} + \cdots + S^{n-1}|b| + S^n$$

or

$$\left(\frac{|b|}{S}\right)^n \leqslant \left(\frac{|b|}{S}\right)^{n-1} + \cdots + \left(\frac{|b|}{S}\right) + 1 \tag{2}$$

or, setting $|b|/S = Z$,

$$Z^n \leqslant Z^{n-1} + \cdots + Z + 1 = \frac{Z^n - 1}{Z - 1}.$$

If $Z \geqslant 2$ this cannot hold, for if $Z \geqslant 2$, $Z - 1 \geqslant 1$, then

$$\frac{Z^n - 1}{Z - 1} \leqslant Z^n - 1 < Z^n.$$

Thus if (2) holds, $Z \leqslant 2$, so $|b| \leqslant 2S$, as was to be proved. ☐

For our earlier example,

$$f(x) = x^5 + 17x^4 - 5x^3 - 277x^2 + 144,$$

these bounds are:

$$R_0 = 1 + 17 + 5 + 277 + 144 = 444;$$

$$R_z = \frac{1}{2^{1/5} - 1} \max\left\{\frac{17}{5}, \left(\frac{5}{10}\right)^{1/2}, \left(\frac{277}{10}\right)^{1/3}, \left(\frac{0}{5}\right)^{1/4}, \left(\frac{144}{1}\right)^{1/5}\right\}$$

$$\leqslant \frac{1}{.15}\left(\frac{277}{10}\right)^{1/3} \leqslant \frac{20}{3}(3.1) \leqslant 21;$$

$$R_k = 2\max\{17, 5^{1/2}, 277^{1/3}, 0^{1/4}, 144^{1/5}\} = 2(17) = 34.$$

Using the bound $R_z = 21$ for the roots of $f(x)$, a bound on the coefficients of possible factors of degree 2 would be $21^2 = 441$. So in factoring $f(x) \bmod M$ we would choose M to be at least 882.

In practice we choose M to be a prime power.

E2. Show that $R_0 \geqslant R_k$ for any polynomial in $\mathbb{Z}[x]$ with at least two nonzero coefficients.

E3. Find R_z, R_k for $f(x) = x^5 - 8x^4 + 7x^3 - 7x^2 - 2x + 6$.

E4. Find R_z, R_k for $f(x) = x^6 + x^5 - 7x^4 + 3x^2 + x - 11$.

E5. Find R_z, R_k for $f(x) = x^4 + 16x^3 + 95x^2 + 256x + 256$.

E6. Find R_z, R_k for $f(x) = x^4 + 16x^3 + 96x^2 + 256x + 255$.

E7. Which is a better bound, R_z or R_k? When is one better than the other?

B. Factoring Modulo High Powers of Primes

The strategy of factoring a polynomial in $\mathbb{Z}[x]$ by factoring mod M described at the beginning of this chapter requires only that M be sufficiently large. In this section we show how to factor mod M where $M = p^{2^N}$. The idea is to "lift" factorizations mod a prime p (which can be small) to factorizations mod p^2, then mod p^4, p^8, p^{16}, ... This lifting rapidly gives factorizations modulo numbers as large as one needs.

To get started, i.e., factor mod p, one can use trial and error, or use Berlekamp's algorithm of Chapter 12.

Henceforth we drop "(x)" in $f(x)$.

First a definition.

If q is any natural number and g, h are polynomials with coefficients in $\mathbb{Z}$, g and h are called *relatively prime* mod q if there are polynomials r and s in $\mathbb{Z}[x]$ such that $rg + sh \equiv 1 \pmod{q}$.

Here is the theorem which tells how to lift factorizations.

Theorem. *Suppose f is a monic polynomial in $\mathbb{Z}[x]$ and $f \equiv g_1 h_1 \pmod{q}$ where g_1 and h_1 are monic and relatively prime mod q. Then there are monic polynomials g_2, h_2 in $\mathbb{Z}[x]$ with $g_2 \equiv g_1 \pmod{q}$, $h_2 \equiv h_1 \pmod{q}$, such that $f \equiv g_2 h_2 \pmod{q^2}$ and g_2 and h_2 are relatively prime mod q^2. The polynomials g_2 and h_2 are unique mod q^2.*

PROOF. We use the following subscript convention: a subscript $(\ )_1$ denotes a polynomial with integer coefficients whose coefficients are defined uniquely only mod q. For example, if $f = 17x^2 + 3x + 8$, then $17x^2 - 3x + 8 \equiv (17x - 3)(x + 2) \pmod{2}$. So $f \equiv g_1 h_1 \pmod{2}$, where we could choose $g_1 = 17x - 3$, $h_1 = x + 2$, but we could also choose $g_1 = x + 1$, $h_1 = x$. On the other hand, a subscript $(\ )_2$ denotes a polynomial which is uniquely defined mod q^2.

To begin the proof we start with g_1, h_1 in $\mathbb{Z}[x]$ with $f \equiv g_1 h_1 \pmod{q}$, and we find g_2, h_2 with $f \equiv g_2 h_2 \pmod{q^2}$.

Let $f \equiv g_1 h_1 + q k_1 \pmod{q^2}$. Since f, g_1, h_1 are all monic we can choose k_1 with $\deg k_1 < \deg f$.

We look for g_2, h_2 in the form $g_2 = g_1 + qG_1$, $h_2 = h_1 + qH_1$ for some G_1, H_1 in $\mathbb{Z}[x]$. If we do, then the equation

$$f \equiv g_2 h_2 \pmod{q^2}$$

becomes

$$g_1 h_1 + qk_1 \equiv (g_1 + qG_1)(h_1 + qH_1) \pmod{q^2}$$

or

$$qk_1 \equiv q(g_1 H_1 + h_1 G_1) \pmod{q^2}$$

or

$$k_1 \equiv g_1 H_1 + h_1 G_1 \pmod{q}. \tag{1}$$

Since g_1 and h_1 are assumed relatively prime mod q, equation (1) is solvable: in fact, if $r_1 g_1 + s_1 h_1 \equiv 1 \pmod{q}$, then $H_1 = k_1 r_1$, $G_1 = k_1 s_1$ solves (1). But we can also choose G_1, H_1 with deg $G_1 <$ deg g_1, deg $H_1 <$ deg h_1, by dividing $k_1 r_1$ by h_1 and $k_1 s_1$ by g_1 and letting H_1, G_1 be the respective remainders. We see this as follows. We have

$$k_1 r_1 = h_1 J_1 + H_1, \quad k_1 s_1 = g_1 K_1 + G_1,$$

with deg $H_1 <$ deg h_1, deg $G_1 <$ deg g_1; substituting, we get

$$k_1 \equiv k_1 r_1 g_1 + k_1 s_1 h_1 \equiv h_1 J_1 g_1 + H_1 g_1 + g_1 K_1 h_1 + h_1 G_1 \pmod{q}$$
$$\equiv h_1 g_1(J_1 + K_1) + H_1 g_1 + h_1 G_1 \pmod{q},$$

or

$$h_1 g_1(J_1 + K_1) \equiv k_1 - H_1 g_1 - h_1 G_1 \pmod{q}.$$

The left side is a multiple of a monic polynomial of degree $=$ deg f, namely, $h_1 g_1$, so if $J_1 + K_1 \not\equiv 0 \pmod{q}$, the left side has degree $\geqslant$ deg f. But the right side is a polynomial of degree $< f$, since deg $k_1 <$ deg $f =$ deg $g_1 h_1$. So $J_1 \equiv K_1 \pmod{q}$ and the remainders H_1, G_1 solve (1).

When we set $g_2 = g_1 + qG_1$, $h_2 = h_1 + qH_1$, then since deg $G_1 <$ deg g_1, etc., g_2 and h_2 are monic; clearly $g_2 \equiv g_1 \pmod{q}$, $h_2 \equiv h_1 \pmod{q}$, and $f \equiv g_2 h_2 \pmod{q^2}$.

Now we show that g_2 and h_2 are relatively prime modulo q^2.

We have $r_1 g_1 + s_1 h_1 \equiv 1 + qW \pmod{q^2}$ for some polynomial W in $\mathbb{Z}[x]$. We look for r_2, s_2 such that $r_2 g_2 + s_2 h_2 \equiv 1 \pmod{q^2}$ by choosing r_2, s_2 of the form $r_2 \equiv r_1 + qT_1 \pmod{q}$, $s_2 \equiv s_1 + qU_1 \pmod{q}$ for some T_1, U_1 to be found. Then

$$r_2 g_2 + s_2 h_2 \equiv (r_1 + qT_1)(g_1 + qG_1) + (s_1 + qU_1)(h_1 + qH_1) \pmod{q^2}$$
$$\equiv r_1 g_1 + s_1 h_1 + q(g_1 T_1 + h_1 U_1 + r_1 G_1 + s_1 H_1) \pmod{q^2}$$
$$\equiv 1 + q(W_1 + g_1 T_1 + h_1 U_1 + r_1 G_1 + s_1 H_1) \pmod{q^2}.$$

It suffices to choose T_1, U_1 so that

$$g_1 T_1 + h_1 U_1 \equiv -W_1 - r_1 G_1 - s_1 H_1 \pmod{q}.$$

But this can be done because g_1 and h_1 are relatively prime mod q. Since

$$r_1 g_1 + s_1 h_1 \equiv 1 \pmod{q}$$

we can choose

$$T_1 = r_1(-W_1 - r_1 G_1 - s_1 H_1),$$
$$U_1 = s_1(-W_1 - r_1 G_1 - s_1 H_1).$$

If we do, then setting $r_2 = r_1 + qT_1$, $s_2 = s_1 + qU_1$, we get

$$r_2 g_2 + s_2 h_2 \equiv 1 \pmod{q^2}$$

and g_2 and h_2 are relatively prime modulo q^2. This completes the existence part of the proof.

For uniqueness, suppose $g_2 \equiv g_1 \pmod{q}$ and $g_2^1 \equiv g_1 \pmod{q}$; also $h_2 \equiv h_1 \pmod{q}$, $h_2^1 \equiv h_1 \pmod{q}$, each of g_2, g_2^1, h_2, h_2^1 is monic, and $f \equiv g_2 h_2 \equiv g_2^1 h_2^1 \pmod{q^2}$. Then

$$g_2^1 = g_2 + qm \quad \text{with deg } m < \text{deg } g_2,$$
$$h_2^1 = h_2 + qn \quad \text{with deg } n < \text{deg } h_2.$$

We show $m, n \equiv 0 \pmod{q}$.

If, say, $n \not\equiv 0 \pmod{q}$, then, since $g_2 h_2 \equiv g_2^1 h_2^1 \pmod{q^2}$, we have

$$g_2 n + h_2 m \equiv 0 \pmod{q}. \tag{2}$$

Since g_2 and h_2 are relatively prime modulo q,

$$g_2 r_2 + h_2 s_2 \equiv 1 \pmod{q}. \tag{3}$$

Multiply (2) by r_2:

$$r_2 g_2 n + r_2 h_2 m \equiv 0 \pmod{q}$$

Substitute for $r_2 g_2$, using (3):

$$(1 - h_2 s_2)n + r_2 h_2 m \equiv 0 \pmod{q}$$

or

$$n \equiv h_2(s_2 n - r_2 m) \pmod{q}.$$

Since h_2 is monic and $n \not\equiv 0$, it must be that $s_2 n - r_2 m \not\equiv 0 \pmod{q}$, and so deg $n \geqslant$ deg h_2, which is a contradiction. Thus n must $\equiv 0$, and, by a similar argument, $m \equiv 0 \pmod{q}$. Thus $g_2 \equiv g_2^1 \pmod{q^2}$, $h_2 \equiv h_2^1 \pmod{q^2}$. The proof is complete. ☐

Notice that once we know a factorization mod q by polynomials which are relatively prime mod q, the proof shows how to construct a factorization mod q^2 by polynomials which are relatively prime mod q^2 and how to exhibit the relative primeness. Thus the proof describes how to proceed from a factorization mod q to a factorization mod q^2, then mod q^4, then mod q^8, etc. In practice, however, we may prefer to use linear algebra to find factorizations, having assurance from the theorem that the equations we set up will have unique solutions. We illustrate this latter approach in the course of developing the following

EXAMPLE. Let $f(x) = x^5 + 2x^4 - 8x^3 + 2x^2 + 3x + 7$. We will detach the coefficients and write $f = (1\ 2\ -8\ 2\ 3\ 7)$.

Here

$$R_z = \frac{1}{2^{1/2} - 1} \max\left\{ \frac{2}{5}, \left(\frac{8}{10}\right)^{1/2}, \left(\frac{2}{10}\right)^{1/3}, \left(\frac{3}{5}\right)^{1/4}, 7^{1/5} \right\}$$

$$= \frac{7^{1/5}}{2^{1/5} - 1} \leqslant \frac{1.50}{.15} = 10,$$

$$R_k = 2 \max\{2, 8^{1/2}, 2^{1/3}, 3^{1/4}, 7^{1/5}\} = 2 \cdot 8^{1/2} = 4\sqrt{2}.$$

Now mod 2, $f \equiv (1\ 0\ 0\ 0\ 1\ 1) = (1\ 1\ 1)(1\ 1\ 0\ 1)$, with both $(1\ 1\ 1)$ and $(1\ 1\ 0\ 1)$ irreducible (mod 2). So if $f(x)$ factors in $\mathbb{Z}[x]$, it must have a factor of degree 2 whose coefficients are in absolute value $\leqslant (4\sqrt{2})^2 = 32$. We can then take $M > 64$: any factorization mod M with $M > 64$ will yield a unique possible factorization in $\mathbb{Z}[x]$.

Starting from $f = (1\ 0\ 0\ 0\ 1\ 1) = (1\ 1\ 1)(1\ 1\ 0\ 1)$ (mod 2), we let

$$g_1 = (1\ 1\ 1),$$
$$h_1 = (1\ 1\ 0\ 1),$$
$$f - g_1 h_1 = (1\ 2\ -8\ 2\ 3\ 7) - (1\ 2\ 2\ 2\ 1\ 1)$$
$$= (0\ 0\ -10\ 0\ 2\ 6) = 2(-5\ 0\ 1\ 3)$$
$$\equiv 2(1\ 0\ 1\ 1)\ (\text{mod } 4).$$

Thus

$$k_1 \equiv (1\ 0\ 1\ 1)\ (\text{mod } 2)$$

We set $g_2 = g_1 + 2G_1$, $h_2 = h_1 + 2H_1$, and solve

$$k_1 \equiv g_1 H_1 + h_1 G_1\ (\text{mod } 2)$$

or

$$(1\ 0\ 1\ 1) \equiv (1\ 1\ 1)H_1 + (1\ 1\ 0\ 1)G_1$$

with deg $G_1 < 2$, deg $H_1 < 3$.

We can solve this by first solving $r_1(1\ 1\ 1) + s_1(1\ 1\ 0\ 1) \equiv 1$ (mod 2) as in the proof. But we prefer to set $H_1 = (a\ b\ c)$, $G_1 = (d\ e)$, and solve directly for the coefficients. If we do, we get from

$$(1\ 0\ 1\ 1) \equiv (1\ 1\ 1)(a\ b\ c) + (1\ 1\ 0\ 1)(d\ e)\ (\text{mod } 2)$$

a system of linear equations mod 2:

$$0 = a + d;$$
$$1 = a + b + d + e;$$
$$0 = a + b + c + e;$$
$$1 = b + c + d;$$
$$1 = c + e.$$

This is easy to solve: $a = 1$, $b = 0$, $c = 0$, $d = 1$, $e = 1$. So

$$H_1 = (1\ 0\ 0), \qquad G_1 = (1\ 1),$$

and

$$g_2 = g_1 + 2G_1 = (1\ 1\ 1) + 2(1\ 1) = (1\ 3\ 3),$$
$$h_2 = h_1 + 2H_1 = (1\ 1\ 0\ 1) + 2(1\ 0\ 0) = (1\ 3\ 0\ 1).$$

Then

$$f \equiv g_2 h_2 \pmod 4.$$

Now we set $q = 4$ and change subscripts, so that

$$f = (1\ 2\ -8\ 2\ 3\ 7),$$
$$g_1 = (1\ 3\ 3),$$
$$h_1 = (1\ 3\ 0\ 1),$$

and try to lift the factorization $f \equiv g_1 h_1 \pmod 4$ to one that is mod 16. Take

$$k_1 = f - g_1 h_1 \equiv (1\ 2\ -8\ 2\ 3\ 7) - (1\ 6\ 12\ 10\ 3\ 3)$$
$$= (0\ -4\ -20\ -8\ 0\ 4) = 4(-1\ -5\ -2\ 0\ 1)$$

Thus, mod 4,

$$k_1 = (-1\ -5\ -2\ 0\ 1) \equiv (3\ 3\ 2\ 0\ 1).$$

We set

$$g_2 = g_1 + 4G_1 = (1\ 3\ 3) + 4(d\ e),$$
$$h_2 = h_1 + 4H_1 = (1\ 3\ 0\ 1) + 4(a\ b\ c),$$

where G_1, H_1 solve

$$k_1 \equiv g_1 H_1 + h_1 G_1 \pmod 4$$

or

$$(3\ 3\ 2\ 0\ 1) \equiv (1\ 3\ 3)(a\ b\ c) + (1\ 3\ 0\ 1)(d\ e)$$

or, mod 4,

$$3 = a + d,$$
$$3 = 3a + b + 3d + e,$$
$$2 = 3a + 3b + c + 3e,$$
$$0 = 3b + 3c + d,$$
$$1 = 3c + e,$$

which we solve mod 4 to get

$$a = 2, \qquad b = 3, \qquad c = 2, \qquad d = 1, \qquad e = 3.$$

Thus

$$g_2 = (1\ 3\ 3) + 4(1\ 3) = (1\ 7\ 15) \equiv (1\ 7\ -1) \pmod{16},$$
$$h_2 = (1\ 3\ 0\ 1) + 4(2\ 3\ 2) = (1\ 11\ 12\ 9) \equiv (1\ -5\ -4\ \ 9) \pmod{16}.$$

Now we try mod $16^2 = 256$. Resubscripting, as before, we have

$$f = (1\ 2 \ -8\ 2\ 3\ 7),$$
$$g_1 = (1\ 7\ -1),$$
$$h_1 = (1\ -5\ -4\quad 9),$$
$$f - g_1 h_1 = (1\ 2 \ -8\ 2\ 3\ 7) - (1\quad 2\ -40\ -14\quad 67\ -9)$$
$$= (0\ 0\ 32\ 16\ -64\ 16)\ = \ 16(2\ 1\ -4\ 1).$$

Thus

$$k_1 = (2\ 1\ -4\ 1).$$

Set

$$g_2 = g_1 + 16G_1 = (1\ 7\ -1) + 16(d\ e),$$
$$h_2 = h_1 + 16H_1 = (1\ -5\ -4\ 9) + 16(a\ b\ c),$$

where $G_1 = (d\ e)$, $H_1 = (a\ b\ c)$. Solve

$$k_1 \equiv g_1 H_1 + h_1 G_1 \pmod{16}$$

or

$$(2\ 1\ -4\ 1) \equiv (1\ 7\ -1)(a\ b\ c) + (1\ -5\ -4\quad 9)(d\ e) \pmod{16}$$

or, mod 16,

$$0 = a + d,$$
$$2 = 7a + b - 5d + e,$$
$$1 = -a + 7b + c - 4d - 5e,$$
$$-4 = -b + 7c + 9d - 4e,$$
$$1 = -c + 9e.$$

After a little effort, we get

$$a = -5, \qquad b = 3, \qquad c = 2, \qquad d = 5, \qquad e = -5.$$

Thus

$$g_2 = (1\quad 7\ -1) + 16(5\ -5) = (1\quad 87\ -81),$$
$$h_2 = (1\ -5\ -4\quad 9) + 16(-5\quad 32) = (1\ -85\quad 44\quad 41),$$

and

$$f - g_2 h_2 = 256(0\quad 0\ -29\quad 42\quad 0 - 13).$$

Now any possible factor of f of degree 2 must be congruent mod 2 to $(1\ 1\ 1)$, hence (by the uniqueness of the theorem) must be congruent mod 256 to $(1\ 87\ -81)$. Since $(1\ 87\ -81)$ has coefficients which are larger in absolute value than 32, and any other polynomial congruent mod 256 to $(1\ 87\ -81)$ is worse, f therefore can have no factor of degree 2. Since a factor of f of degree 1 would yield a factor of degree 1 of f mod 2, and f has no factor of degree 1 mod 2, it follows that f must be irreducible in $\mathbb{Z}[x]$.

The process of getting from a factorization mod p to one that is mod p^{2^N} is one which is quite mechanical, and would not be difficult to program on a computer. The only unpleasant feature of the whole factorization process is that it is rather inefficient if applied to a polynomial $f(x)$ which turns out to be irreducible in $\mathbb{Z}[x]$ but which factors into many factors mod p for all p. Such an $f(x)$ has many factorizations mod p, each of which would have to be lifted to mod $p^{2^N} > M$ and then discarded. In Chapter III-6 we give a class of such bad examples.

I do not know how scarce such examples are.

An interesting article on recent progress in factoring polynomials is Collins (1973).

E8. In the example in the text, in passing from mod 2 to mod 4 show that g_1 and h_1 are relatively prime mod 2 and use that fact to find g_2 and h_2 as in the proof of the theorem. Then show, also as in the proof, that g_2 and h_2 are relatively prime mod 4.

E9. Let $f(x) = x^5 - 8x^4 + 7x^3 - 7x^2 - 2x + 6$, and recall E3.
    (a) Verify that $f(x)$ has no roots in $\mathbb{Z}$.
    (b) Factor $f(x)$ mod 7.
    (c) Factor $f(x)$ mod 7^2.
    (d) Factor $f(x)$ in $\mathbb{Z}[x]$.

E10. Let $f(x) = x^6 + x^5 - 7x^4 + 3x^3 + x - 11$, and recall E4.
    (a) Factor $f(x)$ mod 2.
    (b) Factor $f(x)$ mod 4.
    (c) Factor $f(x)$ mod 16.
    (d) Is $f(x)$ irreducible in $\mathbb{Z}[x]$?

E11. Factor $f(x) = x^4 + 16x^3 + 95x^2 + 256x + 256$ in $\mathbb{Z}[x]$ (cf. E5).

E12. If $f(x)$ factors mod p for a prime p, show that $f(x)$ factors mod p^r for all $r > 1$.

E13. If $f(x)$ is irreducible mod m for some m, show that $f(x)$ must be irreducible mod p for some prime p.

E14. Give a proof, based on the methods of this chapter, of the theorem of Chapter 11, Section B, that any polynomial $f(x)$ in $\mathbb{Q}[x]$ can be factored into a product of irreducible polynomials in $\mathbb{Q}[x]$ in a finite number of steps.

III. FIELDS

This part of the book begins with the primitive element theorem and applications of it, material which continues the development of Part I and could be taken up before Part II. Then the notion of congruence classes modulo a polynomial leads to the invention of many new fields; with the aid of the primitive element theorem we completely classify fields with finitely many elements. After a look at quadratic residues (again a continuation of Part I), the book concludes with a look at algebraic number fields and examples of both rings which do and rings which do not have unique factorization as $\mathbb{Z}$ and $F[x]$ do.

Primitive Elements

1

This last part of the book is mostly about fields. So far, we know that $\mathbb{Q}$, $\mathbb{R}$, and $\mathbb{C}$, with which the book began, are fields, as is $\mathbb{Z}_p$ whenever p is prime, and we also have $GF(9)$, a field with 9 elements we discovered in Chapter I-11.

Using the notion of congruence classes in polynomial rings we shall, starting in III-7, invent many new examples of fields in the same way we invented the fields $\mathbb{Z}_p$. We shall begin this part, however, with the primitive element theorem, a very useful theorem about the structure of fields with a finite number of elements. This theorem is valid both for the examples we shall construct later and for the fields $\mathbb{Z}_p$ we already know about. The five following chapters will be devoted to immediate applications of this theorem.

Recall from Chapter I-11, Section E, that the order of a nonzero element α of a finite field F is the smallest positive integer e for which $\alpha^e = 1$.

Primitive Element Theorem. *Let F be a field with q elements. There is some element α in F such that*

(1) *every non-zero element of F is a power of α, and*
(2) *the order of α is $q - 1$.*

An element α satisfying conditions (1) and (2) is called a *primitive element* of F.

The primitive element theorem is a refinement of Fermat's theorem. Fermat's theorem says that if F is a field with q elements and $\beta \neq 0$ in F, then $\beta^{q-1} = 1$. The primitive element theorem says that there is some element α of F with $\alpha^{q-1} = 1$ and satisfying the extra condition that no smaller positive power of α is equal to 1. Fermat's theorem follows easily from the primitive element theorem. For if α is a primitive element, then any $\beta \neq 0$ is a power of α, say $\beta = \alpha^r$. Since $\alpha^{q-1} = 1$, $\beta^{q-1} = (\alpha^r)^{q-1} = (\alpha^{q-1})^r = 1$.

An example of a primitive element is $\alpha = [2]_{11}$ in $\mathbb{Z}_{11}$. In fact, (1) holds: $[1] = \alpha^{10}$, $[2] = \alpha$, $[3] = \alpha^8$, $[4] = \alpha^2$, $[5] = \alpha^4$, $[6] = \alpha^9$, $[7] = \alpha^7$, $[8] = \alpha^3$, $[9] = \alpha^6$, $[10] = \alpha^5$.

We also showed the existence of a primitive element for $GF(9)$ in Chapter I-11.

E1. Find a primitive element of $\mathbb{Z}_2$; of $\mathbb{Z}_3$; of $\mathbb{Z}_5$; of $\mathbb{Z}_7$; of $\mathbb{Z}_{13}$; of $\mathbb{Z}_{17}$; of $\mathbb{Z}_{19}$; of $\mathbb{Z}_{23}$ (cf. I-11, E24 and E26).

Here is a proof of the primitive element theorem.

PROOF. Let α be any nonzero element of F.

We just observed that by Fermat's theorem the order of α is at most $q - 1$, the number of invertible elements of F; in fact, by Proposition 1, Section I-11E, the order of α divides $q - 1$.

The idea of the proof is to study the set S of orders of elements of F. Since every α in F has order dividing $q - 1$, S is a set of integers $\leqslant q - 1$, and the proof will be essentially complete when we show $q - 1$ is in S. For if $q - 1$ is in S then $q - 1$ is the order of some α in F, so $\alpha^{q-1} = 1$ and *no smaller* positive power of α is equal to 1. That is, part (2) of the theorem is true.

Once we know that an α satisfying part (2) of the theorem exists, then part (1) follows. For if α has order $q - 1$, then $1, \alpha, \alpha^2, \ldots, \alpha^{q-2}$ are all different, by Exercise E2, and so they are all of the $q - 1$ nonzero elements of F.

E2. Show that if $e > 1$ is the order of an element α in a field F, then $1, \alpha, \alpha^2, \ldots, \alpha^{e-1}$ are distinct.

To prove the theorem, then, we need to know about orders of elements of F.

Recall the fact we proved in Chapter I-11 on Fermat's theorem that if e is the order of α, and if $\alpha^f = 1$ for some $f > 0$, then e divides f.

E3. Prove this without looking it up.

Claim 1. *If a, b are in S (that is, a and b are orders of elements of F) and a and b are relatively prime, then ab is in S.*

PROOF. Suppose α has order a, and β has order b. Then $(\alpha\beta)^{ab} = 1$. Suppose c is some positive integer such that $(\alpha\beta)^c = 1$. Then $1 = (\alpha\beta)^{ca} = \alpha^{ca}\beta^{ca} = \beta^{ca}$, so b divides ac. Since b and a are relatively prime, b divides

c. Similarly, from $i = (\alpha\beta)^{cb} = \alpha^{cb}$, we get that a divides c. So $[a, b] = ab$ divides c. Thus ab is the order of $\alpha\beta$, and ab is in S.

Claim 2. *If β has order b and d is any positive divisor of b, then $\beta^{b/d}$ has order d; thus d is in S.*

E4. Prove Claim 2.

Claim 3. *If a, b are in S then $[a, b]$ is in S.*

PROOF. Let $a = p_1^{e_1} \cdots p_r^{e_1}$, $b = p_1^{f_1} \cdots p_r^{f_r}$. By claim 2, if a is in S so is any factor of a. Thus $p_1^{e_1}, \ldots, p_r^{e_r}, p_1^{f_1}, \ldots, p_r^{f_r}$ are all in S. So $p_1^{\max(e_1, f_1)}$, $p_2^{\max(e_2, f_2)}, \ldots$, are all in S. By claim 1,

$$p_1^{\max(e_1, f_1)} \cdot p_2^{\max(e_2, f_2)} \cdots p_r^{\max(e_r, f_r)}$$

is in S, but this is the least common multiple of a and b.

Let e_0 be the largest number in S.

Claim 4. *Every number in S divides e_0.*

For if a is in S and e_0 is in S, then the least common multiple of a and e_0 is in S. But if e_0 is the largest number in S, then that least common multiple, which is always $\geq e_0$, must be equal to e_0. Thus a divides e_0.

Claim 5. $e_0 = q - 1$.

For since every number in S divides e_0, every nonzero element α of F satisfies $\alpha^{e_0} = 1$. There are $q - 1$ nonzero elements of F, so there are $q - 1$ different roots in F of the polynomial $x^{e_0} - 1$. By Corollary 2, Section II-2A, $x^{e_0} - 1$ cannot have more than e_0 roots in the field F, so we must have $e_0 \geq q - 1$. Since every number in S is $\leq q - 1$, therefore $e_0 = q - 1$. This means that $q - 1$ is in S, and therefore there is some element α in F whose order is $q - 1$.

That proves (2). Since we have already proved (1) from (2), the proof of the primitive element theorem is complete. $\qquad\square$

The proof we have just given for the existence of a primitive element gives no simple formula for taking a prime p and finding a primitive element of $\mathbb{Z}_p$. In fact, no one knows of such a formula. About the nicest result known is that if p is a prime of the form $4q + 1$ where q is prime, then 2 is a primitive element of $\mathbb{Z}_p$. Thus 2 is a primitive element of $\mathbb{Z}_p$ for $p = 5, 13, 29, 53$ (among primes < 100). For reasons related to repeating decimals (see below, Chapter 2), it is particularly interesting to know whether 10 is a primitive element of $\mathbb{Z}_p$. It turns out to be so for $p = 7, 17, 19, 23, 29, 47, 59, 61$, and 97 among primes < 100, but it is an unsolved conjecture of Gauss (1801) that there exist infinitely many primes p with 10 a primitive element of $\mathbb{Z}_p$. A generalization of Gauss's conjecture dating from the 1920s by the mathematician E. Artin asserts that if a is any integer not a square, then a is a primitive element of $\mathbb{Z}_p$ for infinitely many primes p. Artin's conjecture has been shown to be true provided that a conjecture called the generalized Riemann hypothesis is true (Hooley (1967)). But the latter, dating from the 19th century and having to do with the values of the complex number s at which a certain function $\zeta(s)$, which for s real and > 1 satisfies

$$\zeta(s) = \sum_{n=1}^{\infty} \frac{1}{n^s},$$

has the value zero, is perhaps the most outstanding unsolved problem in all of mathematics.

Gauss, in his *Disquisitiones Arithemeticae* (1801), discusses the problem of assigning to a prime p a primitive element of $\mathbb{Z}_p$ and quotes Euler as saying that the nature of primitive elements is one of the deepest mysteries of numbers.

For a survey of this mystery, see Goldstein (1973).

Of course, for any given prime p, since $\mathbb{Z}_p$ is finite there are only a finite number of possibilities for a primitive element, and trial and error will always eventually work (see E8).

E5. If F is a field with 32 elements, show that every element of F is a primitive element except 0 and 1.

†**E6.** Show that if F is a field with q elements, α is a primitive element and $\alpha^r = \alpha^s$, then $r \equiv s \pmod{q-1}$.

E7. Let F be a field with q elements, and α be a primitive element of F. If β in F is such that $\alpha^r = \beta$, show that the order of β is $[q-1, r]/r$.

E8. Show that there are $\phi(q-1)$ primitive elements in a field with q elements (Use Exercise E7). Verify this for the fields $\mathbb{Z}_{13}$, $\mathbb{Z}_{17}$.

E9. Let p be an odd prime, and b an integer so that $[b]$ is a primitive element of $\mathbb{Z}_p$. Show: If $(a, p) = 1$, then $x^2 \equiv a \pmod{p}$ has a solution iff $a \equiv b^r \pmod{p}$ with r even.

There is some group theoretic terminology which relates to the primitive element theorem.

Definition. A finite group G is *cyclic* if there is some α in G such that every element of G is obtainable as a product of copies of α. Such an element α is called a *generator* of G.

What the primitive element theorem shows is that the group U of invertible elements of a field F of q elements is a cyclic group. A primitive element of F is the same as a generator of U.

E10. Show that a cyclic group is abelian.

E11. Show that $\mathbb{Z}_n$ under addition is a cyclic group. Find a generator. Find all possible generators for $\mathbb{Z}_n$.

Note that associated with a ring with identity R are two groups: the group of all elements of R under the operation of addition, and the group of all invertible elements of R under the operation of multiplication.

E12. Show that the group of invertible elements of $\mathbb{Z}_8$ is not cyclic.

E13. Show that the group of invertible elements of $\mathbb{Z}_{10}$ is cyclic.

E14. Show that the group of invertible elements of $\mathbb{Z}_{49}$ is cyclic.

E15. Show that the group of invertible elements of $\mathbb{Z}_{15}$ is not cyclic.

E16. Show that $GF(9)$ under addition is not cyclic, but that the group of all invertible elements of $GF(9)$ under multiplication is cyclic. Find all generators of the group of invertible elements of $GF(9)$.

E17. Recall the ring $M_2(\mathbb{Z}_2)$ of 2×2 matrices with entries in $\mathbb{Z}_2$. Is the additive group of $M_2(\mathbb{Z}_2)$ cyclic? Is the group of invertible elements of $M_2(\mathbb{Z}_2)$ cyclic?

2 Repeating Decimals, II

In this chapter we examine the orders of elements of $\mathbb{Z}_t$ for any number t, and complete our study of repeating decimals. These subjects are closely related, for we showed in Chapter I-12 that the base a expansion of u/t, $(t, a) = 1$, $(t, u) = 1$, $0 < u < t$, is repeating:

$$\frac{u}{t} = (.a_1 a_2 \ldots a_d a_1 a_2 \ldots a_d a_1 \ldots)_a = (.\overline{a_1 a_2 \ldots a_d})_a,$$

where d, the length of the period of repetition or, as we shall say for short, the *period*, is the smallest number n such that $a^n \equiv 1 \pmod{t}$. That is, the period of u/t in base a is the *order* of a mod t.

The primitive element theorem gives the maximal possible period for u/t when t is prime.

Theorem 1. *If t is prime, then there are infinitely many bases a for which $1/t$ has a base a expansion with period $\phi(t) = t - 1$.*

PROOF. For if $[a]$ is a primitive element of $\mathbb{Z}_t$, then the smallest number n for which $a^n \equiv 1 \pmod{t}$ is $n = t - 1$. $\square$

As an example, let $t = 7$. Then [3] and [5] are primitive elements. So the expansions of $1/7$ in base a have period 6 whenever $a \equiv 3$ or 5 (mod 7), that is, $a = 3, 5, 10, 12, 17, 19, 24, 26, \ldots$. In fact:

$$\frac{1}{7} = (\overline{.010212})_3$$

$$= (\overline{.032412})_5$$

$$= (\overline{.142857})_{10}$$

$$= (\overline{.2(13)(10)(16)58})_{19},$$

etc.

Here is a nice application of the fact that if a is a primitive element mod t, then every number b, $1 \leqslant b < t$, is congruent mod t to a power of a.

Corollary. *Suppose t is prime and a is a primitive element* mod t. *Suppose*

$$\frac{1}{t} = \left(.\overline{r_0 r_1 r_2 \cdots r_{t-1}} \right)_a.$$

If b is a natural number $< t$ such that $b \equiv a^k$ (mod t), where $0 \leqslant k < t$, then

$$\frac{b}{t} = \left(.\overline{r_k r_{k+1} \cdots r_{t-1} r_0 r_1 \cdots r_{k-1}} \right)_a.$$

PROOF. If

$$\frac{1}{t} = \frac{r_0}{a} + \frac{r_1}{a^2} + \frac{r_2}{a^3} + \cdots + \frac{r_{t-1}}{a^t} + \frac{r_0}{a^{t+1}} + \cdots$$

then

$$\frac{a^k}{t} = r_0 a^{k-1} + r_1 a^{k-2} + \cdots + r_{k-1} + \frac{r_k}{a} + \frac{r_{k+1}}{a^2} + \cdots$$

$$+ \frac{r_{t-1}}{a^{t-k}} + \frac{r_0}{a^{t-k+1}} + \cdots + \frac{r_{k-1}}{a^t} + \frac{r_k}{a^{t+1}} + \cdots .$$

Now $a^k = b + mt$ for some integer m, so $a^k/t = m + (b/t)$. Equating fractional parts of a^k/t gives the result. $\quad\square$

For example, with $a = 10$, $t = 7$,

$$
\begin{aligned}
10^0 &\equiv 1 \ (\text{mod } 7), & 1/7 &= .\overline{142857}, \\
10^1 &\equiv 3 \ (\text{mod } 7), & 3/7 &= .\overline{428571}, \\
10^2 &\equiv 2 \ (\text{mod } 7), & 2/7 &= .\overline{285714}, \\
10^3 &\equiv 6 \ (\text{mod } 7), & 6/7 &= .\overline{857142}, \\
10^4 &\equiv 4 \ (\text{mod } 7), & 4/7 &= .\overline{571428}, \\
10^5 &\equiv 5 \ (\text{mod } 7), & 5/7 &= .\overline{714285}.
\end{aligned}
$$

In order to determine periods of expansions of u/t with t not prime, we have to get an analogue of the primitive element theorem for $\mathbb{Z}_t$. That is, we have to find the maximal order of an invertible element of $\mathbb{Z}_t$. Once we do that, we will have determined all orders, by Exercises E1 and E19.

E1. If a has order e mod t and $f | e$ then $a^{e/f}$ has order f mod t.

E2. What is the period of $1/7$ in base 3? 9? 27?

Let $h(t)$ be the maximal order of an invertible element of $\mathbb{Z}_t$. We know that $h(t)$ divides $\phi(t)$, the number of natural numbers $< t$ which are relatively prime to t, by Chapter I-11. Our aim in the rest of this chapter is to find $h(t)$.

The strategy for finding $h(t)$ is to determine $h(t)$ for $t = p^e$, the power of an odd prime p, then $h(2^e)$, and finally $h(t)$ for any t. We shall do this via a series of exercises.

Proposition 2. *If p is an odd prime and $e \geqslant 1$, then $h(p^e) = \phi(p^e) = p^e(p - 1)$.*

We shall illustrate the proof of this by proving a special case.

Proposition 3. *For each $e \geqslant 1$, 10 has order $\phi(7^e)$ mod 7^e. Thus $1/7^e$ has a repeating decimal expansion in base 10 whose period has length $\phi(7^e) = 7^{e-1} \cdot 6$.*

It follows from Proposition 3 that $\{1/7^e\}_{e > 1}$ forms a sequence of fractions whose periods in base 10 go to ∞ with e. In particular, this gives one way (not the best) to solve Exercise E9 of Chapter I-12.

PROOF OF PROPOSITION 3. We first notice that 10 has order 6 (mod 7), and that $10^6 \equiv (98 + 2)^3 \equiv 8$ (mod 7^2), so that $10^6 \not\equiv 1$ (mod 7^2). We show that the order of 10 (mod 49) is $\phi(49) = 7 \cdot 6$.

Suppose b is the order of 10 (mod 49). Since $10^{7 \cdot 6} \equiv 1$ (mod 49), b divides $7 \cdot 6$. Write $b = a$ or $b = 7a$ where a divides 6. Now if $10^b \equiv 1$ (mod 49), then $10^b \equiv 1$ (mod 7). If $b = a$ then $10^a \equiv 1$ (mod 7); since a divides 6, the order of 10 (mod 7), therefore a must equal 6. If $b = 7a$ then $10^{7a} \equiv 1$ (mod 7), so $10^a \equiv 1$ (mod 7) by Fermat's theorem, so again $a = 6$. Therefore the order b of 10 mod 49 must be $b = 6$ or $b = 7 \cdot 6$. Since $10^6 \not\equiv 1$ (mod 49), therefore $b = 7 \cdot 6$.

Now we show that for any $e \geqslant 2$, the order d of 10 mod 7^e is $7^{e-1} \cdot 6$. Since d divides $\phi(7^e) = 7^{e-1} \cdot 6$, therefore $d = 7^r \cdot c$ with c dividing 6. If $10^d = 10^{7^r \cdot c} \equiv 1$ (mod 7^e) then $10^{7^r \cdot c} \equiv 1$ (mod 7^2), so $7^r \cdot c$ must be divisible by $7 \cdot 6$. Therefore $r \geqslant 1$ and $c = 6$. We need to find the smallest possible r.

Since $10^6 \not\equiv 1$ (mod 49) we can write $10^6 = 1 + 7a$ with $(a, 7) = 1$. Then for any $r > 1$, using the binomial theorem,

$$(10^6)^{7^r} = (1 + 7a)^{7^r} = 1 + 7^{r+1}a + (\text{multiples of } 7^{r+2})$$
$$\equiv 1 + 7^{r+1}a \pmod{7^{r+2}}$$

for any $r \geqslant 0$. In particular,

$$(10^6)^{7^{e-2}} \equiv 1 + 7^{e-1}a \pmod{7^e}.$$

Since $(a, 7) = 1$, it follows that

$$(10^6)^{7^{e-2}} \not\equiv 1 \pmod{7^e}.$$

Thus the smallest number d such that $10^d \equiv 1$ (mod 7^e) is $6 \cdot 7^{e-1} = \phi(7^e)$, proving the theorem.   □

We have shown that $1/7$, $1/49$, $1/343$, ... have repeating decimals of length 6, 42, 294, ...

PROOF OF PROPOSITION 2. This follows the outline of the proof of Proposition 3 and is done in the next two exercises.

E3. Show that for p an odd prime, $h(p^2) = \phi(p^2)$, as follows. You want to show that there is an integer b so that $[b]$ has order $p^2 - p$ in $\mathbb{Z}_{p^2}$. Suppose $[a]_p$ is a primitive element in $\mathbb{Z}_p$. Consider $a + rp$ for r some integer. Show that
   (i) $(a + rp)^{p-1} \equiv 1 \pmod{p}$ for any r,
   (ii) if r happens to have been chosen so that $(a + rp)^{p-1} \equiv 1 \pmod{p^2}$, then for any s with $s \not\equiv r \pmod{p}$, $(a + sp)^{p-1} \not\equiv 1 \pmod{p^2}$,
   (iii) if $(a + rp)^{p-1} \not\equiv 1 \pmod{p^2}$, then $[a + rp]_{p^2}$ has order $p(p - 1)$ in $\mathbb{Z}_{p^2}$.

E4. Show that for p an odd prime and $e \geqslant 2$, $h(p^e) = \phi(p^e)$, as follows. Show that if $[a]_{p^2}$ has order $p(p - 1)$ in $\mathbb{Z}_{p^2}$, then $[a]_{p^e}$ has order $p^{e-1}(p - 1)$ in $\mathbb{Z}_{p^e}$ for any $e \geqslant 2$. (*Hint*: Observe that

$$a^{p-1} \equiv 1 + rp \pmod{p^2} \text{ with } (r, p) = 1.$$

Use the binomial theorem as in the example of $a = 10$, $p = 7$ in the text.)

Exercises E3 and E4 complete the proof of Proposition 2.   □

E5. Show that if a is an element of order $\phi(n)$ mod n, then every integer relatively prime to n is congruent mod n to some power of a.

Because of Exercise E5 we call an integer a a *primitive element* mod n if the order of a mod n is $\phi(n)$. Proposition 3 shows that 10 is a primitive element mod 7^e for any $e \geqslant 1$.

†**E6.** If $[a]_{p^e}$ is a primitive element mod p^e for some $e > 1$, show that $[a]_p$ is a primitive element mod p.

E7. Find another prime p besides $p = 7$ such that in base 10, the periods of $1/p^e$ go to infinity as e goes to infinity.

Having found the maximal orders of invertible elements of $\mathbb{Z}_{p^e}$ for p an odd prime, we now do the same for $p = 2$.

Proposition 4. *For each* $e \geqslant 1$, $h(2^e) = 2^s$, *where* $s = 0$ *if* $e = 1$, $s = 1$ *if* $e = 2$, *and* $s = e - 2$ *if* $e \geqslant 3$.

PROOF. Again, we leave the proof as a sequence of exercises.

E8. Show that the order of 3 in $\mathbb{Z}_{2^e}$ is
   (a) 2 in $\mathbb{Z}_4$,
   (b) 2 in $\mathbb{Z}_8$,
   (c) 4 in $\mathbb{Z}_{16}$,
   (d) 8 in $\mathbb{Z}_{32}$.

E9. Show that for $e \geqslant 5$, $3^{2^{e-3}} = (1 + 2)^{2^{e-3}} \not\equiv 1 \pmod{2^e}$, using the binomial theorem. Conclude that the order of $[3]$ in $\mathbb{Z}_{2^e}$ is 2^{e-2} or 2^{e-1}.

E10. In $\mathbb{Z}_{2^e}$ show -1 is not a square by observing that if -1 were a square mod 2^e, then -1 would be a square mod 4.

E11. Use E9 and E10 to conclude that the maximal orders of invertible elements of $\mathbb{Z}_{2^e}$ are 2^{e-2} for $e \geqslant 3$.

Exercises E8 and E11 complete the proof of Proposition 4.   □

To find $h(t)$ for general t, we have

Proposition 5. *If* $t = mn$ *and* $(m, n) = 1$ *then* $h(t) = \text{l.c.m.}[h(m), h(n)]$.

PROOF. The proof is also an exercise.

E12. Let $t = mn$, and suppose a has order e mod m and b has order f mod n. Using the Chinese remainder theorem, find some c such that c has order e mod m and order f mod n; then show c has order $[e, f]$ mod t.

Combining Propositions 2, 4 and 5, we get

Theorem 6. *Let* $t = 2^e p_1^{e_1} p_2^{e_2} \cdots p_r^{e_r}$ *where* $p_1, \ldots, p_r$ *are odd primes. Let* $h(t)$ *be the largest order of an invertible element of* $\mathbb{Z}_t$. *Then*

$$h(t) = \text{l.c.m.}\left[2^s, \phi(p_1^{e_1}), \phi(p_2^{e_2}), \ldots, \phi(p_r^{e_r})\right]$$

where $s = e - 1$ *if* $e = 1, 2$, *and* $s = e - 2$ *if* $e \geqslant 3$.

Corollary. $\mathbb{Z}_t$ *has a primitive element iff* $t = 2, 4, p^e$, *or* $2p^e$ *for* p *an odd prime.*

†**E13.** Prove the corollary.

Having found $h(t)$ for any t, we now know the possible periods of the fraction $1/t$ in various bases; they are all the possible divisors of $h(t)$.

E14. Find all possible periods of $1/60$ in various bases.

E15. What is $h(48)$? Find some base a such that $1/48$ has an expansion in base a with period $h(48)$.

E16. Suppose $t = rs$ with $(r, s) = 1$. By partial fractions, $1/t = (c/r) + (d/s)$ for some integers c, d. Suppose $(a, t) = 1$ and c/r and d/s have base a expansions with periods m, n respectively. Show that $1/t$ has a base a expansion of period dividing l.c.m.(m, n). Can the period divide l.c.m.(m, n) properly? Can it equal l.c.m.(m, n)?

The question of the orders of elements mod p^2 for p prime relates to a very famous problem in number theory, known as Fermat's last theorem. Fermat claimed (in 1636):

If $n > 2$, then there are no integers x, y, z satisfying

$$x^n + y^n = z^n \tag{1}$$

Fermat provided no proof. No proof has been found since.

If $n \geqslant 3$, one can write $n = ab$ with $a = 4$ or an odd prime. If x, y, z satisfy $x^n + y^n = z^n$, then $(x^b)^a + (y^b)^a = (z^b)^a$. So in order to show that

(1) has no solutions for any $n > 2$, it suffices to show (1) has no solutions for $n = 4$ or an odd prime. Fermat did prove the case $n = 4$.

For odd primes p the problem has been attacked by first trying to show, as a first case, that if $x^p + y^p = z^p$ with $xyz \neq 0$, x, y, z relatively prime, then p divides xyz (so that any solution in the integers reduces to a trivial solution mod p).

In 1909 Wieferich proved: If $2^{p-1} \not\equiv 1 \pmod{p^2}$, then the first case of Fermat's last theorem holds.

By 1914 it had been shown that if the first case of Fermat's last theorem is false (that is, there exist integers x, y, z with p not dividing xyz such that $x^p + y^p = z^p$), then not only is $2^{p-1} \equiv 1 \pmod{p^2}$ but also $q^{p-1} \equiv 1 \pmod{p^2}$ for $q = 2, 3, 5, 7, 11, 13, 17, 19, 23, 29$, and 31.

Primes satisfying such conditions are scarce. The only two primes $p < 3,000,000,000$ for which $2^{p-1} \equiv 1 \pmod{p^2}$ are $p = 1093$ and $p = 3511$. Since $3^{p-1} \not\equiv 1 \pmod{p^2}$ for those two primes, the first case of Fermat's last theorem is true for all primes less than 3,000,000,000.

The largest known prime (as of 1976) is $2^{19937} - 1$. Using Wieferich's result it can be shown that the first case of Fermat's last theorem holds for that prime.

Incidentally, it is now known that if there exist x, y, z with $0 < x < y < z$ and $x^p + y^p = z^p$, then x must have at least a million digits.

The corresponding problem in $\mathbb{R}[x]$ is much easier. It was given as Exercise E6, Chapter II-6.

E17. What about $x^p + y^p = z^p$ when $p = 2$?

E18. What is the order of 2 mod p^2 for $p = 5, 7, 11, 13$?

E19. Show that the order of any invertible element of $\mathbb{Z}_t$ divides $h(t)$. (*Hint*: Show that Claims 1–4 of pages 208–9 hold for the set S of orders of invertible elements of $\mathbb{Z}_t$.)

3

Testing for Primeness

In this chapter we consider how to find large primes for use in the secret codes described in Chapter I-15.

The primitive element theorem says: If p is prime, then there is some natural number $a < p$ whose order mod p is $p - 1$, that is, $a^{p-1} \equiv 1$ (mod p) and no smaller positive power of a is congruent mod p to 1. The converse is

Proposition. *Suppose q is a natural number $\geqslant 2$. If there is some $a < q$ such that the order of a mod q is exactly $q - 1$, then q is a prime number.*

PROOF. If q is not prime, then $\phi(q)$, the number of natural numbers less than q which are relatively prime to q, is less than $q - 1$; if $a < q$, then either a is not relatively prime to q, hence no power of a is congruent to 1 mod q, or a is relatively prime to q, in which case, by Euler's theorem, its order mod q must divide $\phi(q) < q - 1$. □

Thus: q is prime iff there is an element of order exactly $q - 1$ mod q.

How, then, do we decide whether a given number q is prime? We have a negative test, from Fermat's theorem:

$(-)$ *If $a^{q-1} \not\equiv 1$ (mod q) for some $a < q$, then q is not prime.*

We have a positive test, from the proposition:

$(+)$ *If for some a the order of a (mod q) is $q - 1$, then q is prime.*

Unfortunately, test $(+)$ is very impractical. For while it is very easy, using the strategy of Section I-11D, to decide whether $a^{q-1} \equiv 1$ (mod q)

for any a and q, it is much harder to prove that the order of a is $q - 1$ (mod q). It has to be shown that for any prime p dividing $q - 1$, $a^{(q-1)/p} \not\equiv 1$ (mod q). And to do that, one must be able to factor $q - 1$, a problem of almost the same size as that of factoring q.

Here is a small example. Let $q = 11213$. After checking that 11213 is not divisible by 2, 3, 5, 7, 11, and a few other small primes, we suspect that 11213 might be prime. We compute 2^{11212} (mod 11213) and verify that it is $\equiv 1$ (mod 11213). To apply test $(+)$ we must find the order of 2 mod 11213. So we have to factor 11212. Now $11212 = 2^2 \cdot 2803$. Is 2803 prime? This problem is not much less difficult than the original problem. It turns out that 2803 is prime, and that neither $2^{11212/2}$ nor $2^{11212/2803}$ is congruent to 1 mod 11213; hence 2 is a primitive element mod 11213 and 11213 is prime.

If we took, instead of 11213, some 40-digit number q, then to use $(+)$ we would have to factor $q - 1$, and chances are good that after factoring out all the obvious small prime factors one would be left with a factor of $q - 1$ containing 35 or more digits and which we would have to factor—almost as hard a problem as factoring q itself.

So test $(+)$ is not very helpful.

We might ask: Suppose we know only that q is a number for which $2^{q-1} \equiv 1$ (mod q); how likely is it that q is prime?

The answer appears to be: excellent.

Call a number q *fermatian* if $2^{q-1} \equiv 1$ (mod q).

Call a number q a *pseudoprime* if for all integers a, $a^q \equiv a$ (mod q).

Any prime is a pseudoprime. Any odd pseudoprime is fermatian: for if $2^q \equiv 2$ (mod q) and q is odd, we can cancel a factor of 2 from each side.

It turns out that there are infinitely many fermatian numbers which are not primes.

Proposition. *If f is a composite fermatian number, then so is $2^f - 1$.*

PROOF. Suppose $f = ab$, $a > 1$, $b > 1$. Let $g = 2^f - 1$. Then g is composite, for

$$2^{ab} - 1 = (2^a - 1)(1 + 2^a + \cdots + 2^{a(b-1)}). \qquad (*)$$

Suppose $2^{f-1} \equiv 1$ (mod f). We show $2^{g-1} \equiv 1$ (mod g), that is, $2^f - 1$ divides $2^{g-1} - 1$. By $(*)$ it suffices to show that f divides $g - 1$. Now since $2^{f-1} \equiv 1$ (mod f), f divides $2^{f-1} - 1$. Since $g - 1 = 2^f - 1 - 1 = 2(2^{f-1} - 1)$, f divides $g - 1$. So g is a composite fermatian number. $\qquad \square$

Despite this result, fermatians which are not prime are scarce, and pseudoprimes are even scarcer. There are 168 primes < 1000, but only three composite fermatians—341, 561 (the only composite pseudoprime) and 645; there are, according to D. H. Lehmer and P. Poulet, 5,761,455 primes under 100,000,000, but only 2043 composite fermatians and 252 composite pseudoprimes.

Thus if one picks a 40-digit number q and verifies that $2^{q-1} \equiv 1$ (mod q), $3^{q-1} \equiv 1$ (mod q), etc., the likelihood is high that q is prime. (However, I do not know of a result which describes the ratio of composite pseudoprimes $\leqslant n$ (or composite fermatians $\leqslant n$) to all pseudoprimes $\leqslant n$ (or all fermatians $\leqslant n$) for large numbers n.)

In Chapter I-15 we were interested in finding large prime numbers for use in designing secret codes. It turns out that for such codes it is sufficient to use pseudoprimes, and it does not matter whether or not the pseudoprimes are actually prime.

Suppose p and q happen to be pseudoprimes. If we let $N = qp$ and set up a code as in Chapter I-15, it would work provided q and p are relatively prime (which is easy to check by Euclid's algorithm). For then $\phi(N) = \phi(p)\phi(q)$, and since q is a pseudoprime, $q - 1$ is divisible by $\phi(q)$ (by E7(a)), and similarly for p. For any s relatively prime to $(p-1)(q-1)$, we can find t with $st - k(p-1)(q-1) = 1$ for some k; then for any integer a,

$$a^{st} = a^{1+k(p-1)(q-1)}.$$

Now if p and q are relatively prime pseudoprimes, N is squarefree, by E7(b), so by E7(c)

$$a^{k(p-1)(q-1)+1} \equiv a \pmod{pq};$$

thus

$$a^{st} \equiv a \pmod{N}.$$

Thus even if p and q happened to be composite pseudoprimes, the code would work for any code word a with $0 < a < N$.

†**E1.** A *Fermat number* is a number of the form $2^{2^n} + 1$. Fermat conjectured that for any n, $2^{2^n} + 1$ was prime. It was one of his worst conjectures, for no such number has been proved prime for $n > 4$, and some 21 of them have been proved composite. You prove that a Fermat number is fermatian.

E2. A *Mersenne number* is a number of the form $2^p - 1$ where p is prime. It is known that $2^p - 1$ is prime for some 24 primes p, and $2^{19937} - 1$ is the largest known prime of any form. You prove that a Mersenne number is fermatian.

*E3. Test for primeness:
  (a) 22427;
  (b) 32767;
  (c) 65537;
  (d) 274177.

E4. To apply test $(+)$ it is necessary not only to know the primes dividing $q-1$, but also to be lucky enough to pick a primitive element mod q. If q is prime and some number a, $1 \leqslant a \leqslant q-1$, is chosen at random, the likelihood that a is a primitive element is $\phi(q-1)/q-1$. How big can this ratio be? What is it for 11213? For 2803? *How small can it be?

E5. Show that 2803 is prime by showing that 2 is a primitive element mod 2803.

E6. Test $3^{q-1} \equiv 1$ (mod q) for $q = 341$ and 645.

E7. Prove the following.
    (a) If q is pseudoprime, then $\phi(q)$ divides $q - 1$.
    (b) If q is pseudoprime, then q is squarefree (use (a)).
    (c) Conclude from (a), (b), and Exercise E14 of I-14 that if $N = pq$, p, q pseudoprimes with $(p, q) = 1$, then for any integer a,

$$a^{k(p-1)(q-1)+1} \equiv 1 \pmod{N}.$$

There are a number of papers in the journal, *Mathematics of Computation*, Volumes **29** (1975) and **30** (1976), on techniques for factoring large numbers. One of the best is a method in Morrison and Brillhart (1975), which uses continued fractions. Their technique was applied to factor Fermat numbers. Expositions of continued fractions (which in essence are an application of Euclid's algorithm) may be found in many books on number theory, such as Stark (1971), or Chrystal (1898–1900, Chapter XXXII).

The facts about pseudoprimes in this chapter came from Shanks (1962).

4 Fourth Roots of One in $\mathbb{Z}_p$

A. Primes

In Chapter I-4 on primes, we proved that there were infinitely many primes of the form $4k + 3$. We observed that according to a theorem of Dirichlet, for any a, r with $(a, r) = 1$ there are infinitely many primes of the form $ak + r$ (where k varies). In particular (with $a = 4$, $r = 1$) there are infinitely many primes of the form $4k + 1$. Using the existence of primitive elements we can prove this fact without appealing to Dirichlet's theorem.

Here is the main step in the proof.

Theorem. *If p is an odd prime, then -1 is a square mod p iff $p \equiv 1 \pmod 4$.*

PROOF. If $p \equiv 1 \pmod 4$ then $p - 1 = 4r$ for some r. Let $[a]$ be a primitive element of $\mathbb{Z}_p$. The order of $[a]$ in $\mathbb{Z}_p$ is $4r$, so $(a^r)^2 \not\equiv 1 \pmod p$ but is a root of $x^2 - 1 \equiv 0 \pmod p$. Now $x^2 - 1$ in $\mathbb{Z}_p[x]$ has only two roots, $[1]$ and $[-1]$, since $\mathbb{Z}_p$ is a field, so $(a^r)^2 \equiv -1 \pmod p$. Thus -1 is a square mod p.

Conversely, suppose p is odd and $b^2 \equiv -1 \pmod p$. If $[a]$ is a primitive element of $\mathbb{Z}_p$, then $b \equiv a^s \pmod p$ for some s, so $(a^s)^2 \equiv -1 \pmod p$. Since $p - 1$ is the order of a and $(a^{(p-1)/2})^2 \equiv 1 \pmod p$, as in the last paragraph, $a^{(p-1)/2} \equiv -1 \pmod p$. By Exercise E6 of Chapter 1,

$$\frac{p - 1}{2} \equiv 2s \pmod{(p - 1)},$$

so

$$p - 1 = 4s + 2(p - 1)t$$

for some integer t. Since $p - 1$ is even, $p \equiv 1 \pmod 4$. That completes the proof. $\qquad\square$

Using the theorem, we get easily

Corollary 1. *There are infinitely many primes of the form* $4k + 1$.

PROOF. Assume that $p_1, \ldots, p_n$ are all of the primes congruent to 1 mod 4. Consider $s = 4(p_1 \cdots p_n)^2 + 1$. Let q be a prime which divides s. Then clearly q cannot be 2 or any of $p_1, \ldots, p_n$.

Now $s \equiv 0 \pmod{q}$, so $4(p_1 \cdots p_n)^2 \equiv -1 \pmod{q}$. Therefore -1 is a square mod q, and so $q \equiv 1 \pmod 4$. Therefore $p_1, \ldots, p_n$ are not all of the primes congruent to 1 mod 4. This contradicts our assumption and proves the corollary. $\square$

B. Finite Fields of Complex Numbers

A second consequence of the theorem is that if $p \equiv 3 \pmod 4$ then $x^2 + 1$ is irreducible in $\mathbb{Z}_p[x]$. That being so, we can invent "complex numbers mod p" just as we invented $GF(9)$ in Chapter I-11F. Here is how.

Let i be an invented root of $x^2 + 1$. Consider the set of "numbers" of the form $a + bi$, where a, b are in $\mathbb{Z}_p$. Call the set of such numbers $\mathbb{Z}_p[i]$. Add them and multiply them as if they were polynomials in i, or complex numbers:

$$(a + bi) + (c + di) = (a + c) + (b + d)i;$$

$$(a + bi)(c + di) = ac + bdi^2 + (ad + bc)i = ac - bd + (ad + bc)i.$$

(Since i is a root of $x^2 + 1$, $i^2 + 1 = 0$, so $i^2 = -1$.) It is easy to verify that with these operations, $\mathbb{Z}_p[i]$ is a commutative ring.

In fact, $\mathbb{Z}_p[i]$ is a field if $p \equiv 3 \pmod 4$. For then, by the theorem, $x^2 + 1$ is irreducible in $\mathbb{Z}_p[x]$. So if $a + bx$ is any nonzero polynomial in $\mathbb{Z}_p[x]$, Bezout's lemma gives

$$1 = (a + bx)f(x) + (x^2 + 1)g(x)$$

for some $f(x), g(x)$ in $\mathbb{Z}_p[x]$. Setting $x = i$ gives $1 = (a + bi)f(i)$. That shows that any nonzero element of $\mathbb{Z}_p[i]$ is invertible, and therefore $\mathbb{Z}_p[i]$ is a field.

It is easy to see that $\mathbb{Z}_p[i]$ has p^2 elements, for if $a + bi$ is a typical element of $\mathbb{Z}_p[i]$, there are p choices for a and p choices for b.

We have proved

Corollary 2. *If* $p \equiv 3 \pmod 4$ *there is a field with* p^2 *elements.*

Notice that $\mathbb{Z}_{p^2}$ is *not* a field.

Much of Part III of this book is devoted to extensions of Corollary 2.

E1. In $\mathbb{Z}_7[i]$ find the inverse of $3 + 4i$; of $2 + 6i$; of $1 + i$.

E2. Prove: If a or b is relatively prime to 7, then so is $a^2 + b^2$.

E3. Find a primitive element of $\mathbb{Z}_7[i]$.

E4. Verify the associative law for multiplication in $\mathbb{Z}_p[i]$.

E5. If $p \equiv 1 \pmod 4$, we can still invent $\mathbb{Z}_p[i]$, but it will not be a field. Find an element in $\mathbb{Z}_5[i]$ which is nonzero but has no inverse. Do the same for $\mathbb{Z}_p[i]$ for any $p \equiv 1 \pmod 4$.

E6. Prove *Wilson's theorem—if p is an odd prime, then $(p-1)! \equiv -1 \pmod p$—* as follows.
  (a) Show that in $\mathbb{Z}_p[x]$, $f(x) = x^{p-1} - 1$ factors as
$$f(x) = (x-1)(x-2) \cdots (x-(p-1)) \qquad (*)$$
(use Fermat's theorem and the root theorem in $\mathbb{Z}_p[x]$).
  (b) Evaluate (*) at $x = 0$.

†**E7.** Prove that if $p \equiv 1 \pmod 4$, then $((p-1)/2)!^2 \equiv -1 \pmod p$ using Wilson's theorem.

E8. Prove that for any $m > 1$, $(m-1)! \equiv -1 \pmod m$ iff m is prime. (See E6 for the hard implication.)

Telephone Cable Splicing 5

In the 1920s and '30s and perhaps, in some places, yet today, telephone lines were formed out of 1000-foot lengths of wire cables spliced together. These cables, in simplified form, have a crosssection which looks like this:

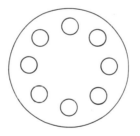

Each little circle is the crosssection of a wire carrying a single conversation.

One problem which arose is that if two conversations are carried on adjacent wires for long distances there is a lot of cross-talk—both conversations are heard on both wires. One way to alleviate this is by splicing lengths of cables together so that two conversations which are adjacent on one length of cable will not be adjacent on the next several lengths of cables. The problem then became how to do this in a simple but effective way.

This problem was considered by H. P. Lawther, of the Southwestern Bell Telephone Company, who, in solving it, used some number theory. What follows is based on part of his article (Lawther, 1935).

It was decided to adopt the following strategy.

(1) The same splicing instructions are to be used for every splice.
(2) At each splice, if the n individual wires are labeled by $1, 2, \ldots, n$, then wire i in the left cable will be spliced to wire $1 + (i - 1)s \pmod{n}$ in

the right cable, where s is some fixed integer, $1 \leqslant s < n$, called the *spacing number*.

E1. Show that this is a viable strategy for any n and any s. That is, show that every wire on the right cable will get attached to some wire on the left cable. Show that two wires on the left cable never get spliced to the same wire on the right cable with this strategy.

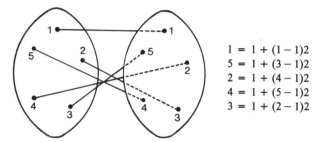

$$1 = 1 + (1-1)2$$
$$5 = 1 + (3-1)2$$
$$2 = 1 + (4-1)2$$
$$4 = 1 + (5-1)2$$
$$3 = 1 + (2-1)2$$

Splicing by the formula, $n = 5$, $s = 2$

Given Exercise E1, the problem becomes: For cables with n wires, which is the best spacing number s to use? To see which s to use, we shall first iterate the formula.

If a given conversation is on wire i_{r-1} in the $(r-1)$st length of cable, it will be on wire $i_r = 1 + (i_{r-1} - 1)s$ on the rth length of cable. So the conversation on wire i_0 at the 0th cable will be on wire i_r on the rth cable, where

$$i_1 = 1 + (i_0 - 1)s,$$
$$i_2 = 1 + (i_1 - 1)s = 1 + ([1 + (i_0 - 1)s] - 1)s,$$
$$= 1 + (i_0 - 1)s^2,$$
$$\vdots$$

and in general, as you can (try to) prove by induction,

$$i_r = 1 + (i_0 - 1)s^r.$$

E2. Prove this.

Now we want two conversations adjacent on the 0th length to not be adjacent for as many lengths as possible after the 0th length. Suppose the conversations on wires i_0 and $i_0 + 1$ are adjacent again on the rth length of cable. Then the location of conversation i_0 on the rth length is $1 + (i_0 - 1)s^r$, the location of conversation $i_0 + 1$ is $1 + ((i_0 + 1) - 1)s^r$, and these are adjacent, so

$$1 + (i_0 - 1)s^r \equiv 1 + ((i_0 + 1) - 1)s^r \pm 1 \pmod{n}.$$

Simplifying, this becomes

$$s^r \equiv \pm 1 \pmod{n}.$$

Thus if we want any two conversations adjacent on the 0th length of cable to be nonadjacent for as long as possible, we should pick s so that the least $r > 0$ with $s^r \equiv \pm 1 \pmod{n}$ is as large as possible.

EXAMPLE. Suppose n is an odd prime p. Then we know that $p - 1$ is even, and every integer s, $1 < s < p - 1$, satisfies

$$s^{p-1} \equiv 1 \pmod{p}.$$

Thus $s^{(p-1)/2}$ is a root of $x^2 - 1$ in $\mathbb{Z}_p[x]$. But $x^2 - 1$ has only the roots $x \equiv 1$, $x \equiv -1 \pmod{p}$ (since $\mathbb{Z}_p$ is a field); therefore

$$s^{(p-1)/2} \equiv 1 \quad \text{or} \quad -1 \pmod{p}.$$

Thus for *any* s, the best possible r is $r = (p - 1)/2$. That is, regardless of the choice of spacing number s, if two conversations are adjacent in the 0th wire, they will necessarily be adjacent in the $((p - 1)/2)$th wire.

If we choose the spacing number s to be an integer so that $[s]_p$ is a primitive element of $\mathbb{Z}_p$, then the best possible r is achieved. For if r is the smallest integer such that $s^r \equiv \pm 1 \pmod{p}$, then $s^{2r} \equiv 1 \pmod{p}$. If $[s]_p$ is a primitive element, $p - 1$ divides $2r$, so $(p - 1)/2$ divides r. Thus $r = (p - 1)/2$.

EXAMPLE. Suppose we have cables with 11 wires. Then $[2]_{11}$ is a primitive element, so we can use $s = 2$. Then two conversations which are adjacent in a given length of cable will again be adjacent only in the fifth subsequent cable length $((11 - 1)/2 = 5)$.

Here is a table which describes the splicing in cables with 11 wires when $s = 2$.

Cable length #	0	1	2	3	4	5
Wire #	1	1	1	1	1	1
	2	3	5	9	6	11
	3	5	9	6	11	10
	4	7	2	3	5	9
	5	9	6	11	10	8
	6	11	10	8	4	7
	7	2	3	5	9	6
	8	4	7	2	3	5
	9	6	11	10	8	4
	10	8	4	7	2	3
	11	10	8	4	7	2

E3. Find the best r and s and describe the splicing for that s when $n = 13$.

E4. Do the same for $n = 12$. (Some of the theory we applied will not work so well for $n = 12$ since $\mathbb{Z}_{12}$ is not a field.)

E5. Lawther shows how to do $n = 12$ in a nice way using the splicing instructions for $n = 11$ and cheating, namely, taking the table for $n = 11$ and altering it by splicing wire #7 to itself always, and replacing all the numbers $r > 7$ in the table by $r + 1$. Thus the first splice becomes

Cable 0	wire #	1	2	3	4	5	6	7	8	9	10	11	12
Cable 1	wire #	1	3	5	8	10	12	7	2	4	6	9	11

Is this a better strategy than that of Exercise E4?

Describe an analogous "cheat" which works for $n = 14$; for $n = p + 1, p$ any prime.

If one adopts the splicing strategy for n not prime, without cheating, the problem of finding the largest r such that calls adjacent on the 0th cable will not again be adjacent until the rth cable turns out to have the following solution.

Let $h(n)$ be the maximum order of an invertible element of $\mathbb{Z}_n$. In Theorem 6 of Chapter 2, $h(n)$ was given as follows: If n factors as $n = 2^e p_1^{e_1} p_2^{e_2} \cdots p_r^{e_r}$, then

$$h(n) = \text{l.c.m.}[2^s \phi(p_1^{e_1}), \phi(p_2^{e_2}), \ldots, \phi(p_r^{e_r})] \quad \text{with } s = \begin{cases} e - 1, & e = 1, 2 \\ e - 2, & e \geqslant 3 \end{cases}.$$

If $\phi(n)$ is the number of invertible congruence classes in $\mathbb{Z}_n$, then (see III-2, E13) $h(n) = \phi(n)$ iff $\mathbb{Z}_n$ has a primitive element iff $n = 2, 4, p^e$, or $2p^e$ for p an odd prime.

Suppose the cables have n wires. If s is chosen to be the best possible spacing number, the corresponding number r will be

$$r = \begin{cases} h(n) & \text{if} \quad h(n) \neq \phi(n), \\ \dfrac{h(n)}{2} & \text{if} \quad h(n) = \phi(n). \end{cases}$$

*E6. Prove this.

It turns out that the problem of cross-talk can be eliminated technologically by using optical fibres to carry telephone signals rather than metal conductors. For an exposition of this rather new (in 1977) advance in telephone communications, see Boyle (1977).

In Chapter II-13, it was claimed that there were polynomials in $\mathbb{Z}[x]$ which were irreducible in $\mathbb{Q}[x]$ but factored mod p for any prime p. Using the existence of a primitive element in $\mathbb{Z}_p$ we can describe a class of such examples.

Proposition. *For any integers a, b and any prime p, $f(x) = x^4 + ax^2 + b^2$ factors mod p.*

PROOF. First suppose $p = 2$. Then any integer is congruent mod 2 to 0 or 1, so for any integers $a, b, f(x)$ is congruent mod 2 to one of the following polynomials:

$$x^4; \qquad x^4 + x^2 = x^2(x^2 + 1);$$

$$x^4 + 1 = (x + 1)^4; \qquad x^4 + x^2 + 1 = (x^2 + x + 1)^2.$$

Each of these is reducible mod 2.

Now suppose p is an odd prime. Choose s so that $a \equiv 2s \pmod{p}$. Then $f(x) \equiv x^4 + 2sx^2 + b^2 \pmod{p}$, which can be written in each of the following three ways:

$$f(x) \equiv (x^2 + s)^2 - (s^2 - b^2)$$

$$(x^2 + b)^2 - (2b - 2s)x^2$$

$$(x^2 - b)^2 - (-2b - 2s)x^2.$$

To show that $f(x)$ factors mod p, it suffices to show that one of $s^2 - b^2$, $2b - 2s$, $-2b - 2s$ is a square mod p, for then $f(x)$ will be the difference of two squares mod p.

Let c be a primitive element mod p. If $2b - 2s$ and $-2b - 2s$ are not squares mod p, then $2b - 2s \equiv c^r$, $-2b - 2s \equiv c^t$ (mod p), where r and t are both odd. Then $(2b - 2s)(-2b - 2s) = 4(b^2 - s^2) \equiv c^{r+t}$, where $r + t = 2u$ is even. So $4(s^2 - b^2)$ is a square mod p. Let $2m \equiv 1$ (mod p). Then $s^2 - b^2 \equiv 4(s^2 - b^2)m^2 \equiv c^{2u}m^2 = (c^u m)^2$. So $s^2 - b^2$ is a square mod p. That completes the proof. $\square$

It is easy to find a, b such that $f(x)$ is irreducible in $\mathbb{Q}[x]$. For example, let $a = 2$, $b = 2$, so that $f(x) = x^4 + 2x^2 + 4$. Then $f(x)$ has no roots. Also $f(x)$ has no factors of degree 2: if you write $f(x) = (x^2 + cx + d)(x^2 + ex + f)$, the equations for the coefficients are

$$df = 4,$$
$$cf + de = 0,$$
$$ce + d + f = 2,$$
$$c + e = 0,$$

with c, d, e, f integers. It is quickly checked that there are no integer solutions. So $f(x)$ is irreducible. The exercises offer other examples.

E1. Show that $x^4 - x^2 + 1$ is irreducible in $\mathbb{Q}[x]$ but factors mod p for all primes p.

E2. Show that $x^4 + 1$ is irreducible in $\mathbb{Q}[x]$ but factors mod p for all primes p.

*E3. Show that if a, b are integers with $0 < b < a$, then $x^4 + 2ax^2 + b^2$ is irreducible in $\mathbb{Q}[x]$.

E4. Can Eisenstein's irreducibility criterion (II-8) be applied to any of the examples of the proposition?

Congruence Classes Modulo a Polynomial: Simple Field Extensions

7

In Chapter I-7 we discovered the rings $\mathbb{Z}_n$ by looking at congruence classes of integers modulo n. For n a prime, $\mathbb{Z}_n$ turned out to be a field.

In this chapter we do the same construction with polynomials. By doing so, we shall discover vast collections of new fields, some of which will turn out to be very useful.

We first look at examples.

EXAMPLE 1. In $\mathbb{Q}[x]$ consider congruence modulo the polynomial $x^2 - 2$. We shall say that

$$a(x) \equiv b(x) \pmod{x^2 - 2}$$

if $x^2 - 2$ divides $a(x) - b(x)$, or if

$$a(x) = b(x) + q(x)(x^2 - 2)$$

for some polynomial $q(x)$ in $\mathbb{Q}[x]$. Thus

$$x^2 \equiv 2 \pmod{x^2 - 2},$$

$$x^3 \equiv 2x \pmod{x^2 - 2},$$

$$x^4 - 2x^2 + 1 \equiv 1 \pmod{x^2 - 2},$$

$$x^9 \equiv 4x^5 \pmod{x^2 - 2},$$

etc.

Write $[f(x)]_{x^2-2}$ for the congruence class mod $x^2 - 2$: two polynomials are in the same congruence class mod $x^2 - 2$ if they are congruent mod $x^2 - 2$. Define addition and multiplication of congruence classes, just as in

$\mathbb{Z}_p$, by

$$[f(x)]_{x^2-2} + [g(x)]_{x^2-2} = [f(x) + g(x)]_{x^2-2},$$
$$[f(x)]_{x^2-2} \cdot [g(x)]_{x^2-2} = [f(x)g(x)]_{x^2-2}.$$

These operations make the set of congruence classes into a ring which we shall call $\mathbb{Q}[x]/(x^2 - 2)$. Each congruence class in $\mathbb{Q}[x]/(x^2 - 2)$ has a nice representative, found by the division theorem. If $f(x)$ is in $\mathbb{Q}[x]$,

$$f(x) = (x^2 - 2)q(x) + r(x)$$

where $r(x)$ is unique and has degree $\leqslant 1$: $r(x) = a + bx$ for some unique a, b, in $\mathbb{Q}$. Then $[f(x)]_{x^2-2} = [r(x)]_{x^2-2} = [a + bx]_{x^2-2} = [a]_{x^2-2} + [b]_{x^2-2} \cdot [x]_{x^2-2}$.

We can think of $\mathbb{Q}$ as being a subset of $\mathbb{Q}[x]/(x^2 - 2)$ by identifying a in $\mathbb{Q}$ with $[a]_{x^2-2}$. (For if $[a]_{x^2-2} = [b]_{x^2-2}$, then $a = b$, for a, b in $\mathbb{Q}$.) Thus every element of $\mathbb{Q}[x]/(x^2 - 2)$ is an element of $\mathbb{Q}$ plus a $\mathbb{Q}$-multiple of $[x]_{x^2-2}$. If we abbreviate $[x]_{x^2-2}$ by α, then every such element of $\mathbb{Q}[x]/(x^2 - 2)$ is of the form $a + b\alpha$, with a, b in $\mathbb{Q}$.

It is convenient to think of elements of $\mathbb{Q}[x]/(x^2 - 2)$ as polynomials in $\mathbb{Q}[x]$ evaluated at $\alpha = [x]_{x^2-2}$ where

$$\alpha^2 = [x]_{x^2-2}^2 = [x^2]_{x^2-2} = [x^2 - 2 + 2]_{x^2-2}$$
$$= [x^2 - 2]_{x^2-2} + [2]_{x^2-2} = 2.$$

(Of course α is not a number in the usual sense, but a congruence class, an invented element satisfying $\alpha^2 = 2$.) Then addition and multiplication are the same as for polynomials: for example,

$$(a + b\alpha)(c + d\alpha) = ac + (ad + bc)\alpha + bd\alpha^2.$$

To express this result as a polynomial in α of degree $\leqslant 1$, we use the fact that $\alpha^2 = 2$ to get

$$(a + b\alpha)(c + d\alpha) = ac + 2bd + (ad + bc)\alpha.$$

We note that $\mathbb{Q}[x]/(x^2 - 2)$ is in fact a field. For if a or $b \neq 0$,

$$(a + b\alpha)\left[(a - b\alpha) \cdot \frac{1}{(a^2 - 2b^2)}\right] = 1$$

provided that the rational numbers a, b do not satify $a^2 - 2b^2 = 0$. But if $a^2 - 2b^2 = 0$, $b \neq 0 \neq a$, then a/b is a rational root of $x^2 - 2$. But $x^2 - 2$ has no rational roots because $x^2 - 2$ is irreducible in $\mathbb{Q}[x]$ (why?).

EXAMPLE 2. In $\mathbb{R}[x]$ the polynomial $x^2 + 1$ is irreducible. Let $\mathbb{R}[x]/(x^2 + 1)$ be the set of congruence classes of polynomials modulo $x^2 + 1$. Then by the division theorem, there are a, b in $\mathbb{R}$ so that

$$[f(x)]_{x^2+1} = [a + bx]_{x^2+1} = a + b[x]$$

where, as in Example 1, we identify a real number with its congruence class modulo $x^2 + 1$. If we abbreviate $[x]_{x^2+1} = i$, then $x^2 \equiv -1 \pmod{x^2 + 1}$, so $i^2 = [x]^2 = [x^2] = [-1] = -1$, and the elements of $\mathbb{R}[x]/(x^2 + 1)$ are of the form $a + bi$, a, b in $\mathbb{R}$. We multiply as if the elements were polynomials in i, and then use the equation $i^2 = -1$.

Then $\mathbb{R}[x]/(x^2 + 1)$ looks just like $\mathbb{C}$.

EXAMPLE 3. We invent a field with 8 elements. Look at $\mathbb{Z}_2[x]/(x^3 + x + 1)$, congruence classes in $\mathbb{Z}_2[x]$ modulo $x^3 + x + 1$.

By the division theorem, namely $f(x) = q(x)(x^3 + x + 1) + r(x)$, any polynomial in $\mathbb{Z}_2[x]$ is congruent modulo $x^3 + x + 1$ to a polynomial of degree $\leqslant 2$. So any congruence class can be represented by a polynomial of degree $\leqslant 2$. There are exactly 8 of these. So $\mathbb{Z}_2[x]/(x^3 + x + 1)$ consists of $[0] = 0$, $[1] = 1$, $[x]$, $[x + 1] = [x] + 1$, $[x^2] = [x]^2$, $[x^2 + 1] = [x]^2 + 1$, $[x^2 + x] = [x]^2 + [x]$, and $[x^2 + x + 1] = [x]^2 + [x] + 1$. These are all the polynomials in $[x]$ with coefficients in $\mathbb{Z}_2$ of degree $\leqslant 2$. We multiply these elements like polynomials and then, to get the product again as a polynomial of degree $\leqslant 2$, use the equation $[x]^3 = [x] + 1$.

Setting $[x] = \alpha$ we get the following multiplication table.

	0	1	α	$\alpha + 1$
0	0	0	0	0
1	0	1	α	$\alpha + 1$
α	0	α	α^2	$\alpha^2 + \alpha$
$\alpha + 1$	0	$\alpha + 1$	$\alpha^2 + \alpha$	$\alpha^2 + 1$
α^2	0	α^2	$\alpha + 1$	$\alpha^2 + \alpha + 1$
$\alpha^2 + 1$	0	$\alpha^2 + 1$	1	α^2
$\alpha^2 + \alpha$	0	$\alpha^2 + \alpha$	$\alpha^2 + \alpha + 1$	1
$\alpha^2 + \alpha + 1$	0	$\alpha^2 + \alpha + 1$	$\alpha^2 + 1$	α

	α^2	$\alpha^2 + 1$	$\alpha^2 + \alpha$	$\alpha^2 + \alpha + 1$
0	0	0	0	0
1	α^2	$\alpha^2 + 1$	$\alpha^2 + \alpha$	$\alpha^2 + \alpha + 1$
α	$\alpha + 1$	1	$\alpha^2 + \alpha + 1$	$\alpha^2 + 1$
$\alpha + 1$	$\alpha^2 + \alpha + 1$	α^2	1	α
α^2	$\alpha^2 + \alpha$	α	$\alpha^2 + 1$	1
$\alpha^2 + 1$	α	$\alpha^2 + \alpha + 1$	$\alpha + 1$	$\alpha^2 + \alpha$
$\alpha^2 + \alpha$	$\alpha^2 + 1$	$\alpha + 1$	α	α^2
$\alpha^2 + \alpha + 1$	1	$\alpha^2 + \alpha$	α^2	$\alpha + 1$

Notice that every nonzero element has an inverse so that (assuming multiplication is associative, which it is) we have the multiplication table for a field.

Here is the general case.

Let F be a field, $p(x)$ be a polynomial in $F[x]$ of degree $\geqslant 1$. Say that $f(x) = g(x) \pmod{p(x)}$ if $p(x)$ divides $f(x) - g(x)$. Write $[f(x)]_{p(x)} =$

$[g(x)]_{p(x)}$ if $f(x) \equiv g(x) \pmod{p(x)}$; $[f(x)]_{p(x)}$ is the congruence class of $f(x)$ and is the set of all those polynomials in $F[x]$ which are congruent to $f(x)$ modulo $p(x)$. Let $F[x]/p(x)$ be the collection of all the congruence classes in $F[x]$ modulo $p(x)$. We shall write $[f(x)]$ instead of $[f(x)]_{p(x)}$ when no confusion can arise.

We first notice that the function which takes a in F to $[a]$ in $F[x]/p(x)$ is 1–1. For if $[a] = [b]$, a, b in F, then $p(x)$ must divide $a - b$. But $\deg(p(x)) \geqslant 1$, so $a - b$ must be the zero polynomial, and thus $a = b$ in F. Therefore we may identify a and $[a]$ in $F[x]/p(x)$.

Addition and multiplication of congruence classes are defined by addition and multiplication of representatives (see E9). Thus

$$[f(x)] + [g(x)] = [f(x) + g(x)],$$

and

$$[f(x)] \cdot [g(x)] = [f(x)g(x)].$$

Using these rules, we can write a typical element of $F[x]/p(x)$ as

$$[a_0 + a_1 x + \cdots + a_r x^r] = [a_0] + [a_1][x] + \cdots + [a_r][x]^r$$

which, with the identification of a and $[a]$, for a in F, can be written

$$a_0 + a_1[x] + \cdots + a_r[x]^r,$$

a polynomial in $[x]_{p(x)}$. If we abbreviate $[x]_{p(x)}$ by α, then this becomes

$$a_0 + a_1\alpha + \cdots + a_r\alpha^r,$$

so a typical element of $F[x]/p(x)$ can be thought of as a polynomial with coefficients in F evaluated at the congruence class $[x]_{p(x)} = \alpha$, where $p(\alpha) = 0$ (since $p(\alpha) = [p(x)]_{p(x)} = [0]_{p(x)}$). With these changes in notation, we shall sometimes denote $F[x]/p(x)$ by $F[\alpha]$, when we are sure no one will forget that $\alpha = [x]_{p(x)}$ and $p(\alpha) = 0$.

If $p(x)$ has degree d, then every element of $F[\alpha]$ is a linear combination of $1, \alpha, \alpha^2, \alpha^3, \ldots, \alpha^{d-1}$. For given any polynomial $f(x)$ in $F[x]$, there is a unique polynomial $r(x)$ of degree $< d$ such that

$$f(x) = p(x)q(x) + r(x).$$

Thus there is a unique polynomial of degree $< d$, namely $r(x)$, to which $f(x)$ is congruent mod $p(x)$. So every polynomial in $F[\alpha]$ is equal to a unique polynomial in α of degree $< d$, thus is a unique F-linear combination of $1, \alpha, \alpha^2, \ldots, \alpha^{d-1}$.

If we think of $F[x]/p(x) = F[\alpha]$ as polynomials in α, then addition and multiplication in $F[\alpha]$ is the same as addition and multiplication of polynomials, evaluated at α. To express $f(\alpha)g(\alpha)$ as a polynomial of degree $< d$ in α, use the division theorem: $f(x)g(x) = p(x)q(x) + r(x)$ with $\deg(r(x)) < d = \deg(p(x))$. Then $f(\alpha)g(\alpha) = r(\alpha)$. Convenient data to have when multiplying in $F[\alpha] = F[x]/p(x)$ is a table expressing powers of $\alpha = [x]$ between d and $2d - 2$ in terms of polynomials in α of degree $< d$.

For example, in $\mathbb{Q}[x]/(x^3 + x + 1)$,

$$\alpha^3 = -\alpha - 1,$$
$$\alpha^4 = -\alpha^2 - \alpha,$$
$$\alpha^5 = -\alpha^3 - \alpha^2 = (\alpha + 1) - \alpha^2.$$

Theorem. $F[x]/p(x)$ *is a commutative ring with unity. It is a field if* $p(x)$ *is irreducible in* $F[x]$.

The converse is also true. See E10.

PROOF. The fact that $F[x]/p(x)$ satisfies the axioms for a commutative ring with unity follows from the fact that $F[x]$ satisfies those same axioms. For example, to show the commutative law for multiplication, take f, g in $F[x]$. Then

$$[f][g] = [f \cdot g] = [g \cdot f] = [g][f],$$

the middle inequality following from the fact that multiplication is commutative in $F[x]$. The other axioms are handled in a similar fashion.

To show that $F[x]/p(x)$ is a field if $p(x)$ is irreducible, we must check the axiom that nonzero elements have inverses.

Suppose $p(x)$ is irreducible. If $f(x)$ is any polynomial, nonzero, of degree $< d$, then $f(x)$ and $p(x)$ are relatively prime, so by Bezout's lemma we can write

$$f(x)a(x) + p(x)b(x) = 1.$$

Then, if $\alpha = [x]_{p(x)}$, evaluating this last equation at α and noting that $p(\alpha) = 0$, we get $f(\alpha)a(\alpha) = 1$ in $F[\alpha] = F[x]/p(x)$, so $f(\alpha)$ has an inverse in $F[\alpha]$. Since every nonzero element of $F[\alpha]$ has an inverse, $F[\alpha]$ is then a field. $\qquad\square$

We now have a way of inventing many new fields. Take an irreducible polynomial $p(x)$ in $F[x]$, and consider $F[x]/p(x)$—it will be a field. Our three examples above were all fields.

A field of the form $F[x]/p(x)$ is called a *simple field extension* of F. Here are some other examples:

$\mathbb{Z}_3[x]/(x^2 + 1)$—called $GF(9)$, it has 9 elements;

$\mathbb{Z}_2[x]/(x^2 + x + 1)$—a field with 4 elements;

$\mathbb{Q}[x]/(x^6 + 2x + 2)$—this looks like complex numbers which are polynomials in α with rational coefficients, where α in $\mathbb{C}$ is a root of $x^6 + 2x + 2$;

$\mathbb{Z}_2[x]/(x^4 + x + 1)$—a field with 16 elements;

$\mathbb{Z}_3[x]/(x^3 + 2x + 1)$—a field with 27 elements;

$\mathbb{Z}_p[x]/(x^2 + 1)$ for $p \equiv 3 \pmod 4$—this looks just like the field $\mathbb{Z}_p[i]$ which we invented in Chapter 4.

We know by Eisenstein's irreducibility criterion that in $\mathbb{Q}[x]$ there are irreducible polynomials of any degree $\geqslant 1$. So given any $d > 1$ there is a field F containing $\mathbb{Q}$ whose dimension as a vector space over $\mathbb{Q}$ is d. We shall see later that the same fact will be true for $\mathbb{Z}_p$ as well.

E1. In $\mathbb{Z}_3[x]/(x^3 + 2x + 1) = \mathbb{Z}[\alpha]$ find the inverse of $\alpha^2 + \alpha + 2$.

E2. Show that $F = \mathbb{Z}_2[x]/(x^5 + x^2 + 1)$ is a field. Find a primitive element of F. How many different primitive elements are there?

E3. Show that each irreducible polynomial in $\mathbb{Z}_2[x]$ of degree 1, 2, or 4 has a root in $\mathbb{Z}_2[\alpha] = \mathbb{Z}_2[x]/(x^4 + x + 1)$. Factor $x^{16} - x$ in $\mathbb{Z}_2[\alpha][x]$; in $\mathbb{Z}_2[x]$.

E4. In $\mathbb{Z}_3[\alpha] = \mathbb{Z}_3[x]/(x^3 + 2x + 1)$:
(a) find the inverse of $\alpha^2 + 1$; of $1 + \alpha + \alpha^2$; of $2 + 2\alpha + \alpha^2$;
(b) write $(\alpha^2 + 2\alpha) \cdot (\alpha^2 + \alpha + 1)$ as a $\mathbb{Z}_3$-linear combination of $1, \alpha, \alpha^2$;
(c) do the same for $(\alpha^2 + \alpha + 2)(\alpha^2 + \alpha + 1)$;
(d) do the same for α^{10};
(e) solve
$$\begin{pmatrix} \alpha^8 & \alpha^{12} \\ \alpha^3 & \alpha^2 \end{pmatrix} \begin{pmatrix} x \\ y \end{pmatrix} = \begin{pmatrix} \alpha^4 \\ \alpha^5 \end{pmatrix};$$
(f) is α a primitive element of $\mathbb{Z}_3[\alpha]$? What are the possibilities for the orders of elements of $\mathbb{Z}_3[\alpha]$?

E5. Write down addition and multiplication tables for $\mathbb{Z}_2[x]/(x^2 + x + 1) = \mathbb{Z}_2[\alpha]$.

E6. Can there be a field F with 6 elements in it?

E7. Describe $F[x]/(x - a)$, for a in F.

E8. Write down the multiplication table for the ring $\mathbb{Z}_3[x]/(x^2 + x + 1)$. (Note that $x^2 + x + 1$ is not irreducible in $\mathbb{Z}_3[x]$.)

E9. Prove that the operations $[f] + [g] = [f + g]$, $[f] \cdot [g] = [f \cdot g]$ in $F[x]/p(x)$ are well defined (see Chapter I-7).

E10. Prove that if $p(x)$ is not irreducible in $F[x]$, then $F[x]/p(x)$ is not a field, by exhibiting nonzero elements of $F[x]/p(x)$ which are zero divisors.

Polynomials and Roots 8

A. Inventing Roots of Polynomials

Perhaps the main reason for the invention of complex numbers was the fact that certain polynomials in $\mathbb{R}[x]$ had no roots in $\mathbb{R}$, for example $x^2 + 1$, $x^2 + x + 1$, $x^2 + 2$, etc. By defining $\mathbb{C}$, the complex numbers, mathematicians provided "imaginary" roots for these polynomials, by, in essence, inventing them.

Now $\mathbb{C}$ looks just like $\mathbb{R}[x]/(x^2 + 1)$. In fact, as we observed in the last section, if you write the elements of $\mathbb{R}[x]/(x^2 + 1)$ as $a + b[x]_{x^2+1}$ and then set $[x]_{x^2+1} = i$, there is no difference between $\mathbb{R}[x]/(x^2 + 1)$ and $\mathbb{C}$: the elements *look* the same, *add* the same way, *multiply* the same way. Of course i is a root of $x^2 + 1$, for

$$i^2 + 1 = [x][x] + 1 = [x^2 + 1]_{x^2+1} = [0]_{x^2+1} = 0.$$

If 16th-century mathematicians can invent $\mathbb{C}$ to provide roots of polynomials with real coefficients we can do the same kind of thing with other polynomials. In fact, we already have the following

Theorem. *Let F be a field, $p(x)$ an irreducible polynomial in $F[x]$. Then there is a field K containing F which has a root of $p(x)$.*

PROOF. The field K is $F[x]/(p(x))$. The root of $p(x)$ is $[x]_{p(x)}$. You verify the details. □

We can do even better:

Corollary. *Let F be a field, $f(x)$ a (monic) polynomial of degree $\geqslant 1$ in $F[x]$. Then there exists a field K containing F such that in $K[x]$, $f(x)$ factors into a product of linear factors.*

PROOF. Induction on $\deg(f(x)) = d$.

The case $\deg(f(x)) = 1$ is trivial.

Let $f(x)$ have degree d. In $F[x]$, $f(x) = p_1(x) \cdots p_s(x)$, a product of irreducible polynomials. If $\deg(p_i(x)) = 1$ for all $i = 1, \ldots, s$, then the field K we are looking for is F itself. Otherwise, suppose (renumbering if necessary) that $\deg(p_1(x)) > 1$. Let $L = F[y]/(p_1(y)) = F[\alpha]$. Then L is a field containing F, and α is a root in L of $p_1(x)$, so in $L[x]$, $p_1(x) = (x - \alpha)q_1(x)$. Thus, in $L[x]$, $f(x) = (x - \alpha)q_1(x)p_2(x) \cdots p_s(x)$. Let $g(x) = q_1(x)p_2(x) \cdots p_s(x)$. Now $\deg(g(x)) = \deg(f(x)) - 1$. By induction there is a field $K \supset L$ such that in $K[x]$, $g(x)$ factors into a product of linear factors. But if $g(x)$ does, so does $f(x) = (x - \alpha)g(x)$. That completes the proof. $\qquad\qquad\qquad\qquad\qquad\qquad\qquad\qquad\qquad\qquad\qquad\qquad\qquad\square$

A field K is called a *splitting field* for $f(x)$ if $f(x)$ factors into linear factors in $K[x]$.

EXAMPLE. In $\mathbb{Q}[x]$, $f(x) = x^3 - 2$ is irreducible. It has a root in $\mathbb{R}$, namely $\sqrt[3]{2}$, but $\mathbb{R}$ is not a splitting field for $f(x)$ because in $\mathbb{R}[x]$,

$$x^3 - 2 = \left(x - \sqrt[3]{2}\right)\left(x^2 + \sqrt[3]{2}\,x + \sqrt[3]{4}\right)$$

and the second factor is irreducible in $\mathbb{R}[x]$. On the other hand, $\mathbb{C}$ is a splitting field for $f(x)$, for if $\omega = (-1 + \sqrt{-3})/2$, then in $\mathbb{C}[x]$,

$$x^3 - 2 = \left(x - \sqrt[3]{2}\right)\left(x - \omega\sqrt[3]{2}\right)\left(x - \omega^2\sqrt[3]{2}\right).$$

Of course, $\mathbb{C}$ is a splitting field for any polynomial in $\mathbb{Q}[x]$ because of the fundamental theorem of algebra.

E1. Prove the theorem.

*E2. Find a field F containing $\mathbb{Z}_2$ which contains all roots of $f(x) = x^{16} - x$. Describe F in the form $\mathbb{Z}_2[x]/(p(x))$ for some polynomial $p(x)$. (*Hint*: recall the generalization of Fermat's theorem in Chapter I-11.)

E3. Show that the field $F[x]/(p(x))$ is in fact a splitting field for $p(x)$ in $F[x]$, in case
  (a) $F = \mathbb{Z}_2, p(x) = x^3 + x + 1$,
  (b) $F = \mathbb{Z}_2, p(x) = x^4 + x + 1$,
  (c) $F = \mathbb{Z}_3, p(x) = x^3 + 2x + 1$,
  (d) F is any field, $p(x)$ has degree 2.

B. Finding Polynomials with Given Roots

Let F be a field, K be a field containing F. Let α be in K. Suppose α is the root of some nonzero polynomial $f(x)$ in $F[x]$. Then we say that α is *algebraic over F*.

Proposition 1. *Let α in K be algebraic over F. Then there is a unique monic irreducible polynomial $p(x)$ in $F[x]$ with α as a root, and every polynomial $f(x)$ in $F[x]$ with α as a root is a multiple of $p(x)$.*

We call $p(x)$ the *minimal polynomial of α over F*, because, as we shall see in the proof, it is a polynomial of minimal degree in $F[x]$ having α as a root.

EXAMPLE. Let $F = \mathbb{R}$, $K = \mathbb{C}$, α be in $\mathbb{C}$. Suppose $\alpha = a + bi$, $b \neq 0$. Then the minimal polynomial of α is

$$(x - \alpha)(x - \bar{\alpha}) = x^2 - 2ax + (a^2 + b^2),$$

as we showed in Chapter II-4.

PROOF OF PROPOSITION 1. Since α is the root of some nonzero polynomial with coefficients in F, there is a nonzero monic polynomial $p(x)$ of smallest degree with α as a root (by well-ordering on the set of degrees of polynomials in $F[x]$ with α as a root). We show that $p(x)$ is irreducible. Otherwise, $p(x) = a(x)b(x)$ with both $\deg(a(x))$ and $\deg(b(x)) < \deg(p(x))$; then $0 = p(\alpha) = a(\alpha)b(\alpha)$. Since K is a field, either $a(\alpha) = 0$ or $b(\alpha) = 0$. Thus α is a root of a polynomial of degree smaller than $\deg(p(x))$, which is a contradiction. Thus $p(x)$ is irreducible. Finally, suppose $f(x)$ is any polynomial in $F[x]$ with α as a root. By the division theorem,

$$f(x) = p(x)q(x) + r(x) \quad \text{with } \deg(r(x)) < \deg(p(x)).$$

Setting $x = \alpha$ yields $0 = r(\alpha)$. By choice of $p(x)$, $r(x)$ must be the zero polynomial, and so $p(x)$ divides $f(x)$. That completes the proof. □

In the next chapter we shall be interested in finding minimal polynomials over $\mathbb{Z}_2$ of elements of $GF(16)$. The next proposition shows that they will exist, and the proof describes how to find them.

Proposition 2. *Let $K = F[\alpha]$ be a simple field extension of F, where the minimal polynomial $p(x)$ of α over F has degree d. Then if β is any element of K, β is algebraic over F, and the minimal polynomial of β over F has degree $\leqslant d$.*

PROOF. To prove this, we need the following fact from linear algebra (see Chapter I-9): Any system of n homogeneous linear equations in $n + 1$ unknowns has a nonzero solution.

Each element of $K = F[\alpha]$ is a polynomial in α of degree $\leqslant d - 1$. In particular, each power of β is such a polynomial in α. Thus

$$1 = 1,$$
$$\beta = a_{1,0} + a_{1,1}\alpha + \cdots + a_{1,d-1}\alpha^{d-1},$$
$$\beta^2 = a_{2,0} + a_{2,1}\alpha + \cdots + a_{2,d-1}\alpha^{d-1},$$
$$\vdots$$
$$\beta^d = a_{d,0} + a_{d,1}\alpha + \cdots + a_{d,d-1}\alpha^{d-1}.$$

We look for a nonzero solution of

$$0 = x_0 + x_1\beta + x_2\beta^2 + \cdots + x_d\beta^d.$$

Substituting, we get

$$0 = x_0 + x_1\big(a_{1,0} + a_{1,1}\alpha + \cdots + a_{1,d-1}\alpha^{d-1}\big) + \cdots$$
$$+ x_d\big(a_{d,0} + a_{d,1}\alpha + \cdots + a_{d,d-1}\alpha^{d-1}\big).$$

Collecting coefficients of powers of α gives

$$x_0 + a_{1,0}x_1 + \cdots + a_{d,0}x_d = 0,$$

$$a_{1,1}x_1 + \cdots + a_{d,1}x_d = 0,$$

$$\vdots$$

$$a_{1,d-1}x_1 + \cdots + a_{d,d-1}x_d = 0.$$

This is a set of d equations in the $d + 1$ unknowns $x_0, \ldots, x_d$. So we can find a solution $x_0 = r_0, x_1 = r_1, \ldots, x_d = r_d$ of these equations with $r_0, r_1, \ldots, r_d$ not all zero. But then

$$0 = r_0 + r_1 x + r_2 x^2 + \cdots + r_d x^d = f(x)$$

is a nonzero polynomial of degree $\leqslant d$ in $F[x]$ with β as a root. That completes the proof. □

While in the proof of Proposition 2 we needed some linear algebra theory, in any case in practice we simply solve a particular set of equations, knowing from the theory invoked in Proposition 2 that the set of equations will have a solution.

EXAMPLE. Let $K = \mathbb{Z}_2[x]/(x^3 + x + 1) = \mathbb{Z}_2[\alpha]$. Then $\alpha^3 + \alpha + 1 = 0$ and K consists of polynomials in α of degree $\leqslant 2$ with coefficients in $\mathbb{Z}_2$. A description of the elements of K is found in Table 1 in Chapter III-9 below.

What is the minimal polynomial over $\mathbb{Z}_2$ of $\alpha^2 + 1$? We write the powers of $\alpha^2 + 1$ in terms of $1, \alpha, \alpha^2$, using Table 1 of Chapter III-9 as follows:

$$1 = 1,$$

$$\alpha^2 + 1 = \alpha^2 + 1;$$

$$(\alpha^2 + 1)^2 = \alpha^4 + 1 = \alpha^2 + \alpha + 1;$$

$$(\alpha^2 + 1)^3 = (\alpha^2 + 1)(\alpha^2 + \alpha + 1) = \alpha^2 + \alpha.$$

We solve

$$x_0 + x_1(\alpha^2 + 1) + x_2(\alpha^2 + \alpha + 1) + x_3(\alpha^2 + \alpha) = 0.$$

Collecting coefficients of powers of α, we get

$$x_0 + x_1 + x_2 = 0,$$
$$x_2 + x_3 = 0,$$
$$x_1 + x_2 + x_3 = 0,$$

which we solve in $\mathbb{Z}_2$. We get $x_1 = 0$, $x_0 = x_2 = x_3$. Thus the minimal polynomial of $\alpha^2 + 1$ over $\mathbb{Z}_2$ is $x^3 + x^2 + 1$.

E4. With $\alpha = [x]_{x^3 + x + 1}$ as in the example, find the minimal polynomial over $\mathbb{Z}_2$ of
(a) $\alpha^2 + \alpha$,
(b) $\alpha^2 + \alpha + 1$.

E5. Find the minimal polynomial over $\mathbb{Q}$ of
(a) $2 + \sqrt{3}$,
(b) $3 - \sqrt[3]{2}$.

E6. Let $K = F[x]/(p(x))$ be a field. What is the minimal polynomial of $[x]_{p(x)}$ over F?

9 Error-Correcting Codes, II

In Chapter I-13, we looked at ways of coding messages so that if in the transmission an error occurred in one of the digits of a coded word, the receiver would be able to correct the error. Those codes, called Hamming codes, were based on defining coded words as vectors of solutions in $\mathbb{Z}_2$ to sets of linear equations.

In this chapter we will use finite fields to describe some multiple-error-correcting codes. These codes, called BCH codes, were discovered in 1960 by Bose, Chaudhuri and Hocquenghem. The coded words in these codes will be vectors which are the coefficients of polynomials in $\mathbb{Z}_2[x]$. The polynomials will have as roots certain powers of a primitive element of some appropriate field extension of $\mathbb{Z}_2$.

The first example uses a field of 8 elements.

Let $m(x) = x^3 + x + 1$ in $\mathbb{Z}_2[x]$. It is easy to see that $m(x)$ is irreducible in $\mathbb{Z}_2[x]$, and so $\mathbb{Z}_2[x]/m(x)$ is a field with 8 elements. Denote the congruence class of x, $[x]_{m(x)}$, by α; then $\mathbb{Z}_2[x]/m(x)$ can be viewed as polynomials in α, where $\alpha^3 + \alpha + 1 = 0$, and so we shall denote $\mathbb{Z}_2[x]/m(x)$ by $\mathbb{Z}_2[\alpha]$. It turns out that α is a primitive element of $\mathbb{Z}_2[\alpha]$. In fact, the elements of $\mathbb{Z}_2[\alpha]$ may be described as powers of α as in Table 1.

Table 1

$GF(8) = \mathbb{Z}_2[x]/(x^3 + x + 1) = \mathbb{Z}_2[\alpha]$	
0	0
1	1
α	α
α^2	$\alpha + 1 = \alpha^3$
$\alpha^3 = \alpha + 1$	α^2
$\alpha^4 = \alpha^2 + \alpha$	$\alpha^2 + 1 = \alpha^6$
$\alpha^5 = \alpha^2 + \alpha + 1$	$\alpha^2 + \alpha = \alpha^4$
$\alpha^6 = \alpha^2 + 1$	$\alpha^2 + \alpha + 1 = \alpha^5$
$\alpha^7 = 1$	

Our first code, like one in Chapter I-13, sends out coded words of length 7 with 4 information digits.

Code III (codes I and II were the Hamming codes of Chapter I-13). For the coding, let (a, b, c, d) be the information digits which we wish to transmit. Form the polynomial

$$C_I(x) = ax^6 + bx^5 + cx^4 + dx^3.$$

Divide $C_I(x)$ by $m(x) = x^3 + x + 1$:

$$C_I(x) = m(x)q(x) + C_R(x),$$

where the remainder $C_R(x)$ has degree $< \deg(m(x))$. Then

$$C_R(x) = rx^2 + sx + t$$

for some r, s, t in $\mathbb{Z}_2$. Since $-1 = 1$ in $\mathbb{Z}_2$, we get

$$m(x)q(x) = C_I(x) + C_R(x)$$

$$= ax^6 + bx^5 + cx^4 + dx^3 + rx^2 + sx + t$$

$$= C(x),$$

and so the polynomial $C(x)$ has the important property that when evaluated at α, a root of $m(x)$, we have

$$C(\alpha) = m(\alpha)q(\alpha) = 0.$$

The coded word is $\mathbf{C} = (a, b, c, d, r, s, t)$, the coefficients of the polynomial $C(x)$. Then $\mathbf{C}$ has 4 information digits. It is characterized by the property that it corresponds to the unique polynomial of degree 6 with given top degree coefficients a, b, c, d and having α as a root.

For the decoding, suppose the receiver receives (A, B, C, D, R, S, T). Form the polynomial

$$R(x) = Ax^6 + Bx^5 + Cx^4 + Dx^3 + Rx^2 + Sx + T.$$

Suppose that at most one error occurred. Then $C(x) - R(x) = E(x)$ is either the zero polynomial or consists of a single term, x^e, whose coefficient in $R(x)$ was erroneous. To decide, look at $R(\alpha)$:

Case 0. If $R(\alpha) = 0$, then, since $C(\alpha) = 0$, $E(\alpha) = 0$ and no errors occurred.

Case 1. If $R(\alpha) = \alpha^e$, then, since $C(\alpha) = 0$, $E(\alpha) = \alpha^e$ and one error occurred, at the coefficient of x^e.

Thus by evaluating $R(x)$ at $x = \alpha$, we can decide whether an error occurred and, if so, where, so that the error can be corrected.

If two or more errors occurred, and we think at most one error occurred, we would be misled. But if more than one error is very unlikely to occur, this is a good code.

EXAMPLE . To code $(1, 1, 0, 1)$, take $x^6 + x^5 + x^3$ and divide it by $x^3 + x + 1$.

$$
\begin{array}{r}
1111 \\
\hline
1011\,)\,1101000 \\
1011 \\
\hline
1100 \\
1011 \\
\hline
1110 \\
1011 \\
\hline
1010 \\
1011 \\
\hline
1
\end{array}
$$

The remainder is $C_R(x) = 1$. So $C(x) = x^6 + x^5 + x^3 + 1$. (Note that $C(\alpha)$ $= \alpha^6 + \alpha^5 + \alpha^3 + 1 = (\alpha^2 + 1) + (\alpha^2 + \alpha + 1) + (\alpha + 1) + 1 = 0$.)
Send $(1, 1, 0, 1, 0, 0, 1)$.
If we receive $(1, 1, 0, 1, 1, 0, 1)$, we find $R(\alpha)$, where

$$R(x) = x^6 + x^5 + x^3 + x^2 + 1.$$

Then, from Table 1, we get $R(\alpha)$ as follows.

$$
\begin{array}{rl}
1 = & 1 \\
+ \alpha^2 = & \alpha^2 \\
+ \alpha^3 = & \alpha + 1 \\
+ \alpha^5 = & \alpha^2 + \alpha + 1 \\
+ \alpha^6 = & \alpha^2 \qquad + 1 \\
\hline
R(\alpha) = & \alpha^2 .
\end{array}
$$

So we change the coefficient of x^2 in $R(x)$, and our corrected word is $(1, 1, 0, 1, \underline{\underline{0}}, 0, 1)$.

E1. Code the messages
    (a) $(1, 0, 0, 0)$,
    (b) $(0, 1, 1, 0)$,
    (c) $(1, 1, 1, 0)$.

E2. Decode the received words
    (a) $(1, 1, 1, 0, 0, 0, 1)$,
    (b) $(1, 0, 1, 1, 0, 1, 1)$,
    (c) $(0, 1, 0, 1, 0, 1, 0)$.

Our second code is set up similarly to the first, but needs a bigger field.
Let $m(x) = x^4 + x + 1$ in $\mathbb{Z}_2[x]$. It is not hard to see that $m(x)$ is irreducible, and so $\mathbb{Z}_2[x]/m(x)$ is a field which we shall call $\mathbb{Z}_2[\alpha]$ or $GF(16)$, the former because we shall let $[x]_{m(x)} = \alpha$, the latter because the

field has 16 elements in it. It turns out that α is a primitive element of $GF(16)$. Thus every element of $\mathbb{Z}_2[\alpha]$ is a power of α. This is exhibited in Table 2, which like Table 1 in the example above, will supply convenient data for decoding.

Table 2

$GF(16) = \mathbb{Z}_2[x]/(x^4 + x + 1) = \mathbb{Z}_2[\alpha]$	
0	0
1	1
α	α
α^2	$\alpha + 1 = \alpha^4$
α^3	α^2
$\alpha^4 = \alpha + 1$	$\alpha^2 + 1 = \alpha^8$
$\alpha^5 = \alpha^2 + \alpha$	$\alpha^2 + \alpha = \alpha^5$
$\alpha^6 = \alpha^3 + \alpha^2$	$\alpha^2 + \alpha + 1 = \alpha^{10}$
$\alpha^7 = \alpha^3 + \alpha + 1$	α^3
$\alpha^8 = \alpha^2 + 1$	$\alpha^3 + 1 = \alpha^{14}$
$\alpha^9 = \alpha^3 + \alpha$	$\alpha^3 + \alpha = \alpha^9$
$\alpha^{10} = \alpha^2 + \alpha + 1$	$\alpha^3 + \alpha^2 = \alpha^6$
$\alpha^{11} = \alpha^3 + \alpha^2 + \alpha$	$\alpha^3 + \alpha + 1 = \alpha^7$
$\alpha^{12} = \alpha^3 + \alpha^2 + \alpha + 1$	$\alpha^3 + \alpha^2 + 1 = \alpha^{13}$
$\alpha^{13} = \alpha^3 + \alpha^2 + 1$	$\alpha^3 + \alpha^2 + \alpha = \alpha^{11}$
$\alpha^{14} = \alpha^3 + 1$	$\alpha^3 + \alpha^2 + \alpha + 1 = \alpha^{12}$
$\alpha^{15} = 1$	

We use $GF(16)$ to construct a code which corrects two errors. The idea for coding is to use for code words vectors of length 15 which are the coefficients of polynomials of degree 14 in $\mathbb{Z}_2[\alpha]$ having α and α^3 as roots.

We know that the polynomial of smallest degree in $\mathbb{Z}_2[x]$ with α as a root is $m(x) = x^4 + x + 1$. We have to find the minimal polynomial over $\mathbb{Z}_2$ of α^3. To do this, observe that every element of $\mathbb{Z}_2[\alpha]$ is a $\mathbb{Z}_2$-linear combination of $1, \alpha, \alpha^2, \alpha^3$. So in the equation

$$0 = a + b\alpha^3 + c\alpha^6 + d\alpha^9 + e\alpha^{12},$$

if we write $\alpha^6, \alpha^9, \alpha^{12}$ in terms of $1, \alpha, \alpha^2, \alpha^3$ by Table 2 we get four equations in the five unknowns a, b, c, d, e in $\mathbb{Z}_2$; any solution will give us the coefficients of a polynomial in $\mathbb{Z}_2[x]$ with α^3 as a root.

Doing so, we get

$$0 = a + b\alpha^3 + c(\alpha^3 + \alpha^2) + d(\alpha^3 + \alpha) + e(\alpha^3 + \alpha^2 + \alpha + 1).$$

Collecting coefficients gives the equations

$$a + e = 0,$$
$$d + e = 0,$$
$$c + e = 0,$$
$$b + c + d + e = 0,$$

so $a = b = c = d = e$, and the only nonzero solution is when they are all equal to 1. The only nonzero polynomial of degree $\leqslant 4$ satisfied by α^3 is therefore

$$1 + x + x^2 + x^3 + x^4 = m_3(x),$$

which must be irreducible.

The polynomial of smallest degree with both α and α^3 as roots is the least common multiple of $m(x)$ and $m_3(x)$; since they are both irreducible in $\mathbb{Z}_2[x]$, that is the same as their product $m_{13}(x) = m(x)m_3(x) = x^8 + x^7 + x^6 + x^4 + 1$.

Since $m_{13}(x)$ has degree 8 we are going to send out words of length 15 with 7 information digits.

Note the important fact (see Theorem 1 of Chapter II-10) that if $p(x)$ is a polynomial in $\mathbb{Z}_2[x]$, then $p(x^2) = (p(x))^2$. Thus if α is a root of $m(x)$, so are α^2 and α^4.

E3. Prove (using the binomial theorem and induction on the degree) that any polynomial $p(x)$ with coefficients in $\mathbb{Z}_2$ has the property that $(p(x))^2 = p(x^2)$.

Code IV. We encode as follows. Let $(a_{14}, a_{13}, \ldots, a_8)$ be the information word. Let $C_I(x) = a_{14}x^{14} + a_{13}x^{13} + \cdots + a_8x^8$. Divide $C_I(x)$ by $m_{13}(x)$:

$$C_I(x) = m_{13}(x)q(x) - C_R(x),$$

where $C_R(x)$, the remainder, has degree $\leqslant 7$,

$$C_R(x) = a_7x^7 + \cdots + a_1x + a_0.$$

Then

$$C(x) = a_{14}x^{14} + \cdots + a_1x + a_0 = C_R(x) + C_I(x) = m_{13}(x)q(x),$$

has α and α^3 as roots, and (since the remainder in the division algorithm is unique) is the unique polynomial of degree $\leqslant 14$ with given coefficients $a_{14}, \ldots, a_8$ and with α and α^3 as roots.

Let $\mathbf{C} = (a_{14}, a_{13}, \ldots, a_0)$. Send $\mathbf{C}$. [Note the interplay between a polynomial as a formal sum of powers of x, and as an n-tuple of coefficients. Recall Chapter II-1.]

For decoding, suppose we receive $\mathbf{R}$. Set $\mathbf{R} = \mathbf{C} + \mathbf{E}$, where $\mathbf{E}$ is the error vector. Think of $\mathbf{R}, \mathbf{C}, \mathbf{E}$ as polynomials. Then, since $m_{13}(x)$ divides $C(x)$:

$R(\alpha) = E(\alpha)$, because $m_{13}(\alpha) = 0$;
$R(\alpha^2) = E(\alpha^2)$, because $m_{13}(\alpha) = 0$, hence $m_{13}(\alpha^2) = (m_{13}(\alpha))^2 = 0$;
$R(\alpha^3) = E(\alpha^3)$, because $m_{13}(\alpha^3) = 0$.

Consider the polynomial

$$R(\alpha)x^2 + R(\alpha^2)x + (R(\alpha^3) + R(\alpha)R(\alpha^2)) = P(x).$$

We assume $E(x) = R(x) - C(x)$ has at most two nonzero terms in it.

Case 0. If $E(x) = 0$, then $p(x) = 0$.
Case 1. If $E(x) = x^e$, then

$$P(x) = \alpha^e x^2 + \alpha^{2e} x + (\alpha^{3e} + \alpha^{2e}\alpha) = (\alpha^e x)(x + \alpha^e).$$

Case 2. If $E(x) = x^e + x^f$, then

$$P(x) = (\alpha^e + \alpha^f)x^2 + (\alpha^{2e} + \alpha^{2f})x + (\alpha^{3e} + \alpha^{3f}) + (\alpha^{2e} + \alpha^{2f})(\alpha^e + \alpha^f)$$
$$= (\alpha^e + \alpha^f)\left[x^2 + (\alpha^e + \alpha^f)x + \alpha^e\alpha^f \right]$$
$$= (\alpha^e + \alpha^f)\left[(x + \alpha^e)(x + \alpha^f) \right].$$

Thus, if there are roots of $P(x)$, then the roots are the powers of α corresponding to the errors. So we receive **R**, compute $P(x)$, and find its roots.

This code therefore corrects two errors.

EXAMPLE. We want to send $(1, 1, 0, 1, 1, 0, 1)$. So we set

$$C_I(x) = x^{14} + x^{13} + x^{11} + x^{10} + x^8.$$

Divide $C_I(x)$ by $m_{13}(x) = x^8 + x^7 + x^6 + x^4 + 1$.

```
                        1010110
            111010001 ) 110110100000000
                        111010001
                        ─────────
                         110010100
                         111010001
                         ─────────
                          100010100
                          111010001
                          ─────────
                           110001010
                           111010001
                           ─────────
                            10110110
```

So $C_R(x) = x^7 + x^5 + x^4 + x^2 + x$ and the coded word is

$$(1, 1, 0, 1, 1, 0, 1, 1, 0, 1, 1, 0, 1, 1, 0).$$

Suppose we receive

$$(1, 1, 0, 1, 1, 1, 1, 0, 0, 1, 1, 0, 1, 1, 0).$$

We compute $R(\alpha)$, $R(\alpha^2)$, $R(\alpha^3)$:

$R(\alpha) = \alpha^{14} + \alpha^{13} + \alpha^{11} + \alpha^{10} + \alpha^9 + \alpha^8 + \alpha^5 + \alpha^4 + \alpha^2 + \alpha$;

$R(\alpha^2) = (R(\alpha))^2$ (since the field has characteristic 2);

$R(\alpha^3) = \alpha^{42} + \alpha^{39} + \alpha^{33} + \alpha^{30} + \alpha^{27} + \alpha^{24} + \alpha^{15} + \alpha^{12} + \alpha^6 + \alpha^3.$

Using Table 2, we replace all these terms by polynomials in α of degree ≤ 3, first in $R(\alpha)$.

Coefficients of	α^3	α^2	α	1
α^{14}	1			1
$+\alpha^{13}$	1	1		1
α^{11}	1	1	1	
α^{10}		1	1	1
α^9	1		1	
α^8		1		1
α^5		1	1	
α^4			1	1
α^2		1		
α			1	
$R(\alpha)$	0	0	0	1

$R(\alpha) = 1$. Then $R(\alpha)^2 = R(\alpha)^2 = 1$. Since $\alpha^{15} = 1$,

$$R(\alpha^3) = \alpha^{42} + \alpha^{39} + \alpha^{33} + \alpha^{30} + \alpha^{27} + \alpha^{24} + \alpha^{15} + \alpha^{12} + \alpha^6 + \alpha^3$$
$$= \alpha^{12} + \alpha^9 + \alpha^3 + 1 + \alpha^{12} + \alpha^9 + 1 + \alpha^{12} + \alpha^6 + \alpha^3$$
$$= \alpha^{12} + \alpha^6 = (\alpha^3 + \alpha^2 + \alpha + 1) + (\alpha^3 + \alpha^2) = \alpha + 1.$$

So

$$P(x) = x^2 + x + \left(\frac{\alpha+1}{1} + 1\right) = x^2 + x + \alpha.$$

To find the roots of $P(x)$ we use trial and error, using Table 2 to express everything in terms of $1, \alpha, \alpha^2, \alpha^3$.

if $x =$	$x^2 =$	(By Table 2) $x =$	$x^2 =$	$x^2 + x + \alpha =$
0	0	0	0	α
1	1	1	1	α
α	α^2	α	α^2	α^2
α^2	α^4	α	$\alpha + 1$	$\alpha + 1$
α^3	α^6	α^3	$\alpha^3 + \alpha^2$	$\alpha^2 + \alpha$
α^4	α^8	$\alpha + 1$	$\alpha^2 + 1$	α^2
α^5	α^{10}	$\alpha^2 + \alpha$	$\alpha^2 + \alpha + 1$	$\alpha + 1$
α^6	α^{12}	$\alpha^3 + \alpha^2$	$\alpha^3 + \alpha^2 + \alpha + 1$	1
α^7	α^{14}	$\alpha^3 + \alpha + 1$	$\alpha^3 + 1$	0

(When we hit 0 in the last column, we stop.) Then $(x^2 + x + \alpha) = (x + \alpha^7)(x + \alpha^e)$ for some e. But then $\alpha^7\alpha^e = \alpha^{16} = \alpha$, and $e = 9$, so the two errors are at x^7 and x^9.

E4. Code
    (a) (1, 1, 1, 0, 0, 1, 1),
    (b) (0, 0, 1, 1, 0, 1, 1),
    (c) (1, 0, 1, 0, 1, 0, 1).

E5. Decode
  (a) (011, 001, 011, 101, 100),
  (b) (011, 110, 101, 110, 110),
  (c) (100, 100, 100, 100, 100).

To generalize the ideas of these codes in order to get codes which correct more than two errors, it is useful to use the notion of row rank of a matrix (Section I-9F). So for the remainder of this chapter we shall assume some acquaintance with Chapter I-9.

To introduce the matrix techniques, we describe anew how to decode code IV. We decoded above by using the polynomial $P(x)$, which appeared in a rather mysterious manner. We shall here use, instead, a matrix.

Let $R(\alpha^i) = S_i$ for $i = 1, 2, 3, 4$.

Let

$$S = \begin{pmatrix} S_1 & S_2 \\ S_2 & S_3 \end{pmatrix}.$$

Using this 2×2 matrix with entries in $GF(16)$ we can correct 0, 1, or 2 errors as follows:

Case 0. No errors. Then $E(x) = 0$, so $S = 0$ (and in particular, the row rank of S is 0).

Case 1. One error. Then $E(x) = x^i$ for some i, $0 \le i \le 14$; thus

$$S = \begin{pmatrix} \alpha^i & \alpha^{2i} \\ \alpha^{2i} & \alpha^{3i} \end{pmatrix}.$$

The second row is α^i times the first row. So S has (row) rank 1. We can correct the received vector R by changing the digit corresponding to x^i (where i is known from $R(\alpha) = \alpha^i$).

Case 2. Two errors. Then

$$S = \begin{pmatrix} \alpha^i + \alpha^j & \alpha^{2i} + \alpha^{2j} \\ \alpha^{2i} + \alpha^{2j} & \alpha^{3i} + \alpha^{3j} \end{pmatrix}$$

$$= \begin{pmatrix} 1 & 1 \\ \alpha^i & \alpha^j \end{pmatrix} \begin{pmatrix} \alpha^i & 0 \\ 0 & \alpha^j \end{pmatrix} \begin{pmatrix} 1 & \alpha^i \\ 1 & \alpha^j \end{pmatrix}.$$

so is nonsingular, hence has row rank 2.

We see that *the rank of S = the number of errors.*

With two errors, $E(x) = x^i + x^j$. We find i and j in two steps, as follows.

(a) Solve

$$\begin{pmatrix} S_1 & S_2 \\ S_2 & S_3 \end{pmatrix} \begin{pmatrix} \sigma_2 \\ \sigma_1 \end{pmatrix} = \begin{pmatrix} S_3 \\ S_4 \end{pmatrix},$$

in $GF(16)$, for σ_1 and σ_2.

(b) Solve $x^2 + \sigma_1 x + \sigma_2 = 0 = p_2(x)$ in $GF(16)$, for x.

We can solve (a) because S is nonsingular, so in fact we can use the inverse of S, which for 2×2 matrices is easy to write down. There will be a unique solution.

E6. Show that in fact $\sigma_1 = R(\alpha)$, $\sigma_2 = (R(\alpha^3)/R(\alpha)) + R(\alpha^2)$, so that $R(\alpha)p_2(x) = P(x)$, the mysterious polynomial used in decoding code IV.

The reason we want to solve (b) is that the roots in $GF(16)$ of $x^2 + \sigma_1 x + \sigma_2 = 0$ turn out to be α^i and α^j, the powers of α corresponding to where the errors occur in $R(x)$. To see this, we examine the coefficients of $(x - \alpha^i)(x - \alpha^j) = x^2 + \tau_1 x + \tau_2$ in $GF(16)[x]$. They satisfy

$$\alpha^{2i} + \tau_1 \alpha^i + \tau_2 = 0,$$

$$\alpha^{2j} + \tau_1 \alpha^j + \tau_2 = 0. \tag{3}$$

Multiplying the equations (3) by α^i and α^j respectively gives

$$\alpha^{3i} + \tau_1 \alpha^{2i} + \tau_2 \alpha^i = 0,$$

and

$$\alpha^{3j} + \tau_1 \alpha^{2j} + \tau_2 \alpha^j = 0.$$

Adding, we get

$$S_3 + \tau_1 S_2 + \tau_2 S_1 = 0.$$

Similarly, multiplying the two equations (3) by α^{2i} and α^{2j} respectively and adding, we get

$$S_4 + \tau_1 S_3 + \tau_2 S_2 = 0.$$

Putting these last two equations together in matrix form (noting $+ = -$ in $\mathbb{Z}_2$), gives

$$\begin{pmatrix} S_1 & S_2 \\ S_2 & S_3 \end{pmatrix} \begin{pmatrix} \tau_2 \\ \tau_1 \end{pmatrix} = \begin{pmatrix} S_3 \\ S_4 \end{pmatrix}.$$

Since there was a unique solution of this equation, therefore $\tau_1 = \sigma_1$, $\tau_2 = \sigma_2$.

The way we solve $x^2 + \sigma_1 x + \sigma_2 = 0$ is by "brute force," as we did before. There are only two roots and only 15 candidates, so start trying $x = 1, \alpha, \alpha^2, \alpha^3, \ldots, \alpha^{14}$ until two of them solve the equation.

To sum up: this code sends words of length 15, of which 7 are information digits. The receiver can decide whether 0, 1, or 2 errors occurred, and correct them.

E7. Using the matrix S, decode
    (a) (110, 001, 110, 010, 110),
    (b) (101, 011, 110, 010, 110),
    (c) (110, 010, 111, 110, 110).

E8. (This and the following exercise assume understanding of the Hamming codes I and II of Chapter I-13.) Is the two-error-correcting code, code IV, a Hamming code of the kind described in Chapter I-13? Is there a coding matrix **H** of the type used in codes I and II such that if **C** is a 15-tuple, with given first 7 digits, then **C** is a code word if and only if $\mathbf{HC} = \mathbf{0}$? If there is such an **H**, write it down.

E9. Find a matrix **H** with the property that given a vector **v** which is the sum of two columns of **H**, **v** can be written as such a sum in exactly one way.

Our next example is a code which corrects three errors in words of length 15. We shall use $GF(16)$ as in our last example, and some matrix theory.

We find the polynomial in $Z_2[x]$ of smallest degree which has α, α^3, and α^5 as roots. It turns out to be $m_{135}(x) = m_{13}(x)m_5(x)$, where $m_5(x) = x^2 + x + 1$ is the minimal polynomial of α^5. So m_{135} has degree 10, and we are allowed 5 information digits, $(a_{14}, \ldots, a_{10})$. Note that $m_{135}(x)$ has α^2, α^4, and α^6 as roots also, by Exercise E3.

Code V. For coding, let $C_I(x) = a_{14}x^{14} + \cdots + a_{10}x^{10}$. Divide $C_I(x)$ by $m_{135}(x)$ to get a remainder $C_R(x)$ of degree $\leqslant 9$; then $C(x) = C_I(x) + C_R(x)$ is a multiple of $m_{135}(x)$ and is the unique polynomial of degree $\leqslant 14$ with given $a_{14}, \ldots, a_{10}$ and having α, α^2, α^3, α^4, α^5, α^6 as roots.
We send **C**.

For decoding, the receiver receives **R**, and evaluates $R(x)$ at α^i, $i = 1, \ldots, 6$. Since $C(\alpha^i) = 0$, $i = 1, \ldots, 6$, $R(\alpha^i) = E(\alpha^i) = S_i$, $i = 1, \ldots, 6$. We consider the matrix

$$\mathbf{S} = \begin{bmatrix} S_1 & S_2 & S_3 \\ S_2 & S_3 & S_4 \\ S_3 & S_4 & S_5 \end{bmatrix}.$$

Case 0. No errors. $E(x) = 0$. Then $\mathbf{S} = \mathbf{0}$.
Case 1. One error. $E(x) = x^i$. Then

$$\mathbf{S} = \begin{bmatrix} \alpha^2 & \alpha^{2i} & \alpha^{3i} \\ \alpha^{2i} & \alpha^{3i} & \alpha^{4i} \\ \alpha^{3i} & \alpha^{4i} & \alpha^{5i} \end{bmatrix}$$

has (row) rank 1, and the single error is at the ith spot.
Case 2. Two errors. $E(x) = x^i + x^j$. Then

$$\mathbf{S} = \begin{bmatrix} \alpha^i + \alpha^j & \alpha^{2i} + \alpha^{2j} & \alpha^{3i} + \alpha^{3j} \\ \alpha^{2i} + \alpha^{2j} & \alpha^{3i} + \alpha^{3j} & \alpha^{4i} + \alpha^{4j} \\ \alpha^{3i} + \alpha^{3j} & \alpha^{4i} + \alpha^{4j} & \alpha^{5i} + \alpha^{5j} \end{bmatrix}$$

and the row space of S is the two-dimensional space over $GF(16)$ spanned by $(\alpha^i, \alpha^{2i}, \alpha^{3i})$ and $(\alpha^j, \alpha^{2j}, \alpha^{3j})$. We can let $S(2)$ be the invertible matrix

$$S(2) = \begin{pmatrix} S_1 & S_2 \\ S_2 & S_3 \end{pmatrix}$$

and decode as in code IV.

Case 3. Three errors. $E(x) = x^i + x^j + x^k$. Then

$$S = \begin{bmatrix} \alpha^2 + \alpha^j + \alpha^k & \alpha^{2i} + \alpha^{2j} + \alpha^{2k} & \alpha^{3i} + \alpha^{3j} + \alpha^{3k} \\ \alpha^{2i} + \alpha^{2j} + \alpha^{2k} & \alpha^{3i} + \alpha^{3j} + \alpha^{3k} & \alpha^{4i} + \alpha^{4j} + \alpha^{4k} \\ \alpha^{3i} + \alpha^{3j} + \alpha^{3k} & \alpha^{4i} + \alpha^{4j} + \alpha^{4k} & \alpha^{5i} + \alpha^{5j} + \alpha^{5k} \end{bmatrix}$$

is invertible, so has rank 3.

For these matrices S, the rank and the number of errors can be computed by looking at the square matrices in the upper left-hand corners. Let

$$S(1) = (S_1), \qquad S(2) = \begin{pmatrix} S_1 & S_2 \\ S_2 & S_3 \end{pmatrix}, \qquad S(3) = S.$$

Then

$$\text{rank } S = 0 \quad \text{if det } S(1) = 0,$$
$$\text{rank } S = 1 \quad \text{if det } S(1) \neq 0 \quad \text{but det } S(2) = 0,$$
$$\text{rank } S = 2 \quad \text{if det } S(2) \neq 0 \quad \text{but det } S(3) = 0,$$
$$\text{rank } S = 3 \quad \text{if det } S(3) \neq 0.$$

E10. If $\alpha^i, \alpha^j, \alpha^k$ are all distinct, show that S is nonsingular. Generalize to $n \times n$ S with $\alpha^{i_1}, \alpha^{i_2}, \ldots, \alpha^{i_n}$ all distinct.

Since S is invertible, we can solve uniquely the equation

$$S \begin{bmatrix} \sigma_3 \\ \sigma_2 \\ \sigma_1 \end{bmatrix} = \begin{bmatrix} S_4 \\ S_5 \\ S_6 \end{bmatrix} \quad \text{for} \quad \sigma_1, \sigma_2, \sigma_3,$$

and then find the roots in $GF(16)$ of

$$x^3 + \sigma_1 x^2 + \sigma_2 x + \sigma_3 = 0.$$

The three roots of this equation will be $\alpha^i, \alpha^j, \alpha^k$, the powers of α corresponding to where the three errors are.

E11. Fill in the details of this last sentence.

So the receiver can decode by

(1) finding the row rank of the matrix S, which tells the number of errors (up to 3),
(2) finding where the errors are by the techniques described above.

These codes generalize. Unfortunately, when working with words of length 15, if a code which corrects 4 errors is desired, one is allowed only 1 information digit, so the repetition code described in Chapter I-13 would work as well. But if longer words are permitted, $GF(32)$ can be used to send out coded words of length 31, with 10 information digits and with up to 5 errors correctable, or with 6 information digits and with up to 7 errors correctible. Or $GF(64)$ can be used to send out coded words of length 63 as follows:

30 information digits, up to 6 errors correctable;
24 information digits, up to 7 errors correctable;
18 information digits, up to 10 errors correctable;
16 information digits, up to 11 errors correctable;
10 information digits, up to 13 errors correctable;
 7 information digits, up to 15 errors correctable.

All of these codes are designed the same way as our last code, which corrected 3 errors.

E12. How would you go about designing one of these codes with $GF(32)$ or $GF(64)$?

Good expositions of algebraic coding theory include Berlekamp (1968), Peterson and Weldon (1972), and Blake and Mullen (1975).

E13. The coder, using code IV, is sending out 15-bit words, coefficients of a polynomial $C(x)$ in $\mathbb{Z}_2[x]$ of degree 14 where, if $\alpha^4 + \alpha + 1 = 0$, then $C(\alpha) = C(\alpha^2) = C(\alpha^3) = 0$. You receive $R(x) = x^{14} + x^{11} + x^9 + x^8 + x^4 + x^2 + x + 1$, and assume that at most two errors were made. What was $C(x)$?

E14. In $\mathbb{Z}_2[x]/(x^4 + x + 1) = \mathbb{Z}_2[\alpha] = GF(16)$, solve

$$\begin{pmatrix} \alpha^3 + \alpha^4 & \alpha^6 + \alpha^8 \\ \alpha^6 + \alpha^8 & \alpha^9 + \alpha^{12} \end{pmatrix}\begin{pmatrix} \sigma_2 \\ \sigma_1 \end{pmatrix} = \begin{pmatrix} \alpha^9 & \alpha^{12} \\ \alpha^{12} & \alpha^{16} \end{pmatrix}$$

for α_1, σ_2; find the roots in $\mathbb{Z}_2[\alpha]$ of $x^2 + \sigma_1 x + \sigma_2 = 0$

E15. Solve in $GF(16)$

$$\begin{bmatrix} \alpha^2 + \alpha^5 + \alpha^{10} & \alpha^4 + \alpha^{10} + \alpha^{20} & \alpha^6 + \alpha^{15} + \alpha^{30} \\ \alpha^4 + \alpha^{10} + \alpha^{20} & \alpha^6 + \alpha^{15} + \alpha^{30} & \alpha^8 + \alpha^{20} + \alpha^{40} \\ \alpha^6 + \alpha^{15} + \alpha^{30} & \alpha^8 + \alpha^{20} + \alpha^{40} & \alpha^{10} + \alpha^{25} + \alpha^{50} \end{bmatrix}\begin{pmatrix} \sigma_3 \\ \sigma_2 \\ \sigma_1 \end{pmatrix}$$

$$= \begin{bmatrix} \alpha^8 + \alpha^{20} + \alpha^{40} \\ \alpha^{10} + \alpha^{25} + \alpha^{50} \\ \alpha^{12} + \alpha^{30} + \alpha^{60} \end{bmatrix}$$

for $\sigma_1, \sigma_2, \sigma_3$. Verify that

$$x^3 + \sigma_1 x^2 + \sigma_2 x + \sigma_3 = (x - \alpha^2)(x - \alpha^5)(x - \alpha^{10}).$$

E16. Find the minimal polynomial over $\mathbb{Z}_2$ of α^7, where $\alpha^4 + \alpha + 1 = 0$.

E17. Does there exist an example of an irreducible polynomial $p(x)$ in $\mathbb{Z}_2[x]$ such that in $\mathbb{Z}_2[x]/p(x) = \mathbb{Z}_2[\alpha]$, α is not a primitive element?

E18. Using code IV, code $\mathbf{C}_i = (1110001)$

E19. Using code V,
  (a) code $\mathbf{C}_I = (1011)$,
  (b) decode (101100110011100),
  (c) decode (101000010011110).

E20. Do there exist received vectors $\mathbf{R}$ in code IV for which it can be determined that more than two errors occurred?

E21. Using code IV, decode
  (a) (110110000101101),
  (b) (111110000101101),
  (c) (111110000111101),
  (d) (111110000111111),
  (e) (011110000111111),
  (f) (001110000111111).

E22. Define, by analogy to code IV, a double error-correcting code using $GF(8)$. How many information digits will it have?

Isomorphisms, I 10

A. Definitions

This chapter is about functions from one ring to another.

When we define a function in algebra, we specify three things: a set R, the *domain* of the function; a set S, the *range* of the function; and the function f, which assigns to each element r of R a unique element $f(r)$ of S. When we call S the range, we mean only that S acts as a target for f, *not* that every element s of S has the form $s = f(r)$ for some r in R. The set of elements s of S with $s = f(r)$ for some r in R is called the *image* of f.

If R and S are commutative rings, then we are particularly interested in functions f from R to S which preserve the ring properties of R and S. How do we say this more precisely?

A ring is a set with operations $+$ (addition), $\cdot$ (multiplication), and $-$ (negation), and with distinguished elements 0 and 1, satisfying a certain set of axioms (see Chapter I-8). A function f from a ring R to a ring S is a *ring homomorphism* if

(i) $f(a + b) = f(a) + f(b)$ in S for all a, b in R,
(ii) $f(a \cdot b) = f(a) \cdot f(b)$ in S for all a, b in R, and
(iii) $f(1_R) = 1_S$, where 1_R, 1_S are the multiplicative identities of R, S, respectively.

It turns out that if f satisfies the conditions (i)–(iii), then

(iv) $f(0_R) = 0_S$:

for

$$f(b) = f(0_R + b) = f(0_R) + f(b),$$

and adding $-f(b)$ to both sides gives

$$0_S = f(0_R) + 0_S = f(0_R).$$

Also,

(v) $f(-a) = -f(a)$:

for

$$0_S = f(0_R) = f(a + (-a)) = f(a) + f(-a),$$

so that $f(-a)$ is a negative for $f(a)$ in S, and since negatives are unique, $f(-a) = -f(a)$. Also

(vi) if a has an inverse a^{-1} in R, so that $a \cdot a^{-1} = 1$, then $f(a)^{-1} = f(a^{-1})$:

the proof is an easy consequence of the fact that inverses in a commutative ring are unique.

E1. Prove that if f is a ring homomorphism, then $f(a^{-1}) = f(a)^{-1}$.

Thus a function which preserves addition and multiplication and sends identity to identity also sends zero to zero, negatives to negatives, and inverses to inverses.

A ring homomorphism f is 1–1 if it is 1–1 as a function, that is, $f(a) = f(b)$ implies $a = b$. It is not hard to show that f is 1–1 if 0_R, the zero element of R, is the only element a in R with $f(a) = 0_S$.

E2. Prove this.

A ring homomorphism f from R to S is *onto* (or surjective) if it is onto as a function, that is, the image of f in S is all of S, that is, for every s in S there is some r in R with $f(r) = s$. Note that for the concept "onto" to make sense, the range S of the function f must be included as part of the definition. See Example 2 below.

A ring homomorphism f is called an *isomorphism* if f is 1–1 and onto.

If $f : R \to S$ is an isomorphism, then R and S are *isomorphic*, and we shall write $R \cong S$. $R \cong S$ means that even if R and S look different, nonetheless in terms of their algebraic structure they are not (see Example 12, below).

The rest of this chapter is devoted to examples of ring homomorphisms.

EXAMPLE 1. The most trivial example is the *identity homomorphism*. Let R be a commutative ring, and define a function $i : R \to R$ by $i(r) = r$ for all r in R. It is very easy to check that i is a ring homomorphism and is 1–1 and onto.

EXAMPLE 2. Let S be a commutative ring, and let R be a subset of S which is also a commutative ring with the same addition and multiplication that

S has, and such that the identity and zero elements of S are the identity and zero elements for R respectively. For example, let S be the complex numbers and R be the real numbers, or the rational numbers. We can then define a homomorphism i from R to S, called the *inclusion map*, by $i(r) =$ the element r thought of as being in S.

The only difference between Example 2 and Example 1 is that in Example 1 the range of i is the ring R, whereas in Example 2 the range of i is the ring S. Thus if S is a bigger set than R, then the function of Example 1 is onto, while the function of Example 2 is not onto.

These two examples illustrate a difference between functions in algebra and functions in beginning calculus. In beginning calculus, all the functions have the same range, namely the real numbers (depicted geometrically as the y-axis); in algebra there are many different rings and hence many different sets for functions to map into, so when we introduce a function it is important to specify both the domain and the range of the function as part of the definition of the function.

B. Examples Involving $\mathbb{Z}$

Here is a special case of Example 2:

EXAMPLE 3. Let $f : \mathbb{Z} \to \mathbb{Q}$ by $f(a) = a/1$. Since $(a/1) + (b/1) = (a + b)/1$, $(a/1) \cdot (b/1) = (ab)/1$, and $1 = (1/1)$, f is a ring homomorphism, which is 1–1 but not onto.

Here is a function which is not 1–1:

EXAMPLE 4. Let $f : \mathbb{Z} \to \mathbb{Z}_n$ by $f(a) = [a]$. Since $[a]_n + [b]_n = [a + b]_n$, $[a]_n \cdot [b]_n = [ab]_n$, and $[1]_n$ is the multiplicative identity of $\mathbb{Z}_n$, f is a homomorphism. (In fact, addition and multiplication were defined in $\mathbb{Z}_n$ so as to make f a homomorphism!) One sees easily that f is onto but not 1–1.

The next example arose in Chapter II-10, when we defined the characteristic of a field.

EXAMPLE 5. Let F be a field, and let $f : \mathbb{Z} \to F$ be defined as follows:

$f(0) = 0_F = 0 \cdot 1_F$;
if $n > 0$, $f(n) = 1_F + 1_F + 1_F + \cdots + 1_F$ (n times), $= n \cdot 1_F$;
if $n < 0$, $f(n) = - f(-n) = -(-n) \cdot 1_F = n \cdot 1_F$.

This function is easily seen to be a homomorphism.

In Chapter II-10 we defined the characteristic of a field F to be 0 if this function f is 1–1; if f is not 1–1, then the smallest natural number p such that $f(p) = 0_F$ is the characteristic of F. We showed that if F is a field, then p must be prime.

As an exercise in properties of homomorphisms, we prove:

Proposition 1. *Let F be a field, and $f : \mathbb{Z} \to F$ the function of Example 5. If f is onto, then f cannot be 1–1; in fact, F has characteristic p for some prime p, and F has exactly p elements.*

PROOF. We first assume that f is onto, and show f cannot be 1–1. Suppose f were 1–1. By property (iv), $f(0) = 0_F$, the zero element of F, and by property (iii), $f(1) = 1_F$, the identity element of F. Let $f(2) = \alpha$. If f were 1–1, $\alpha \neq 1_F$ and $\alpha \neq 0_F$. Since F is a field, there is some β in F with $\alpha\beta = 1_F$. Since f is onto, there is some integer n such that $f(n) = \beta$. Then by property (ii), $f(2n) = f(2)f(n) = \alpha\beta = 1_F = f(1)$. If f were 1–1, $2n = 1$. But that is impossible: there is no integer n such that $2n = 1$. Thus f cannot be 1–1.

This means that F must have characteristic p for some prime p, by Chapter II-10.

Now f is onto; it follows that all elements of F have the form 0_F, 1_F, $2 \cdot 1_F, \ldots, (p-1) \cdot 1_F$. For since F has characteristic p, $0_F = p \cdot 1_F$. Given any integer n, $n = qp + r$ where $0 \leqslant r < p$; hence $n \cdot 1_F = qp \cdot 1_F + r \cdot 1_F = r \cdot 1_F$ where $0 \leqslant r < p$. So F has p elements. That completes the proof. □

EXAMPLE 6. Extending the ideas of the proof of Proposition 1 gives an example of an isomorphism which will be useful in other chapters.

Proposition 2. *Let F be a field of characteristic p, a prime. Let F_0 be the set of all integer multiples of 1_F in F, that is, the image of the function F of Example 5. Then F_0 is isomorphic to $\mathbb{Z}_p$.*

PROOF. We define a function $\bar{f}$ from $\mathbb{Z}_p$ to F_0 as follows: If $n = qp + r$, $0 \leqslant r < p$, then $\bar{f}([n]_p) = r \cdot 1_F$. We show that $\bar{f}$ is an isomorphism by showing that it is well defined, a homomorphism, 1–1, and onto.

$\bar{f}$ is well defined because $[n]_p = [m]_p$ iff n and m leave the same remainder when divided by p.

To show that $\bar{f}$ is a homomorphism we must check properties (i)–(iii).

(i) Given integers n_1, n_2, let $n_1 = q_1 p + r_1$, $n_2 = q_2 p + r_2$, with $0 \leqslant r_1 < p$, $0 \leqslant r_2 < p$. Then $n_1 + n_2 = (q_1 + q_2)p + r_1 + r_2$, and also $= q_3 p + r_3$ where $0 \leqslant r_3 < p$. Thus $r_3 + sp = r_1 + r_2$ for some $s \geqslant 0$. So $r_1 \cdot 1_F + r_2 \cdot 1_F = (r_1 + r_2) \cdot 1_F = r_3 \cdot 1_F + sp \cdot 1_F = r_3 \cdot 1_F$. That means that $\bar{f}([n_1]_p + [n_2]_p) = \bar{f}([n_1 + n_2]_p) = r_3 \cdot 1_F = r_1 \cdot 1_F + r_2 \cdot 1_F = \bar{f}([n_1]_p) + \bar{f}([n_2]_p)$, which proves (i).

(ii) is similar to (i) and is left as an exercise (E12).

(iii) $\bar{f}([1]_p) = 1_F$ by definition.

As to f being 1–1, if $\bar{f}([n]_p) = 0_F$, then p divides n, so $[n]_p = 0$ in $\mathbb{Z}_p$. So $\bar{f}$ is 1–1 by Exercise E2.

Finally, $\bar{f}$ is onto as follows. Since F has characteristic p, F_0 consists of $0_F = p \cdot 1_F, 1_F, 2 \cdot 1_F, \ldots (p-1) \cdot 1_F$. For if m is any integer, $m = qp + r$, $0 \leqslant r < p$, so $m \cdot 1_F = qp \cdot 1_F + r \cdot 1_F$. Since $p \cdot_F = 0$, so is $qp \cdot 1_F$ for any integer q. So $m \cdot 1_F = r \cdot 1_F$. Since $0 \leqslant r < p$, $r \cdot 1_F = \bar{f}([r]_p)$. So $\bar{f}$ is onto.

That completes the proof of Proposition 2. $\qquad\qquad\square$

C. Examples Involving $F[x]$

From Chapter 7 we have the next example.

EXAMPLE 7. Define $f : F \to F[x]/p(x)$ where F is any field and $p(x)$ is a polynomial of degree $\geqslant 1$, by $f(r) = [r]_{p(x)}$. This is easily seen to be a homomorphism, and we showed in Chapter 7 that f is 1–1.

EXAMPLE 8. Let $f : F[x] \to F[x]/p(x)$ be defined by $f(r(x)) = [r(x)]_{p(x)}$, where $F, p(x)$ are as above . Then f is an onto homomorphism. As with the homomorphism $\mathbb{Z} \to \mathbb{Z}_n$ of Example 4, this is a homomorphism because of the way we defined addition and multiplication in $F[x]/p(x)$.

EXAMPLE 9. We wish to extend Example 8, and consider homomorphisms from $F[x]$, F a field, into other rings R.

Proposition 3. *If a function $\phi : F[x] \to R$ is a homomorphism, then the values of ϕ are completely determined by specifying $\phi(a)$ for all a in F, and $\phi(x)$.*

PROOF. A typical element of $F[x]$ is a polynomial $p(x)$, and since ϕ is a ring homomorphism we may write $\phi(p(x))$ as follows:

$$\phi(p(x)) = \phi(a_0 + a_1 x + \cdots + a_n x^n)$$

$$= \phi(a_0) + \phi(a_1 x) + \phi(a_2 x^2) + \cdots + \phi(a_n x^n)$$

$$= \phi(a_0) + \phi(a_1)\phi(x) + \phi(a_2)\phi(x)^2 + \cdots + \phi(a_n)\phi(x)^n.$$

Thus $\phi(p(x))$ is described in terms of $\phi(a_i)$, a_i in F, and $\phi(x)$, as was to be shown. $\qquad\qquad\square$

A special case of this last situation is if F is a subset of R and $\phi(a) = a$ for a in F. Then ϕ is completely determined by its value on x; in fact, $\phi(p(x)) = p(\phi(x))$.

EXAMPLE 10. Let a be any element of F. We get a homomorphism $\phi_a : F[x] \to F$ by setting $\phi_a(r) = r$ for r in F and $\phi_a(x) = a$. This gives the "evaluation at a" map: if $p(x)$ is any polynomial, $\phi_a(p(x)) = p(a)$.

EXAMPLE 11. Suppose now we want to define a homomorphism f from $F[x]/p(x)$ to R, where $F \subseteq R$ and $f(a) = a$ for a in F. We may not let $f([x]_{p(x)})$ be just any element of R. In fact, if $f([x]_{p(x)}) = b$, then $0 = f([p(x)]_{p(x)}) = p(f[x]) = p(b)$, so b must be a root of $p(x)$ in R. Conversely, if b is a root of $p(x)$, then the function $f : F[x]/p(x) \to R$ given by $f(q([x]_{p(x)})) = q(b)$ is well defined (that is, if $[q_1(x)]_{p(x)} = [q_2(x)]_{p(x)}$, then $q_1(b) = q_2(b)$) and is a ring homomorphism). To see that f is well defined, suppose $[q_1(x)]_{p(x)} = [q_2(x)]_{p(x)}$. Then $q_1(x) = q_2(x) + k(x)p(x)$ for some polynomial $k(x)$. So $q_1(b) = q_2(b) + k(b)p(b)$. Since $p(b) = 0$, $q_1(b) = q_2(b)$.

We sum up the above as

Proposition 4. *Let F be a field, R a ring containing F, $p(x)$ a polynomial in $F[x]$. The function $f : F[x]/p(x) \to R$ defined by $f(a) = a$ for a in F, $f([x]) = b$ is a well-defined homomorphism iff b is an element of R with $p(b) = 0$.*

EXAMPLE 12. If we wish to define a homomorphism f from $\mathbb{R}[x]/(x^2 + 1)$ to $\mathbb{C}$, we can do it by letting $f(r) = r$ for r in $\mathbb{R}$ and letting $f([x]) = i$ or $-i$, one of the two roots of $x^2 + 1$ in $\mathbb{C}$. We may not let $f([x])$ be anything else for, if f is a homomorphism, $0 = f([0]) = f([x^2 + 1]) = f([x])^2 + 1$, so $f([x])^2$ must be a root in $\mathbb{C}$ of $x^2 + 1$.

Observe that the function $f : \mathbb{R}[x]/(x^2 + 1) \to \mathbb{C}$ given by $f(r) = r$, $f([x]) = i$ is an isomorphism of rings. We essentially already observed that. For in Chapter 7 we described $\mathbb{R}[x]/(x^2 + 1)$ as the set of elements of the form $[a + bx] = [a] + [b][x]$, a, b real, where $[x]^2 = -1$. Then we "relabeled": $[x]$ was called i, $[a]$ became a, $[b]$ became b. Having done that, $\mathbb{R}[x]/(x^2 + 1)$ became indistinguishable from $\mathbb{C}$. In fact what we were doing was implicitly describing the isomorphism

$$f : \mathbb{R}[x]/(x^2 + 1) \to \mathbb{C}$$

given by $f([a + bx]) = a + bi$.

In general, the existence of an isomorphism between a ring R and a ring S implies that anything algebraic which we know about one ring we then know about the other. For example, in $\mathbb{C}$ we know there are five different fifth roots of 1. Let u_1, u_2, u_3, u_4, u_5 be the five fifth roots of 1 in $\mathbb{C}$. Since $f : \mathbb{R}[x]/(x^2 + 1) \to \mathbb{C}$ is 1–1 and onto, there are five elements h_1, h_2, h_3, h_4, h_5 in $\mathbb{R}[x]/(x^2 + 1)$ such that $f(h_i) = u_i$. Each h_i is then a fifth root of 1 in $\mathbb{R}[x]/(x^2 + 1)$, for $f(h_i^5) = u_i^5 = 1$ in $\mathbb{C}$ and since $f(1) = 1$ and f is 1–1, $1 = h_i^5$ in $\mathbb{R}[x]/(x^2 + 1)$. As another example, in the proof of Proposition 1, we showed that if f is onto a field F, where $f : \mathbb{Z} \to F$ is as in

Example 5, then f could not be 1–1: for if f were 1–1, then f would be an isomorphism, and so F would be isomorphic to $\mathbb{Z}$. Then, since F is a field, $\mathbb{Z}$ would have to be a field also, which it is not.

D. Automorphisms

One special kind of homomorphism is that of an isomorphism from a ring R onto itself. Such a function is called an *automorphism* of the ring R.

One such example is Example 1, the function $i : R \to R$ defined by $i(r) = r$, called the *identity automorphism*. Here are two other examples.

EXAMPLE 13. Let $j : \mathbb{C} \to \mathbb{C}$ be defined by $j(r + si) = r - si$ for r, s real. The function j is called *complex conjugation*; if α is a complex number, $j(\alpha)$ is usually denoted $\bar{\alpha}$. Geometrically, j is the map which takes a complex number, viewed as a point in the plane, and sends it to its reflection across the real axis.

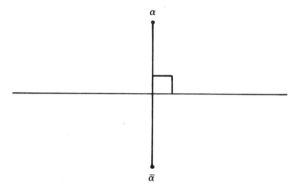

EXAMPLE 14. Let F be a field of characteristic p. Then the map $\phi_p : F \to F$ given by $\phi_p(a) = a^p$ is a 1–1 ring homomorphism. If F is a finite field, then ϕ_p is onto, and is therefore an automorphism of F.

All of this is easily shown using Theorem 1 and the proof of Theorem 2 of Chapter II-10.

E3. Verify the claims of Example 14.

E4. Let F be the field of rational functions f/g with f, g in $\mathbb{Z}_p[x]$. Show that $\phi_p : F \to F$ is not surjective.

The map ϕ_p has the following nice property.

Proposition 5. *Let F be a field of characteristic p, and let $\mathbb{Z}_p$ be the subset of F consisting of $0, 1, 1 + 1, \ldots, 1 + 1 + 1 + \ldots + 1, \ldots$, where $0, 1$ are* *respectively the zero and identity elements of F. Let* $\phi_p : F \to F$ *be as in Example* 14. *Then for a in F,* $\phi_p(a) = a$ *iff a is in* $\mathbb{Z}_p$.

PROOF. $\phi_p(a) = a$ iff $a^p = a$ iff a is a root in F of the polynomial $x^p - x$ in $F[x]$. Now $x^p - x$ has at most p roots in the field F, but by Fermat's theorem each of the p elements of $\mathbb{Z}_p$ is a root of $x^p - x$. Thus $a^p = a$ iff a is in $\mathbb{Z}_p$.

We shall use this proposition subsequently in Chapter 15.

If f, g are two automorphisms of a ring R, so is their composition fg, $fg(r) = f(g(r))$. It is not hard to see that the set of all automorphisms of R is a group with respect to the operation of composition. The study of groups of automorphisms of fields is a beautiful theory, known as *Galois theory* after its inventor. Expositions of Galois theory may be found in numerous more advanced books in abstract algebra, such as Herstein (1964).

E5. Let F be a field with 4 elements. Show that the only automorphisms of F are ϕ_2 and the identity automorphism.

E6. Find all automorphisms of $\mathbb{Z}$.

†**E7.** Two fields with 8 elements are $\mathbb{Z}_2[x]/(x^3 + x + 1)$ and $\mathbb{Z}_2[x]/(x^3 + x^2 + 1)$. Find an isomorphism between them. Find all isomorphisms between them.

E8. Suppose $f : R \to S$ is an isomorphism of rings. Define a homomorphism of rings $g : S \to R$ such that $gf : R \to R$ is the identity function, $gf(r) = r$ for all r in R, and also fg is the identity function on S. Conclude that if $R \cong S$, then $S \cong R$.

E9. Let F be a field and consider $E(F)$, the set of functions from F to F. We can add and multiply in $E(F)$ by $(f + g)(a) = f(a) + g(a)$, $(f \cdot g)(a) = (f(a))(g(a))$. There is a function 0, $0(a) = 0$, and a function 1, $1(a) = 1$ for all a in F. Then $E(F)$ is a commutative ring with unity. Consider the function

$$\phi : F[x] \to E(F)$$

which treats a polynomial as a function on F. Show ϕ is a ring homomorphism. When is it 1–1? When is it onto?

E10. Show that complex conjugation is an automorphism of $\mathbb{C}$ and is the only automorphism of $\mathbb{C}$, except for the identity automorphism, which is the identity on $\mathbb{R} \subset \mathbb{C}$.

E11. Prove that f is as in Example 5 and if $\bar{f}$ is as in Example 6, then $\bar{f}([n]_p) = f(n)$ for any integer n.

E12. Verify that property (ii) holds for the function of Example 6.

†**E13.** Prove that if $R \cong S$ and $S \cong T$, then $R \cong T$.

†**E14.** Let $f : F \to S$ be a ring homomorphism where F is a field. Show that f must be 1–1. (*Hint: use* property (iii).)

E15. Find all automorphisms of $\mathbb{Z}_2[x]/(x^4 + x + 1)$. Show that all are obtained by composing ϕ_2 (Example 14) with itself n times, $n \leqslant 4$.

E16. Suppose $m = rs$ is a factorization of the natural number m, where $r, x > 1$. Define $\phi : \mathbb{Z}_m \to \mathbb{Z}_r \times \mathbb{Z}_s$ by $\phi([a]_m) = ([a]_r, [a]_s)$ (I-14, E12).
(a) Show that ϕ is a well-defined ring homomorphism.
(b) Show that ϕ is an isomorphism iff r and s are relatively prime.

11 Finite Fields are Simple

In Chapter 7 we constructed a collection of simple field extensions, that is, fields of the form $K = F[x]/(p(x))$, by starting with a field F and an irreducible polynomial $p(x)$ in $F[x]$, and letting K be the set of congruence classes of polynomials modulo $p(x)$. We observed that we could also think of K as $K = F[\beta]$, polynomials with coefficients in F evaluated at β, where $\beta = [x]_{p(x)}$ is an "invented root" of $p(x)$.

If $q(x)$ is an irreducible polynomial of degree d with coefficients in $\mathbb{Z}_p$, then $\mathbb{Z}_p[x]/(q(x))$ is a field with p^d elements. For there is a 1–1 correspondence between elements of $\mathbb{Z}_p[x]/(q(x))$ and polynomials with coefficients in $\mathbb{Z}_p$ of degree $\leq d - 1$, and there are p^d such polynomials (since there are d coefficients $a_0, a_1, \ldots, a_{d-1}$ and p choices for each). In this way we can construct many *finite fields* (i.e., fields with finitely many elements) as simple field extensions of $\mathbb{Z}_p$ for some p.

Are there finite fields which are not simple field extensions of $\mathbb{Z}_p$ for some p?

Surprisingly, the answer is no.

Theorem. *Any finite field is isomorphic to a simple field extension of $\mathbb{Z}_p$ for some prime p.*

We shall prove this theorem in this chapter.

The theorem says that given the finite field F, there is a prime p, a polynomial $q(x)$ in $\mathbb{Z}_p[x]$ with invented root β, and an isomorphism $\phi : \mathbb{Z}_p[\beta] \to F$.

If the theorem is true, then, thinking of $\mathbb{Z}_p$ as polynomials of degree 0 in $\mathbb{Z}_p[\beta]$, $\phi(\mathbb{Z}_p)$ is a subfield of F isomorphic to $\mathbb{Z}_p$. We begin the proof of the theorem by finding the prime p and the subfield of F isomorphic to $\mathbb{Z}_p$.

Lemma 1. *If F is a finite field, then F has characteristic p for some prime $p > 0$ and so contains a subfield isomorphic to $\mathbb{Z}_p$.*

PROOF. If 1 is the multiplicative identity of F, let $m \cdot 1 = 1 + 1 + \cdots + (m$ times) and consider the set $\{1, 2 \cdot 1, 3 \cdot 1, \ldots, m \cdot 1, \ldots\}$. Since F is finite this set has only a finite number of distinct elements of F in it. So for some m and n, $m < n$, $m \cdot 1 = n \cdot 1$. Subtracting $m \cdot 1$ from both sides, $(n - m) \cdot 1 = 0$. Thus F has characteristic p for some prime p (see Chapter II-10). Define a function f from $\mathbb{Z}_p$ to F by $f([n]) = n \cdot 1$, $n \geqslant 0$ (cf. Proposition 2, Chapter III-10). Then f is well defined and 1–1, for $[n] = [m]$ iff $n = m + rp$ iff $n \cdot 1 = m \cdot 1 + rp \cdot 1$ iff $n \cdot 1 = m \cdot 1$ (since $p \cdot 1 = 0$); also, f is easily seen to be a ring homomorphism. Thus the image of f, namely, $\{n \cdot 1 | n = 0, 1, 2, \ldots, p - 1\}$ in F, is isomorphic to $\mathbb{Z}_p$. That proves the lemma. $\square$

We can now proceed to the proof of the theorem.

PROOF. Suppose F has n elements and has characteristic p. Then F has a primitive element (Chapter 1), that is, an element α such that every nonzero element of F is a power of α.

By Fermat's theorem, $\alpha^{n-1} = 1$, so α is a root of the polynomial $x^{n-1} - 1$, thought of as having coefficients in $\mathbb{Z}_p$ (where we regard $\mathbb{Z}_p$ as being a subfield of F via the map $f : [n] \to n \cdot 1$ of the lemma above). Therefore, the set of polynomials in $\mathbb{Z}_p[x]$ with α as a root is nonempty. Let $q(x)$ be the minimal polynomial over $\mathbb{Z}_p$ of α. By Proposition 1, Chapter III-8, $q(x)$ is irreducible in $\mathbb{Z}_p[x]$, and so $\mathbb{Z}_p[x]/(q(x))$ is a field.

Consider the function

$$\phi : \mathbb{Z}_p[x]/(a(x)) \to F$$

defined by

$$\phi([f(x)]) = f(\alpha).$$

Then ϕ is a well-defined ring homomorphism since α is a root of $q(x)$ (by Proposition 4 of Chapter 10). ϕ is onto because $\phi([0]) = 0$, $\phi([x^r]) = \alpha^r$, and every nonzero element of F is a power of α. It remains only to show that ϕ is 1–1. Since $\mathbb{Z}_p[x]/(q(x))$ is a field, that follows from Exercise E14, Chapter 10, and completes the proof. $\square$

One important consequence of this theorem is that it shows that there are severe restrictions on the number n of elements in a finite field.

Corollary. *If F is a finite field then F has p^d elements for some prime p and natural number d. Thus if n is not a power of a prime, there is no field with n elements.*

PROOF. If F is a finite field, $F \cong \mathbb{Z}_p[x]/(q(x))$ for some irreducible polynomial $q(x)$ of degree d for some d. Then $\mathbb{Z}_p[x]/(q(x))$ has p^d elements (as we observed in the second paragraph of this chapter), and so F does also. $\square$

Table 1 gives a list of the fields with n elements for n small, based on what we know so far. Table 1 will be of interest when we turn in the next chapter to Latin squares.

In Chapter 13 we shall prove the converse of the corollary, namely, if $n = p^d$, p prime, then there is a field with n elements.

Table 1 A list of fields with n elements for n small

$n =$	fields
2	$\mathbb{Z}_2$
3	$\mathbb{Z}_3$
4	$\mathbb{Z}_2[x]/(x^2 + x + 1)$
5	$\mathbb{Z}_5$
6	none
7	$\mathbb{Z}_7$
8	$\mathbb{Z}_2[x]/(x^3 + x + 1)$ and $\mathbb{Z}_2[x]/(x^3 + x^2 + 1)$ (which are isomorphic)
9	$\mathbb{Z}_3[x]/(x^2 + 1)$ (are there others?)
10	none
11	$\mathbb{Z}_{11}$
12	none
13	$\mathbb{Z}_{13}$
14	none
15	none
16	$\mathbb{Z}_2[x]/(x^4 + x + 1)$ (others??)

[†]E1. (a) Find a primitive element β of $\mathbb{Z}_2[x]/(x^4 + x^3 + x^2 + x + 1)$.
　　(b) Find the minimal polynomial $q(x)$ in $\mathbb{Z}_2[x]$ of β.
　　(c) Show that $\mathbb{Z}_2[x]/(x^4 + x^3 + x^2 + x + 1)$ is isomorphic to $\mathbb{Z}_2[x]/(q(x))$.

[†]E2. (a) Find a primitive element β of $\mathbb{Z}_3[x]/(x^2 + 1)$.
　　(b) Find the minimal polynomial $q(x)$ in $\mathbb{Z}_3[x]$ of β.
　　(c) Show that $\mathbb{Z}_3[x]/(x^2 + 1)$ is isomorphic to $\mathbb{Z}_3[x]/q(x)$.

E3. (a) Find a root in $K = \mathbb{Z}_2[x]/(x^4 + x + 1)$ of $x^2 + x + 1$.
　　(b) Describe a homomorphism from $\mathbb{Z}_2[x]/(x^2 + x + 1)$ into K.

E4. Continue Table 1 for $17 \leqslant n \leqslant 50$.

Latin Squares

12

An $n \times n$ *Latin square* is a square matrix in which each of the numbers from 1 to n occurs once in each row and once in each column. Here is an example:

$$
\begin{matrix}
4 & 1 & 2 & 3 \\
1 & 2 & 3 & 4 \\
2 & 3 & 4 & 1 \\
3 & 4 & 1 & 2
\end{matrix}.
$$

You may recognize this example as the table for addition in $\mathbb{Z}_4 = \{1, 2, 3, 4\}$. Similarly the addition table for $\mathbb{Z}_n$ is an $n \times n$ Latin square for any $n \geqslant 2$. More generally, if G is any group with operation $*$ and elements $a_1, a_2, \ldots, a_n$, then the multiplication table for G is a table whose subscripts form a Latin square. For example, if we let G be the set of invertible elements of $\mathbb{Z}_8$, namely $G = \{1, 3, 5, 7\}$ under multiplication, then the multiplication table is

	1	3	5	7
1	1	3	5	7
3	3	1	7	5
5	5	7	1	3
7	7	5	3	1

if we now replace 1, 3, 5, 7 by 1, 2, 3, 4 we get a Latin square:

$$
\begin{matrix}
1 & 2 & 3 & 4 \\
2 & 1 & 4 & 3 \\
3 & 4 & 1 & 2 \\
4 & 3 & 2 & 1
\end{matrix}.
$$

Latin squares are of interest in agricultural experiments (see Fisher (1935)). Here are two examples.

EXAMPLE 1. Suppose five strains of wheat are to be tested for yield on a rectangular field (= plot of ground). The yield depends not only on the strain of wheat but also on the fertility of the soil, which may vary around the field. Suppose, for example, that the north side of the field happens to be more fertile than the south side. Suppose the experimenters did not know how the fertility varied around the field, and planted the five strains of wheat (labeled 1–5) as follows.

North

| 1 |
| 2 |
| 3 |
| 4 |
| 5 |

South

If the yield of strain 1 were higher than that of strain 5 the experimenters would not know whether the result was caused by the fertility of the soil or the difference in the strains.

Fertility tends to be more uniform along strips parallel to the edges of the field, because of the mixing effect of plowing parallel to the edges. So in doing the wheat yield experiment, the problem is to plant the wheat in such a way that variations in fertility of the soil along strips parallel to the edges can be neglected. A nice way to do this is to plant the strains of wheat in a Latin square arrangement, like so:

$$
\begin{array}{ccccc}
1 & 2 & 3 & 4 & 5 \\
2 & 4 & 5 & 3 & 1 \\
4 & 3 & 1 & 5 & 2 \\
5 & 1 & 4 & 2 & 3 \\
3 & 5 & 2 & 1 & 4
\end{array}.
$$

EXAMPLE 2. Three diets—all hay, half hay and half corn, all corn—are to be tested on three dairy cows, to see the effect of diet on milk yield. Different cows have different milk yields, and the same cow's milk yield varies over time. To try to test diet independent of these variations, a Latin square is a useful design.

Week	Cow 1	2	3
1	Corn	1/2	Hay
2	1/2	Hay	Corn
3	Hay	Corn	1/2

Returning to Example 1, suppose in addition to testing five strains of wheat, five kinds of fertilizer are also to be tested. We would like to use a Latin square arrangement for the fertilizer in such a way that each kind of fertilizer is used with each strain of wheat. What is needed, therefore, are two *orthogonal* Latin squares, that is, two 5×5 Latin squares such that each ordered pair (r, s) of (wheat, fertilizer) occurs exactly once on a plot. Here is such a pair.

$$
\text{I:} \quad
\begin{array}{ccccc}
1 & 2 & 3 & 4 & 5 \\
2 & 4 & 5 & 3 & 1 \\
4 & 3 & 1 & 5 & 2 \\
5 & 1 & 4 & 2 & 3 \\
3 & 5 & 2 & 1 & 4 \\
\end{array}
\quad \text{and II:} \quad
\begin{array}{ccccc}
1 & 2 & 3 & 4 & 5 \\
4 & 3 & 1 & 5 & 2 \\
5 & 1 & 4 & 2 & 3 \\
3 & 5 & 2 & 1 & 4 \\
2 & 4 & 5 & 3 & 1 \\
\end{array}
$$

(Wheat)   (Fertilizer)

Suppose in addition we wish simultaneously to test the effect on yield of five kinds of fungicides. For that we would like to test each fungicide with each fertilizer, and with each strain of wheat, so we need to find another Latin square orthogonal to each of the two above. Here is one.

$$
\text{III:} \quad
\begin{array}{ccccc}
1 & 2 & 3 & 4 & 5 \\
5 & 1 & 4 & 2 & 3 \\
3 & 5 & 2 & 1 & 4 \\
2 & 4 & 5 & 3 & 1 \\
4 & 3 & 1 & 5 & 2 \\
\end{array}
$$

(Fungicide)

Suppose we wish also to test five kinds of herbicides; we would like yet another Latin square orthogonal to the previous three. Here is one.

$$
\text{IV:} \quad
\begin{array}{ccccc}
1 & 2 & 3 & 4 & 5 \\
3 & 5 & 2 & 1 & 4 \\
2 & 4 & 5 & 3 & 1 \\
4 & 3 & 1 & 5 & 2 \\
5 & 1 & 4 & 2 & 3 \\
\end{array}
$$

(Herbicide)

Suppose we wished also to test five levels of soil acidity; we would like one more Latin square orthogonal to the previous four. But there is none. For if we had such a square, we could number the five levels of acidity appearing on the top row by 12345 and then the new square would start

$$
\text{V:} \quad \begin{pmatrix} 1 & 2 & 3 & 4 & 5 \\ a & & & & \end{pmatrix}.
$$

But if V is to be orthogonal to all of the other squares, then the number $a \neq 1$ cannot coincide with the corresponding number in any other square. For example, $a \neq 2$, for otherwise the pair $(2, 2)$ occurs twice in the pair of squares (V, I), once at the second entry of the top row, and once in the first entry of the second row, so V and I are not orthogonal. The same argument prevents a from being 3 or 4 or 5; $a \neq 1$, since 1 already occurs in the first column of V.

This leads to the following problem: Given m, how many pairwise orthogonal $m \times m$ Latin squares can be constructed?
Here are two facts:

Theorem. (1) *There cannot be more than $m - 1$ pairwise orthogonal $m \times m$ Latin squares.*
(2) *If there is a field with m elements then there are $m - 1$ pairwise orthogonal $m \times m$ Latin squares.*

We leave the proof of the first statement of the theorem for Exercise E8.

PROOF OF (2). Suppose we have a field F with m elements. Let α be a primitive element. Then $\alpha^{m-1} = 1$ and every nonzero element of F is a power of α. Consider the addition table for F set up as follows:

+	α	α^2	$\cdots$	α^s	$\cdots$	α^{m-1}	0
0	α	α^2		α^s		α^{m-1}	0
α^i	$\alpha^i + \alpha$	$\alpha^i + \alpha^2$		$\alpha^i + \alpha^s$		$\alpha^i + \alpha^{m-1}$	α^i
α^{i+1}	$\alpha^{i+1} + \alpha$	$\cdots$		$\alpha^{i+1} + \alpha^s$			
$\vdots$				$\vdots$			$\vdots$
α^{i+r}	$\alpha^{i+r} + \alpha$	$\cdots$		$\alpha^{i+r} + \alpha^s$			
$\vdots$				$\vdots$			
α^{i+m-1}	$\alpha^{i+m-1} + \alpha$	$\cdots$		$\alpha^{i+m-1} + \alpha^s$	$\cdots$		α^{i+m-1}

Examining the entries of the table, we see that each element of F occurs once in each row and once in each column (Exercise E3). If we write the nonzero entries of the table as powers of α (possible because α is a primitive element of F) and then replace the elements of F by the numbers 1 to m, using the correspondence

$$\begin{array}{cccccc} \alpha & \alpha^2 & \alpha^3 & & \alpha^{m-1} & 0 \\ 1 & 2 & 3 & & m-1 & m \end{array}$$

we get a Latin square; call it L_i.

If i, j are two different integers between 1 and $m - 1$, then L_i and L_j are orthogonal Latin squares. For example, with $m = 5$, $\alpha = 2$, $i = 1$, we get L_1.

+	2	2^2	2^3	2^4	0
0	2	2^2	2^3	2^4	0
2	2^2	2^4	0	2^3	2
2^2	2^4	2^3	2	0	2^2
2^3	0	2	2^4	2^2	2^3
2^4	2^3	0	2^2	2	2^4

or

1	2	3	4	5
2	4	5	3	1
4	3	1	5	2
5	1	4	2	3
3	5	2	1	4

This is the Latin square I we gave in the wheat example above. With $i = 3$ we get L_3.

+	2	2^2	2^3	2^4	0
0	2	2^2	2^3	2^4	0
2^3	0	2	2^4	2^2	2^3
2^4	2^3	0	2^2	2	2^4
2	2^2	2^4	0	2^3	2
2^2	2^4	2^3	2	0	2^2

or

1	2	3	4	5
5	1	4	2	3
3	5	2	1	4
2	4	5	3	1
4	3	1	5	2

This was Example 3 above.

This construction gives $m - 1$ pairwise orthogonal Latin squares. For the pair of entries in L_i and L_j at the (r, s)th position is $(\alpha^{i+r} + \alpha^s, \alpha^{j+r} + \alpha^s)$. Suppose $i \neq j$. If the pair of entries at the (r, s)th position is equal to the pair of entries at the (p, q)th position, then the pairs

$$(\alpha^{i+r} + \alpha^s, \alpha^{j+r} + \alpha^s) \quad \text{and} \quad (\alpha^{i+p} + \alpha^q, \alpha^{j+p} + \alpha^q)$$

are the same, so

$$\alpha^{i+r} + \alpha^s = \alpha^{i+p} + \alpha^q$$

and

$$\alpha^{j+r} + \alpha^s = \alpha^{j+p} + \alpha^q.$$

Then

$$\alpha^{i+r} - \alpha^{i+p} = \alpha^q - \alpha^s = \alpha^{j+r} - \alpha^{j+p},$$

so

$$\alpha^i(\alpha^r - \alpha^p) = \alpha^j(\alpha^r - \alpha^p).$$

Since $i \neq j$, we must have $\alpha^r - \alpha^p = 0$, $r = p$, hence $\alpha^q = \alpha^s$ and $q = s$. Thus L_i and L_j are orthogonal if $i \neq j$. That completes the proof. $\square$

We shall prove in the next chapter that if n is any number which is a power of a prime, $n = p^e$, then there is a field with n elements. For such numbers n the theorem says that there are $n - 1$ pairwise orthogonal $n \times n$ Latin squares, but not n pairwise orthogonal $n \times n$ Latin squares.

In the case that n is not a prime power, the question remains, how many pairwise orthogonal $n \times n$ Latin squares can there be? For many n this is unknown. The smallest nonprime power is $n = 6$, and that question was the content of a famous problem of Euler, called the problem of 36 officers. It goes as follows. 36 officers are to be placed in review in a square, 6 rows deep with 6 men in each row. The officers come from 6 different regiments, and each regiment is represented by 6 officers, each of different ranks. For reasons of protocol it is desired that each row and column is to have one officer from each regiment and one officer of each rank. Can this be done?

If it could be done, then one would have a pair of orthogonal 6×6 Latin squares. Euler believed that it could not be done, but it was not proved impossible for well over 100 years, until a proof was finally achieved by M. G. Tarry in 1901. Thus the situation for nonprime powers is apparently much different than for prime powers. There are 4 pairwise orthogonal 5×5 Latin squares and 6 pairwise orthogonal 7×7 Latin squares, but no two 6×6 Latin squares are orthogonal.

The construction of orthogonal Latin squares is due to R. C. Bose. See Mann (1949).

E1. Find three pairwise orthogonal 4×4 Latin squares.

E2. Use the construction in the proof of part (2) of the theorem with $\alpha = 3$, to find 4 pairwise orthogonal 5×5 Latin squares.

E3. In the proof of part (2) of the theorem, verify that each L_i is a Latin square.

E4. Find three pairwise orthogonal 8×8 Latin squares.

E5. Prove that there is no field with 6 elements.

E6. Show that if G is a group under multiplication, with n elements, then the multiplication table for G yields a Latin square.

E7. Find a Latin square which cannot be viewed as the multiplication table for a group.

E8. Prove part (1) of the theorem: that is, show that there cannot be m pairwise orthogonal $m \times m$ Latin squares.

Irreducible Polynomials in $\mathbb{Z}_p[x]$ 13

A. Factoring $x^{p^n} - x$

In this chapter we show that there are irreducible polynomials of any degree over $\mathbb{Z}_p$, and get a formula for the number of monic irreducible polynomials of each degree.

We begin with a theorem about the number of elements there can be in a finite field. We showed in Chapter 11 that if F is a field with n elements, then n has to be a power of a prime. Here is the converse.

Theorem 1. *Given any prime p and any $n > 0$ there is a field with exactly p^n elements.*

PROOF. Consider $f(x) = x^{p^n} - x$ in $\mathbb{Z}_p[x]$. By the corollary of Chapter 8 there is a splitting field K for $f(x)$, that is, a field K such that in $K[x]$, $f(x)$ factors into a product of linear factors.

Let F be the subset of K consisting of all roots of $x^{p^n} - x$ in K. We shall show F is the desired field.

Claim. F contains p^n distinct elements of K.

To prove this claim, recall (Chapter II-10, Theorem 2) that the derivative $f'(x)$ of a polynomial $f(x)$ has the property that if $f(x)$ and $f'(x)$ are relatively prime in $K[x]$, then $f(x)$ has no multiple roots in K. Computing the derivative of $x^{p^n} - x$, we get $(d/dx)(x^{p^n} - x) = p^n x^{p^n-1} - 1 = -1$. Thus $x^{p^n} - x$ has no multiple roots. That means that when $x^{p^n} - x$ factors in $K[x]$ into a product of linear factors, they are p^n distinct linear factors. So $x^{p^n} - x$ has p^n distinct roots in K, as claimed.

Claim. F is a field.

For F is the set of elements a of K which satisfy $a^{p^n} = a$. Thus, if a, b are in F, then:

(i) so is $a + b$: $(a + b)^{p^n} = a^{p^n} + b^{p^n} = a + b$ (the first equality is by Example 14, Chapter 10);

(ii) so is $a \cdot b$: $(ab)^{p^n} = a^{p^n} b^{p^n} = ab$;

(iii) so is $-a$: $(-a)^{p^n} = (-1)^{p^n} a^{p^n} = -a$ $((-1)^{p^n} = -1$ by Fermat's theorem);

(iv) so is a^{-1}: $(a^{-1})^{p^n} = (a^{p^n})^{-1} = a^{-1}$.

Since 0 and 1 are in F, and addition and multiplication in F is the same as that in K, therefore F is a field. That completes the proof. $\square$

Corollary. *There is an irreducible polynomial in $\mathbb{Z}_p[x]$ of degree n for each n.*

PROOF. Let F be a field with p^n elements. By the theorem of Chapter 11, F is isomorphic to $\mathbb{Z}_p[x]/(q(x))$ for some irreducible polynomial $q(x)$ in $\mathbb{Z}_p[x]$. Since F has p^n elements, $\mathbb{Z}_p[x]/(q(x))$ must have p^n elements, so $q(x)$ must have degree n, and is the desired polynomial. $\square$

Observe that the irreducible polynomial $q(x)$ is a divisor of $x^{p^n} - x$. For otherwise we could write

$$1 = q(x)r(x) + (x^{p^n} - x)s(x) \quad \text{in} \quad \mathbb{Z}_p[x]. \tag{*}$$

Let β be a root of $q(x)$ in F. By Fermat's theorem, β is also a root of $x^{p^n} - x$. Setting $x = \beta$ in (*) we would get $1 = 0$. Thus $q(x)$ must divide $x^{p^n} - x$.

This observation leads to

Theorem 2. $x^{p^n} - x$ *is the product of all irreducible polynomials in $\mathbb{Z}_p[x]$ of degree d for all d dividing n.*

We prove this in two parts.

Theorem 2(a). *If $q(x)$ is an irreducible polynomial of degree d and $d \mid n$, then $q(x)$ divides $x^{p^n} - x$.*

PROOF. Let $F = \mathbb{Z}_p[x]/q(x) = \mathbb{Z}_p[\alpha]$, where $\alpha = [x]_{q(x)}$. Then $q(x)$ is the minimal polynomial over $\mathbb{Z}_p$ of α. Now F is a field with p^d elements. So by Fermat's theorem,

$$\alpha^{p^d} = \alpha.$$

Since $d \mid n$, $\alpha^{p^n} = \alpha^{p^{de}} = (\ldots ((\alpha^{p^d})^{p^d}) \ldots)^{p^d} = \alpha$, so α is a root of $x^{p^n} - x$. By Proposition 1, Chapter III-8 (or by the observation just preceding the theorem), $q(x)$ divides $x^{p^n} - x$. $\square$

Theorem 2(b). *If $q(x)$ is an irreducible factor of $x^{p^n} - x$ and has degree d, then $d \mid n$.*

PROOF. Suppose that $q(x)$ is an irreducible factor of $x^{p^n} - x$. Consider $\mathbb{Z}_p[x]/(q(x)) = K$. If $q(x)$ has degree d, K has p^d elements. Let F be a field with p^n elements. Since $q(x)$ divides $x^{p^n} - x$ and $x^{p^n} - x$ has p^n roots in F, there is a root β of $q(x)$ in F. Thus there is a homomorphism ϕ (Chapter 10, Proposition 4) from K to F which is 1–1 by Chapter 10, Exercise E14. Let K' be the image of K in F; K' is then a subfield of F isomorphic to K.

Let α be a primitive element of F. We now copy the proof of the theorem of Chapter 11. Let $s(x)$ be the minimal polynomial of α over K'. Then sending $[x]$ to α defines a homomorphism ϕ' from $K'[x]/s(x)$ to F (Chapter 10, Proposition 4) which is 1–1 since $s(x)$ is irreducible, and onto since every nonzero element of F is a power of α. So ϕ' is an isomorphism from $K'[x]/(s(x))$ onto F. So $K'[x]/(s(x))$ and F have the same number of elements.

How many elements are in $K'[x]/(s(x))$? If $s(x)$ has degree e, and K has m elements, then $K'[x]/(s(x))$ has m^e elements. But $m = p^d$. Now F has p^n elements. So $(p^d)^e = p^n$. So $de = n$ and d, the degree of $q(x)$, divides n. That completes the proof. $\qquad\square$

E1. Factor $x^{16} - x$ in $\mathbb{Z}_2[x]$.

E2. Factor $x^9 - x$ in $\mathbb{Z}_3[x]$.

E3. Show that if p, q are primes, then $x^{p^q} - x = (x^p - x)h(x)$ in $\mathbb{Z}_p[x]$, where $h(x)$ is the product of all monic irreducible polynomials in $\mathbb{Z}_p[x]$ of degree q.

E4. Factor $x^{25} - x$ in $\mathbb{Z}_5[x]$.

E5. Show that $GF(16)$ is a splitting field for $x^4 - x$ in $\mathbb{Z}_2[x]$. If $GF(16) = \mathbb{Z}_2[\alpha]$, $\alpha^4 + \alpha + 1 = 0$ (as in Table 2 of Chapter III-9), which are the roots in $GF(16)$ of $x^4 - x$?

B. Counting Irreducible Polynomials

Let N_n^p be the number of irreducible polynomials of degree n in $\mathbb{Z}_p[x]$. Using Theorem 2 we will find an explicit formula for N_n^p.

To obtain such a formula, we use the Möbius function, a classical tool in number theory which has attracted new interest in combinatorics in recent years (see Bender and Goldman (1975)).

Definition. The *Möbius function* $\mu(n)$ is defined for $n \geq 1$ by

$$\mu(n) = \begin{cases} 1 & \text{if } n = 1, \\ 0 & \text{if } p^e \mid n \text{ for some prime } p \text{ and some } e > 1, \\ (-1)^r & \text{if } n \text{ is the product of } r \text{ distinct primes.} \end{cases}$$

The formula we want is

$$N_n^p = \frac{1}{n} \sum_{d|n} \mu\left(\frac{n}{d}\right) p^d. \tag{1}$$

Formula (1) is a special case of the Möbius inversion formula, which we now derive. We begin with two facts about the Möbius function.

Proposition 1. *If* $(m, n) = 1$, *then* $\mu(mn) = \mu(m)\mu(n)$.

This is easy to verify.

Such a function is called *multiplicative*. Another example of a multiplicative function is Euler's ϕ function.

Proposition 2. $\sum_{d|n}\mu(d) = 0$ *unless* $n = 1$.

The proof of this is an exercise in manipulating sums. Before doing the proof in general we illustrate with $n = 36 = 2^2 3^2$:

$$\sum_{d|36} \mu(d) = \left[\mu(1) + \mu(2) + \mu(2^2) \right] + \left[\mu(3) + \mu(2\cdot 3) + \mu(2^2\cdot 3) \right]$$

$$+ \left[\mu(3^2) + \mu(3^2\cdot 2) + \mu(3^2\cdot 2^2) \right].$$

Now $\mu(d) = 0$ if d is divisible by the square of a prime, so this sum reduces to

$$\left[\mu(1) + \mu(2) \right] + \left[\mu(3) + \mu(2\cdot 3) \right]$$

$$= \left[\mu(1) + \mu(2) \right] + \left[\mu(1)\mu(3) + \mu(2)\mu(3) \right] \qquad \text{(by Proposition 1)}$$

$$= \left[\mu(1) + \mu(2) \right] + \left[\mu(1) + \mu(2) \right] \mu(3)$$

$$= \left[\mu(1) + \mu(2) \right]\left[\mu(1) + \mu(3) \right].$$

Now $\mu(1) = 1$, $\mu(3) = -1$, so $\mu(1) + \mu(3) = 0$. Hence $\sum_{d|36} \mu(d) = 0$.

The proof in general works in exactly the same way; instead of writing it all out, however, we have to use summation notation.

PROOF OF PROPOSITION 2. Write $n = p_1^{e_1} \cdots p_r^{e_r}, e_1, \ldots, e_r \geq 1$. Then any d dividing n has the form

$$d = p_1^{f_1} \cdots p_1^{f_r} \quad \text{with} \quad 0 \leq f_1 \leq e_i.$$

So

$$\sum_{d|n} \mu(d) = \sum \mu(p_1^{f_1} \cdots p_1^{f_r}) = \sum \left(\mu(p_1^{f_1}) \cdots \mu(p_r^{f_r}) \right)$$

where the sum runs through all $(f_1 \cdots f_r)$ with $0 \leqslant f_i \leqslant e_i$. Since $\mu(p^f) = 0$ if $f > 1$, this sum equals

$$\sum_{f_r=0}^{1} \sum_{f_{r-1}=0}^{1} \cdots \sum_{f_1=0}^{1} \mu(p_1^{f_1})\mu(p_2^{f_2}) \cdots \mu(p_r^{f_r})$$

$$= \sum_{f_r=0}^{1} \left[\sum_{f_{r-1}=0}^{1} \cdots \sum_{f_1=0}^{1} \mu(p_1^{f_1}) \cdots \mu(p_{r-1}^{f_{r-1}}) \right] \mu(p_r^{f_r})$$

$$= \left[\sum_{f_{r-1}=0}^{1} \cdots \sum_{f_1=0}^{1} \mu(p_1^{f_1}) \cdots \mu(p_{r-1}^{f_{r-1}}) \right] \left[\mu(p_r^0) + \mu(p_r^1) \right].$$

But the sum in the right brackets is $1 + (-1) = 0$. That completes the proof of Proposition 2. □

Proposition 3 (Möbius inversion formula). *For any function f defined on natural numbers, if we set*

$$F(n) = \sum_{d|n} f(d) \quad \text{for every} \quad n \geqslant 1,$$

then

$$f(n) = \sum_{d|n} \mu\left(\frac{n}{d}\right)F(d) = \sum_{e|n} \mu(e)F\left(\frac{n}{e}\right).$$

PROOF. If we substitute $e = n/d$, $d = n/e$, then as d runs through all divisors of n, so does e. Hence the last two sums are equal.

Now by definition of F,

$$\sum_{e|n} \mu(e)F\left(\frac{n}{e}\right) = \sum_{e|n} \mu(e)\left(\sum_{d|(n/e)} f(d) \right) = \sum_{e|n} \left(\sum_{d|(n/e)} \mu(e)f(d) \right).$$

Interchanging the order of summation, (if $d|(n/e)$, then $de|n$ so $e|(n/d)$), we get

$$= \sum_{d|n} \left(\sum_{e|(n/d)} \mu(e)f(d) \right) = \sum_{d|n} \left(\sum_{e|(n/d)} \mu(e) \right)f(d). \tag{2}$$

Now by Proposition 2, for each d,

$$\sum_{e|(n/d)} \mu(e) = 0 \quad \text{unless} \quad \frac{n}{d} = 1.$$

So the coefficient of $f(d)$ is 0 unless $n/d = 1$, $d = n$. Hence the sum (2) reduces to the single term $f(n)$, as was to be shown. □

E6. If F is a function defined on natural numbers and f is defined by $f(n) = \sum_{d|n} \mu(d)F(n/d)$, prove that $F(n) = \sum_{d|n} f(d)$.

E7. If f is a function defined on natural numbers and $F(n) = \sum_{d|n} f(d)$, prove that F is multiplicative.

E8. Prove Proposition 1.

With these generalities out of the way, we can get formula (1) for N_n^p, the number of irreducible polynomials of degree n in $\mathbb{Z}_p[x]$. We shall write N_n^p as N_n if p is understood.

Theorem 2 describes the complete factorization of $x^{p^n} - x$ in $\mathbb{Z}_p$ for any n. Theorem 2 readily gives the formula

$$p^n = \sum_{d|n} dN_d,$$

since $x^{p^n} - x$, a polynomial of degree p^n, has as factors N_d irreducible polynomials of degree d for each d dividing n. Applying the Möbius inversion formula with $F(n) = p^n$, $f(d) = dN_d$, we get

$$nN_n = \sum \mu\left(\frac{n}{d}\right)p^d,$$

or our desired formula:

$$N_n = \frac{1}{n}\sum \mu\left(\frac{n}{d}\right)p^d.$$

We can derive from the formula the corollary to Theorem 1, namely, that $N_n > 0$ for all n. Indeed, since $\mu(n/n) = 1$, we have that $N_n > (1/n)(p^n - \sum_{d<n} p^d)$ and

$$\sum_{d=0}^{n-1} p^d = \frac{1-p^n}{1-p} = \frac{p^n-1}{p-1} < p^n,$$

so

$$\frac{1}{n}\left(p^n - \sum_{d<n} p^d\right) > 0.$$

EXAMPLE. Let $p = 3$, $n = 4$. Then the number of irreducible monic polynomials of degree 4 over $\mathbb{Z}_3$, N_4^3, is

$$N_4^3 = \frac{1}{4}\left(\mu(1)3^4 + \mu(2)3^2 + \mu(4)3\right) = \frac{1}{4}(3^4 - 3^2 + 0) = 18.$$

E9. What are the 18 monic irreducible polynomials of degree 4 in $\mathbb{Z}_3[x]$?

E10. Show that $N_6^7 = 19544$.

It is interesting to see how rapidly N_n grows: for example, if $p = 7$, we have the following set of values (from Simmons (1970)).

n	N_n	Total number of monic polynomials of degree $n(= 7^n)$
1	7	7
2	21	49
3	112	343
4	588	2401
5	3,360	16,807
6	19,544	117,649
7	117,648	823,543
8	720,300	5,764,801
9	4,483,696	40,353,607

Of course, every irreducible polynomial in $\mathbb{Z}_7[x]$ of degree n gives rise to infinitely many different irreducible polynomials of degree n in $\mathbb{Q}[x]$. So there are many irreducible polynomials in $\mathbb{Q}[x]$!

*E11. Can you prove without using the Möbius inversion formula that for any n and any prime p, n divides $\sum_{d|n}\mu(d)p^{n/d}$?

E12. Compute N_n when $p = 2$, $n \leqslant 9$.

*E13. Can you make any guess as to the following limit: $\lim_{n\to\infty}(N_n^p/p^n)$, where $N_n^p =$ number of monic irreducible polynomials of degree n in $\mathbb{Z}_p[x]$, and $p^n =$ number of monic polynomials of degree n in $\mathbb{Z}_p[x]$? (Asking how $N_n \to \infty$ is the analogue in $\mathbb{Z}_p[x]$ of the prime number theorem discussed in Chapter I-4.)

14 Finite Fields

In the last chapter we showed that there are, for example, 588 irreducible monic polynomials of degree 4 in $\mathbb{Z}_7[x]$. Thus it is conceivable that there could be 588 different fields with $7^4 = 2401$ elements. But that is not really the case. For in Exercise E7 of Chapter 10 you showed, for example, that the two different irreducible polynomials of degree 3 in $\mathbb{Z}_2[x]$ gave simple field extensions which are isomorphic to each other, and you showed in Exercises E1 and E2 of Chapter 11 that the same is true for certain fields with 9 or 16 elements.

The remarkable fact is that if F_1 and F_2 are *any* two fields with p^n elements, then F_1 and F_2 are isomorphic. Thus, rather than 588 different fields of 7^4 elements, there is, up to isomorphism, only *one*, which can be presented as a simple field extension of Z_7 in 588 different ways.

The following theorem was proved by E. H. Moore in 1903.

Theorem. *Any two fields with p^n elements are isomorphic.*

PROOF. We prove this theorem by showing that any field with p^n elements is isomorphic to the field F consisting of all roots of $x^{p^n} - x$ which we constructed in Theorem 1 of Chapter 13.

We know that if F_1 is any field with p^n elements, then F_1 is isomorphic to $\mathbb{Z}_p[x]/(q(x))$ for $q(x)$ some irreducible polynomial in $\mathbb{Z}_p[x]$ of degree n, by the theorem of Chapter 11. So we choose any such irreducible polynomial $q(x)$ and show that $\mathbb{Z}_p[x]/(q(x))$ is isomorphic to F.

We know that $q(x)$ divides $x^{p^n} - x$ by Theorem 2 of Chapter 13. Since $x^{p^n} - x$ has p^n different roots in F, $q(x)$ has a root in F; call it β. We define a homomorphism ϕ from $\mathbb{Z}_p[x]/(q(x))$ to F by $\phi(f(x)) = f(\beta)$. Since $q(\beta) = 0$, ϕ is well defined (Proposition 4, Chapter 10). Since $\mathbb{Z}_p[x]/(q(x))$

is a field, ϕ is 1–1 by Exercise E14 of Chapter 10. Since both $\mathbb{Z}_p[x]/(q(x))$ and F have p^n elements and ϕ is 1–1, ϕ must be onto as well, so ϕ is an isomorphism.

Since two fields which are isomorphic to F must be isomorphic to each other (Exercise E13 of Chapter 10), the proof is complete.   □

Since there is essentially only one field with p^n elements, it is customary to give it a special name, namely $GF(p^n)$. GF stands for Galois field, after Galois, who first developed the theory of finite fields.

Corollary. *Any irreducible polynomial of degree n in $\mathbb{Z}_p[x]$ has a root in any field with p^n elements.*

PROOF. For if $q(x)$ is the polynomial, and F_2 is the field, then $F_2 \cong \mathbb{Z}_p[x]/(q(x))$, by the theorem. Since $q(x)$ has a root in $\mathbb{Z}_p[x]/(q(x))$, namely $[x]_{q(x)}$, it has a root in F_2.   □

E1. If d divides n, prove that any polynomial of degree d in $\mathbb{Z}_p[x]$ has a root in any field F with p^n elements.

The theorem of this chapter completes a development which is quite remarkable. We have, starting from nothing, given a complete description of all finite fields, up to isomorphism. For a mathematician who does "algebra" this is a very satisfying outcome. To ask analogous questions, such as, "Describe all commutative rings with unity, up to isomorphism," or, "Describe all finite groups, up to isomorphism" is to raise unsolved questions which have motivated the mathematical research of hundreds of mathematicians over the past several generations.

E2. Show that if $q(x)$ in $\mathbb{Z}_p[x]$ has degree d and F is a field with p^n elements, where $d|n$, then F is a splitting field (Chapter III-8) of $q(x)$.

15 The Discriminant and Stickelberger's Theorem

In this chapter we describe a technique for deciding whether a given polynomial in $\mathbb{Z}_p[x]$, p an odd prime, has an odd or an even number of distinct irreducible factors. The theorem is this:

Stickelberger's Theorem. *Let p be an odd prime, f a monic polynomial of degree d with coefficients in $\mathbb{Z}_p[x]$, without repeated roots in any splitting field. Let r be the number of irreducible factors of f in $\mathbb{Z}_p[x]$. Then $r \equiv d$ (mod 2) iff the discriminant $D(f)$ is a square in $\mathbb{Z}_p$.*

Of course you do not know yet what the discriminant $D(f)$ is. The rest of this chapter is devoted to explaining what $D(f)$ is and how to compute it, and then to proving Stickelberger's theorem.

The condition in the theorem that f have no repeated roots in any splitting field is one that can be checked by finding the greatest common divisor of $f(x)$ and $f'(x)$. The greatest common divisor of f' and f is a constant iff f has no repeated roots, by Theorem 2, Chapter II-10. We shall see in fact that the discriminant $D(f)$ is also computed by finding the greatest common divisor of $f(x)$ and $f'(x)$.

A. The Discriminant

The discriminant is defined in terms of something called the resultant. Let f, g be two polynomials with coefficients in a field F. Suppose K is a splitting field for fg, so that in $K[x]$,

$$f(x) = a(x - \alpha_1)(x - \alpha_2) \cdots (x - \alpha_n),$$
$$\text{and } g(x) = b(x - \beta_1)(x - \beta_2) \cdots (x - \beta_m)$$

for some $\alpha_1, \ldots, \alpha_n, \beta_1, \ldots, \beta_m$ in K. We define the *resultant* of f and g, $R(f, g)$, by

$$R(f, g) = a^m b^n \prod_{i=1}^{n} \prod_{j=1}^{m} (\alpha_i - \beta_j) \qquad (n = \deg f, \ m = \deg g).$$

That is, $R(f, g) = a^m b^n$ times the product of all possible differences of the form (root of f − root of g).

Here are some facts about the resultant:

Fact 1. $R(g, f) = (-1)^{mn} R(f, g)$.

PROOF. For obtaining $R(g, f)$ from $R(f, g)$ involves making mn changes of sign: (root of g − root of f) = − (root of f − root of g). ☐

Fact 2. $R(f, g) = 0$ *if f and g have any common factors.*

PROOF. For then in K, f and g would have a common root, so $R(f, g)$ would have a factor of 0. ☐

Fact 3. $R(f, g) = a^{\deg g} \prod_{i=1}^{n} g(\alpha_i)$, *where a is the leading coefficient of f and α_i are the various roots of f.*

PROOF. Since $g(x) = b \prod_{j=1}^{n} (x - \beta_j)$, $g(\alpha_i) = b \prod_{j=1}^{n} (\alpha_i - \beta_j)$ and $\prod_{i=1}^{n} g(\alpha_i) = b^n \prod_{i=1}^{n} \prod_{j=1}^{n} (\alpha_i - \beta_j)$. Multiplying both sides by a^m gives the result. ☐

Fact 4. *If $g = fq + r$, then $R(f, g) = a^{(\deg g - \deg r)} R(f, r)$ where a is the leading coefficient of f.*

PROOF. By Fact 3, $R(f, g) = a^{\deg g} \prod_{i=1}^{n} g(\alpha_i) = a^{\deg g} \prod_{i=1}^{n} [f(\alpha_i) q(\alpha_i) + r(\alpha_i)]$. Since the α_i are roots of f, we get $R(f, g) = a^{\deg g} \prod_{i=1}^{n} r(\alpha_i)$. On the other hand, $R(f, r) = a^{\deg r} \prod_{i=1}^{n} r(\alpha_i)$, also by Fact 3. That gives Fact 4 immediately. ☐

Fact 5. $R(f, b) = b^{\deg f}$ *if b is a scalar.*

PROOF. This follows directly from Fact 3 with $g(x) = b$. ☐

Facts 1, 4, and 5 permit the computation of the resultant of any two polynomials by Euclid's algorithm. It is then easy to show that the resultant $R(f, g)$ is an element of the field F, even though it is defined in terms of elements of the bigger field K.

E1. Prove that $R(f, g)$ is in F by induction on the number of steps in Euclid's algorithm for f and g.

The *discriminant* of f is $D(f) = (-1)^{n(n-1)/2} R(f, f')$, where f' is the derivative of f and $n = $ degree of f.

By Fact 2, $D(f) \neq 0$ iff f and f' have no common factors.

We compute $D(f)$ by using Euclid's algorithm on f and f'. Here are some examples.

EXAMPLE 1. Let $f(x) = x - a$. Then $f'(x) = 1$, so $D(f) = (-1)^{(1 \cdot 0)/2} R(f, 1)$ $= R(f, 1) = 1^{\deg f} = 1$, by 5.

EXAMPLE 2. Let $f(x) = x^2 + ax + b$. Then $D(f) = - R(f, f')$. Now $f'(x) = 2x + a$, and

$$x^2 + ax + b = (2x + a)\left(\frac{x}{2} + \frac{a}{4}\right) + \left(b - \frac{a^2}{4}\right).$$

Call $b - (a^2/4) = r$. Then

$$\begin{aligned} D(f) &= (-1)^{2(1)/2} R(f, f') = (-1)R(f, f') \\ &= -R(f', f) \qquad \text{(by Fact 1)} \\ &= 2^{\deg f - \deg r}(-1)\big(R(f', r) \qquad \text{(where } r = b - (a^2/4), \text{ by Fact 4)} \\ &= -2^{(2-0)} R(f', b - (a^2/4)) \\ &= a^2 - 4b. \end{aligned}$$

Note that the roots of f are

$$-\frac{a}{2} \pm \frac{1}{2}\sqrt{a^2 - 4b}\ .$$

So f has two irreducible factors iff $a^2 - 4b = D(f)$ is a square in F. Stickelberger's theorem is true in this case over any field F.

EXAMPLE 3. Let $f(x) = x^3 + qx + r$. Then $f'(x) = 3x^2 + q$ and, doing Euclid's algorithm,

$$x^3 + qx + r = (3x^2 + q)\left(\frac{x}{3}\right) + \left(\frac{2}{3}qx + r\right),$$

$$3x^2 + q = \left(\frac{2qx}{3} + r\right)\left(\frac{9x}{2q} - \frac{27r}{4q^2}\right) + \left(q + \frac{27r^2}{4q^2}\right).$$

Then $D(f)$, using these calculations, is $-4q^3 - 27r^2$.

Now that we know what the discriminant is, Stickelberger's theorem makes sense. We try it out on a few cubic polynomials of the form in Example 3, with $F = \mathbb{Z}_5$.

f	$D(f)$	Information from Stickelberger's theorem	Actual factorization of f in $\mathbb{Z}_5[x]$
$x^3 + 3x + 1$	0	repeated roots	$(x - 1)(x + 3)^2$
$x^3 + 3x + 2$	4	1 or 3 factors	irreducible (no roots)
$x^3 + x$	2	2 factors	$(x + 2)(x^2 + 3x + 3)$
$x^3 + x$	1	1 or 3 factors	$x(x + 1)(x - 1)$

EXAMPLE 4. Let p, q be distinct odd primes, and let $f(x) = x^q - 1$ in $\mathbb{Z}_p[x]$. We need a splitting field K for $x^q - 1$. Let K be a field containing $\mathbb{Z}_p$ with p^n elements so that q divides $p^n - 1$ (for example, let n be the order of p mod q). If β is a primitive element of K, $\beta^{(p^n - 1)/q} = \alpha$ will satisfy $\alpha^q = 1$, with $\alpha \neq 1$. Moreover, q will be the order of α in K (since $\beta^{(p^n - 1)/q} = \alpha$ and β has order $p^n - 1$ in K), so $1, \alpha, \alpha^2, \alpha^3, \cdots \alpha^{q-1}$ are distinct elements of K. Since they are roots in K of $x^q - 1$, they are all the roots of $x^q - 1$. Thus K is a splitting field for $x^q - 1$.

Then we may compute $D(x^q - 1)$ by using Fact 3 of resultants:

$$D(x^q - 1) = (-1)^{(q-1)/2} \prod_{i=1}^{q} q(\alpha^i)^{q-1}.$$

Collecting terms, this becomes

$$(-1)^{(q-1)/2} q^q (1 \cdot \alpha \cdot \alpha^2 \cdots \alpha^{q-1})^{q-1} = (-1)^{(q-1)/2} q^q (\alpha^{q(q-1)/2})^{q-1}$$

$$= (-1)^{(q-1)/2} q^q$$

in $\mathbb{Z}_p$ since $\alpha^q = 1$, and q is odd.

For a special case, suppose $q = 11$, $p = 7$. Then in $\mathbb{Z}_7[x]$,

$$D(x^{11} - 1) = (-1)^{(11-1)/2} 11^{11} \qquad \text{(in } \mathbb{Z}_7)$$

$$= -11^{11}$$

$$= 5$$

which is not a square in $\mathbb{Z}_7$. Applying Stickelberger's theorem, if $r =$ number of irreducible factors of $x^{11} - 1$ in $\mathbb{Z}_7$,

$$r \equiv 11 \pmod{2} \quad \text{iff } 5 = D(x^{11} - 1) \text{ is a square.}$$

Hence $r \not\equiv 11 \pmod 2$ so r is even: $x^{11} - 1$ has an even number of factors in $\mathbb{Z}_7[x]$.

To find the actual factorization of $x^{11} - 1$ in $\mathbb{Z}_7[x]$ would be harder.

We shall use Example 4 in the next chapter.

To prove Stickelberger's theorem we shall need the following two facts about the discriminant.

Proposition 1. *If f is a monic polynomial and has roots $\alpha_1, \ldots, \alpha_n$ in K, then*

$$D(f) = \left(\prod_{j=2}^{n} \prod_{i=1}^{j-1} (\alpha_i - \alpha_j) \right)^2 = \left(\prod_{j=1}^{n} \prod_{i<j} (\alpha_i - \alpha_j) \right)^2$$

(We write this last expression for short as $(\prod_{1 \leqslant i < j \leqslant n}(\alpha_i - \alpha_j))^2$.)

PROOF. If in K, $f(x) = (x - \alpha_1)(x - \alpha_2) \cdots (x - \alpha_n)$, then

$$f'(x) = (x - \alpha_2)(x - \alpha_3) \cdots (x - \alpha_n)$$

$$+ (x - \alpha_1)(x - \alpha_3) \cdots (x - \alpha_n) + \cdots + (x - \alpha_1) \cdots (x - \alpha_{n-1})$$

$$= \sum_{i=1}^{n} \prod_{j=1, j \neq i}^{n} (x - \alpha_j)$$

Thus $f'(\alpha_i) = \prod_{j=1, j\neq i}^n (\alpha_i - \alpha_j)$, so by Fact 3 of resultants,

$$D(f) = (-1)^{n(n-1)/2} \prod_{i=1}^n f'(\alpha_i) = (-1)^{n(n-1)/2} \prod_{i=1}^n \prod_{j=1, j\neq i}^n (\alpha_i - \alpha_j).$$

Now this last product contains $n(n-1)$ terms, of which half involve $\alpha_i - \alpha_j$ with $i < j$, and half involve $\alpha_i - \alpha_j$ with $i > j$. Multiply each of the $n(n-1)/2$ terms of the second type by -1. That uses up all of the $n(n-1)/2$ factors of -1 and gives the result. $\square$

Proposition 2. *Let* $f_1, \ldots, f_r$ *be monic polynomials in* $F[x]$. *Then* $D(f_1 \cdots f_r) = D(f_1) \cdots D(f_r) \cdot R^2$ *for some* R *in* F.

PROOF. We show $D(fg) = D(f)D(g)R(f, g)^2$; the general result will follow easily by induction on r.

Let $f(x)$ have roots $\alpha_1, \alpha_2, \ldots, \alpha_n$ and $g(x)$ have roots $\beta_1, \beta_2, \ldots, \beta_m$. Relabel $\alpha_1, \alpha_2, \ldots, \alpha_n, \beta_1, \ldots, \beta_m$ as $\gamma_1, \ldots, \gamma_n, \gamma_{n+1}, \ldots, \gamma_{n+m}$. Then by Proposition 1,

$$D(fg) = \prod_{j=2}^{n+m} \prod_{i=1}^{j-1} (\gamma_i - \gamma_j)^2,$$

which we split up into three factors:

$$D(fg) = \left[\prod_{j=2}^n \prod_{i=1}^{j-1} (\gamma_i - \gamma_j)^2 \cdot \prod_{j=n+1}^{n+m} \prod_{i=1}^n (\gamma_i - \gamma_j)^2 \right.$$

$$\left. \cdot \prod_{j=n+2}^{n+m} \prod_{i=n+1}^{j-1} (\gamma_i - \gamma_j)^2 \right]$$

$$= \left[\prod_{j=2}^n \prod_{i=1}^{j-1} (\alpha_i - \alpha_j) \right]^2 \cdot \left[\prod_{k=1}^m \prod_{n=1}^n (\alpha_i - \beta_k) \right]^2$$

$$\cdot \left[\prod_{k=1}^m \prod_{i=1}^{k-1} (\beta_i - \beta_k) \right]^2$$

$$= D(f)R(f, g)^2 D(g),$$

as was to be shown. $\square$

E2. Derive information about the cubic polynomials $x^3 + ax + b$ in $\mathbb{Z}_3[x]$ using Stickelberger's theorem, above, when
   (i) $(a, b) = (1, 1)$,
   (ii) $(a, b) = (1, 2)$,
   (iii) $(a, b) = (2, 2)$.

E3. Compute the discriminant of $x^4 + x^2 + x + 1$ in $\mathbb{Z}_3[x]$.

E4. Verify Proposition 2 for $(x^2 + x + 1)(x^2 + 1)$ in $\mathbb{Z}_3[x]$.

B. Roots of Irreducible Polynomials in $\mathbb{Z}_p[x]$

Let K be a field of characteristic p, and let $\mathbb{Z}_p$ be the subfield $\{n \cdot 1 | n = 1, \ldots, p\}$ of K. Let $\phi_p : K \to K$ be the automorphism given by $\phi_p(a) = a^p$. Then, as we observed in Chapter 10, ϕ_p gives a means to check whether an element of K is actually in $\mathbb{Z}_p$, namely, a is in $\mathbb{Z}_p$ iff $\phi_p(a) = a^p = a$. The reason for this being true is that if $a^p = a$, then a is a root in K of the polynomial $x^p - x$. But the p elements of $\mathbb{Z}_p$ are all roots of $x^p - x$, by Fermat's theorem, and $x^p - x$ cannot have more than p roots in the field K. So if $a^p = a$, then a must be in $\mathbb{Z}_p$.

We extend ϕ_p to a function from $K[x]$ to itself by defining $\phi_p(x) = x$; hence

$$\phi_p(a_n x^n + a_{n-1} x^{n-1} + \cdots + a_1 x + a_0) = a_n^p x^n + a_{n-1}^p x^{n-1} + \cdots + a_0^p.$$

Then we have easily

Proposition 3. $\phi_p(f(x)) = f(x)$ iff $f(x)$ is in $\mathbb{Z}_p[x]$.

Suppose now that $f(x)$ is in $\mathbb{Z}_p[x]$ and $f(x) = (x - \alpha_1) \cdots (x - \alpha_m)$ in $K[x]$. Since $f(x)$ is in $\mathbb{Z}_p[x]$, $\phi_p(f(x)) = f(x)$. So

$$f(x) = \phi_p(f(x)) = (x - \alpha_1^p) \cdots (x - \alpha_m^p).$$

By uniqueness of factorization of polynomials in $K[x]$, the set of roots $\{\alpha_1^p, \ldots, \alpha_m^p\}$ is the same as the set $\{\alpha_1, \ldots, \alpha_m\}$. Thus we have

Corollary. *The pth power of any root of a polynomial $f(x)$ in $\mathbb{Z}_p[x]$ is a root of $f(x)$.*

For irreducible polynomials in $\mathbb{Z}_p[x]$ it turns out that once you have one root in K you have them all:

Proposition 4. *If $f(x)$ in $\mathbb{Z}_p[x]$ is irreducible of degree d, K is a field containing $\mathbb{Z}_p$, and α in K is a root of $f(x)$, then in $K[x]$, $f(x) = (x - \alpha)(x - \alpha^p)(x - \alpha^{p^2}) \cdots (x - \alpha^{p^{d-1}})$.*

We have already noted this fact when $p = 2$ in connection with roots of irreducible polynomials used in error-correcting codes.

PROOF. Certainly $\alpha, \alpha^p, \alpha^{p^2}, \ldots$ are all roots of $f(x)$ by the last corollary.

Suppose r is the smallest natural number such that $\alpha^{p^r} = \alpha$. Then $\alpha, \alpha^p, \alpha^{p^2}, \ldots, \alpha^{p^{r-1}}$ are all distinct (otherwise $\alpha^{p^k} = \alpha^{p^h}$ for some h, k, $1 \leqslant k < h < r$; so $(\alpha^{p^k})^{p^{r-h}} = (\alpha^{p^h})^{p^{r-h}} = \alpha$, giving $\alpha^{p^{r-(h-k)}} = \alpha$, contradicting minimality of r). So f has at least r roots in K. Thus r must be $\leqslant$ the degree of f, namely d.

Let $g(x) = (x - \alpha)(x - \alpha^p) \cdots (x - \alpha^{p^{r-1}})$. Since $(\alpha^{p^{r-1}})^p = \alpha$, $\phi_p(g(x)) = g(x)$, so $g(x)$ is in $\mathbb{Z}_p[x]$. Evidently α is a root of $g(x)$. Now $f(x)$ is irreducible and has α as a root, so $f(x)$ must divide $g(x)$. But since the degree of $g(x)$ is $\leqslant$ the degree of $f(x)$, $f(x) = g(x)$. That, plus Exercise E5, completes the proof. □

E5. How do you know in this last proof that there exists a number r such that $\alpha^{p^r} = \alpha$?

From Proposition 4 we get a new proof of a corollary which we know already from Chapter II-10.

Corollary. *An irreducible polynomial $f(x)$ in $\mathbb{Z}_p[x]$ has no multiple roots in any extension field.*

PROOF. We just showed that if $f(x)$ has degree d and α is a root in some field K containing $\mathbb{Z}_p$, then

$$f(x) = (x - \alpha)(x - \alpha^p) \cdots (x - \alpha^{p^{d-1}})$$

and all those roots $\alpha, \alpha^p, \ldots, \alpha^{p^{d-1}}$ are distinct. □

E6. In $GF(16)$ what are the roots of $x^4 + x^3 + 1$? Write them in terms of α where α is a root of $x^4 + x + 1$ in $GF(16)$.

C. Stickelberger's Theorem

We are now ready to prove

Theorem. *Let p be an odd prime, $f(x)$ a monic polynomial of degree m with coefficients in $\mathbb{Z}_p[x]$ without multiple factors (so that its discriminant $D(f) \neq 0$). Let r be the number of irreducible factors of $f(x)$ in $\mathbb{Z}_p[x]$. Then $r \equiv m \pmod 2$ iff $D(f)$ is a square in $\mathbb{Z}_p$.*

PROOF. We prove it first for $r = 1$. The result for general r will follow easily, as we shall see.

Case $r = 1$. We assume $f(x)$ is irreducible of degree m. Let K be a field containing $\mathbb{Z}_p$ in which $f(x)$ splits into linear factors. Let α be a root of f. Then the roots of $f(x)$ are $\alpha, \alpha^p, \alpha^{p^2}, \ldots, \alpha^{p^{m-1}}$ by Proposition 4 of the last section.

Let $\delta(f) = \prod_{j=1}^{m-1}\prod_{i=0}^{j-1}(\alpha^{p^i} - \alpha^{p^j})$ in K. Then $D(f) = (\delta(f))^2$, in K, so $D(f)$ is a square in $\mathbb{Z}_p$ iff $\delta(f)$ is in $\mathbb{Z}_p$.

We test to see whether $\delta(f)$ is in $\mathbb{Z}_p$ by examining $\phi_p(\delta(f))$:

$$\phi_p(\delta(f)) = \prod_{0 \leqslant i < j \leqslant m-1} \left(\phi_p(\alpha^{p^i}) - \phi_p(\alpha^{p^j})\right)$$

$$= \prod_{0 \leqslant i < j \leqslant m-1} (\alpha^{p^{i+1}} - \alpha^{p^{j+1}});$$

changing indices, this becomes

$$\phi_p(\delta(f)) = \prod_{i \leqslant i < j \leqslant m} (\alpha^{p^i} - \alpha^{p^j})$$

$$= \prod_{1 \leqslant i < j \leqslant m-1} (\alpha^{p^i} - \alpha^{p^j}) \cdot \prod_{i=1}^{m-1} (\alpha^{p^i} - \alpha^{p^m})$$

$$= (-1)^{m-1} \prod_{1 \leqslant i < j \leqslant m-1} (\alpha^{p^i} - \alpha^{p^j}) \cdot \prod_{j=1}^{m-1} (\alpha - \alpha^{p^j})$$

$$= (-1)^{m-1} \delta(f).$$

Thus $\delta(f)$ is in $\mathbb{Z}_p$ iff $(-1)^{m-1}\delta(f) = \delta(f)$ iff $(-1)^{m-1} = 1$ iff $m \equiv 1$ (mod 2). Since we have assumed $r = 1$, the theorem is true for $r = 1$.

Suppose now $f = f_1 f_2 \cdots f_r$, a product of r distinct irreducible factors. Then by Proposition 2, above,

$$D(f) = D(f_1 f_2 \cdots f_r) = D(f_1)D(f_2) \cdots D(f_r)R^2$$

for some R in $\mathbb{Z}_p$. Thus in K,

$$\delta(f) = \delta(f_1)\delta(f_2) \cdots \delta(f_r)R$$

where R is in $\mathbb{Z}_p$. To see whether $\delta(f)$ is in $\mathbb{Z}_p$ we apply ϕ_p. By the case $r = 1$, $\phi_p\delta(f_i) = (-1)^{d_i-1}\delta(f_i)$, where $d_i = \deg f_i$. So

$$\phi_p\delta(f) = \delta(f_1)\delta(f_2) \cdots \delta(f_r)R \cdot (-1)^{d_1-1}(-1)^{d_2-1} \cdots (-1)^{d_r-1}$$

$$= \delta(f) \cdot (-1)^{d_1+d_2+ \cdots +d_r-r}.$$

Now $d_1 + d_2 + \cdots + d_r = m$, the degree of $f(x)$. Thus $D(f)$ is a square in $\mathbb{Z}_p$ iff $\delta(f)$ is in $\mathbb{Z}_p$ iff $(-1)^{d_1+ \cdots +d_r-r} = 1$ iff $(-1)^{m-r} = 1$ iff $m \equiv r$ (mod 2), which was what we wanted to prove.

To apply Stickelberger's theorem we need to be able to decide whether $D(f)$ is a square mod p. We shall find a nice way to determine this when we do quadratic reciprocity. It will also turn out that the law of quadratic reciprocity will be a consequence of Stickelberger's theorem.

The proof of Stickelberger's theorem given here was adapted from Swan (1962) and Berlekamp (1968).

E7. Using Stickelberger's theorem, decide how each of these polynomials in $\mathbb{Z}_{11}[x]$ factors:
(a) $x^3 + 3x + 10$;
(b) $x^3 + 4x + 7$;
(c) $x^3 + 5x + 6$;
(d) $x^3 + 5x + 5$;
(e) $x^3 + 4x + 6$;
(f) $x^3 + 3x + 7$.

E8. Do the same as in the above exercise for
(a) $x^3 + 2x^3 + 3x + 5$,
(b) $x^3 + 4x^2 + 9x + 8$.

E9. In $\mathbb{Z}_3[x]$ apply Stickelberger's theorem to
   (a) $x^6 + 2x^5 + x^4 + x^3 + x + 2$,
   (b) $x^5 + x^4 + 2x^3 + 1$,
   (c) $x^7 - 1$.

E10. Consider $x^5 + 4x^4 + 2x^2 + 5x + 9 = f(x)$ in $\mathbb{Z}[x]$. Does the information obtained by applying Stickelberger's theorem to $f(x) \bmod p$ for $p = 3$ and/or 5 and/or 7 . . . give any useful information on how to factor $f(x)$ in $\mathbb{Z}[x]$?

E11. Show that Stickelberger's theorem holds for monic polynomials of degree 3 with real coefficients. One way to do this is to use Sturm's theorem.

*E12. (a) Using complex conjugation instead of ϕ_q, prove the analogue of Stickelberger's theorem for polynomials with real coefficients.
   (b) Assuming (a), determine whether the following is true: If $f(x)$ is a polynomial with real coefficients of degree d, and $D(f) > 0$, then f has $d - 4k$ real roots for some $k \geqslant 0$; if $D(f) < 0$, then f has $d - 2 - 4k$ real roots for some $k \geqslant 0$.

E13. What happens to Stickelberger's theorem for polynomials in $\mathbb{Z}_2[x]$?

Quadratic Residues 16

A. Reduction to the Odd Prime Case

In Chapter I-6 we mentioned that the question of when it is possible to solve $x^2 \equiv a \pmod{m}$ was one that involved a certain amount of nontrivial mathematics. In this chapter we describe a procedure for determining efficiently whether or not a is a square mod m. The main result, known as the law of quadratic reciprocity, was first proved by Gauss and is a cornerstone of number theory.

We begin this chapter with results which show that to decide whether a number a is a square mod m for any m, it suffices to be able to decide whether a is a square mod p when p is an odd prime.

Theorem 1. *Let $m = p_1^{e_1} p_2^{e_2} \cdots p_r^{e_r}$. Then the number a is a square mod m iff there are numbers $x_1, \ldots, x_r$ such that*

$$x_1^2 \equiv a \pmod{p_1^{e_1}},$$
$$x_2^2 \equiv a \pmod{p_2^{e_2}},$$
$$\vdots$$
$$x_r^2 \equiv a \pmod{p_r^{e_r}}.$$

PROOF. If $x_1^2 \equiv a \pmod{p_1^{e_1}}, \ldots, x_r^2 \equiv a \pmod{p_r^{e_r}}$, then by the Chinese remainder theorem there is some x_0 satisfying $x_0 \equiv x_1 \pmod{p_1^{e_1}}, \ldots, x_0 \equiv x_r \pmod{p_r^{e_r}}$. Such an x_0 satisfies $x_0^2 \equiv a \pmod{m}$. Conversely, if $x_0^2 \equiv a \pmod{m}$, then certainly $x_0^2 \equiv a \pmod{p_i^{e_i}}$ for each $i = 1, \ldots, r$. $\square$

Thus to decide whether a number a is a square mod m, it suffices to decide it mod powers of primes dividing m. To do that we must consider separately p odd, and $p = 2$.

Theorem 2. *Let p be an odd prime, and $(a, p) = 1$. Then there is a solution of $x^2 \equiv a \pmod{p^e}$, $e \geq 1$, iff there is a solution of $x^2 \equiv a \pmod{p}$.*

PROOF. If $x_0^2 \equiv a \pmod{p^e}$, then $x_0^2 \equiv a \pmod{p}$. Conversely, suppose a is a square mod p. Let b be a primitive element mod p^e: b exists by Proposition 2 of Chapter III-2. Then $b^r \equiv a \pmod{p^e}$ for some r. We must show that r is even. If $b^r \equiv a \pmod{p^e}$, then $b^r \equiv a \pmod{p}$. Since b is also a primitive element mod p, by Exercise E6 of Chapter III-2, and a is a square mod p, there exists s such that $b^{2s} \equiv a \pmod{p}$, so $b^r \equiv b^{2s} \pmod{p}$. But this implies that $r \equiv 2s \pmod{p-1}$ by Exercise E6 of Chapter III-1. Since p is odd, r must be even, so $r = 2t$ for some number t. But then $b^r = (b^t)^2 \equiv a \pmod{p^e}$ and a is a square mod p^e. $\square$

Theorem 3. *Suppose a is odd. Then*

(a) *a is always a square mod 2,*
(b) *a is a square mod 4 iff $a \equiv 1 \pmod 4$,*
(c) *a is a square mod 2^e, $e \geq 3$, iff $a \equiv 1 \pmod 8$.*

PROOF. (a) and (b) are easy. For (c), suppose first that $b^2 \equiv a \pmod{2^e}$, $e \geq 3$. Then $b^2 \equiv a \pmod 8$, so a must be $\equiv 1 \pmod 8$ by an easy computation.

Conversely, suppose $a = 1 + 8n$ for some fixed integer n. We show that for any e there is some x_e with

$$x_e^2 \equiv 1 + 8n \pmod{2^e}$$

by induction on e.

For $e = 3$, $2^e = 8$, let $x_3 = 1$. Then $x_3^2 \equiv 1 + 8n \pmod 8$, and the case $e = 3$ is solved. Write $x_3^2 = 1 + 8n + 2^3 u_3$ for some integer u_3 (here of course $u_3 = -n$).

Assume $e > 3$ and $x_{e-1}^2 = 1 + 8n + 2^{e-1}u_{e-1}$ for some integer u_{e-1}. Then x_{e-1} is odd. Let $x_e = x_{e-1} + 2^{e-2}u_{e-1}$. Then since $2(e-2) \geq e$,

$$x_e^2 \equiv x_{e-1}^2 + 2^{e-1}x_{e-1}u_{e-1}$$

$$\equiv 1 + 8n + 2^{e-1}u_{e-1} + 2^{e-1}x_{e-1}u_{e-1} \pmod{2^e}.$$

Since x_{e-1} is odd, $x_e^2 \equiv 1 + 8n \pmod{2^e}$. Thus for any $e > 3$ we can solve $x^2 \equiv 1 + 8n \pmod{2^e}$ by induction on e. $\square$

E1. What happens in Theorems 2 and 3 if a is not relatively prime to p?

B. The Legendre Symbol

From Theorems 2 and 3 the only case where we have to find criteria to decide when $x^2 \equiv a \pmod{m}$ is solvable is when $m = p$, an odd prime.

Say that a is a *quadratic residue* (mod p) if $x^2 \equiv a \pmod{p}$ has a solution; otherwise a is a *quadratic nonresidue* (mod p).

To formulate the criteria, we introduce the *Legendre symbol*.

Definition. Let p be an odd prime, and a any number not divisible by p. Then the Legendre symbol $\left(\frac{a}{p}\right)$ is defined by

$$\left(\frac{a}{p}\right) = \begin{cases} 1 & \text{if } a \text{ is a quadratic residue mod } p \\ -1 & \text{if } a \text{ is a quadratic nonresidue mod } p. \end{cases}$$

To decide whether a is a quadratic residue mod p, we manipulate Legendre symbols. Here are the rules.

Theorem 4. *Assume a and b are integers, p is an odd prime, and $(p, ab) = 1$. Then*

(1) $\left(\dfrac{a^2}{p}\right) = 1,$

(2) *if* $a \equiv b \pmod{p}$, $\left(\dfrac{a}{p}\right) = \left(\dfrac{b}{p}\right),$

(3) $\left(\dfrac{ab}{p}\right) = \left(\dfrac{a}{p}\right)\left(\dfrac{b}{p}\right),$

(4) $\left(\dfrac{-1}{p}\right) = (-1)^{(p-1)/2},$

(5) $\left(\dfrac{2}{p}\right) = (-1)^{(p^2-1)/8},$

(6) $\left(\dfrac{p}{q}\right)\left(\dfrac{q}{p}\right) = (-1)^{[(p-1)/2][(q-1)/2]}$ *if p and q are both odd primes.*

Of these, (6) is the *law of quadratic reciprocity*. It says, for example, that if either p or q is $\equiv 1 \pmod 4$, then p is a square mod q iff q is a square mod p.

EXAMPLE. Is -42 a square mod 97? We ask, is $\left(\frac{-42}{97}\right) = 1$ or -1? Using the rules of Theorem 4, we manipulate as follows:

$$\left(\frac{-42}{97}\right) = \left(\frac{-1}{97}\right)\left(\frac{2}{97}\right)\left(\frac{3}{97}\right)\left(\frac{7}{97}\right) \quad \text{(by rule (3))}$$

$$= \left(\frac{3}{97}\right)\left(\frac{7}{97}\right) \quad \text{(by rules (4) and (5))}$$

$$= \left(\frac{97}{3}\right)\left(\frac{97}{7}\right) \quad \text{(by rule (6))}$$

$$= \left(\frac{1}{3}\right)\left(\frac{6}{7}\right) \quad \text{(by rule (2))}$$

$$= \left(\frac{6}{7}\right) \quad \text{(by rule (1))}$$

$$= \left(\frac{2}{7}\right)\left(\frac{3}{7}\right) = \left(\frac{3}{7}\right) \quad \text{(by rules (3) and (5))}$$

$$= (-1)\left(\frac{7}{3}\right) = (-1)\left(\frac{1}{3}\right) = -1 \quad \text{(by rules (6), (2) and (1)).}$$

Thus -42 is not a square mod 97.

It is apparent that the law of quadratic reciprocity, rule (6), is the rule which gives the most striking results. It permits one to decide whether 3 is a square mod 97 by seeing whether 97 is a square mod 3. The former question is not so easy to decide, but the latter is very easy.

E2. Is 45 a quadratic residue mod 47?

E3. Is -13 a quadratic residue mod 37?

E4. Is 8 a quadratic residue mod 37?

E5. Is 99 a quadratic residue mod 101? (*Hint*: Use Theorem 1.)

E6. Is 31 a quadratic residue mod 65?

E7. Is 31 a quadratic residue mod 200?

E8. Is 311 a quadratic residue mod 1001?

The rest of this chapter is devoted to proofs of the various properties of the Legendre symbol collected as Theorem 4. The proofs range from trivial to very clever.

PROOF OF THEOREM 4, RULE (1). Trivial.   □

PROOF OF RULE (2). $\left(\frac{a}{p}\right) = 1$ iff there is some c with $c^2 \equiv a \pmod{p}$ iff there is some c with $c^2 \equiv b \pmod{p}$ (since $a \equiv b \pmod{p}$) iff $\left(\frac{b}{p}\right) = 1$.   □

PROOF OF RULE (3). Let e be a primitive element mod p. Then $a \equiv e^r$, $b \equiv e^s$ for some r, s, so $ab \equiv e^{r+s}$. Now e^t is a square mod p iff t is even.

Thus $\left(\frac{ab}{p}\right) = 1$ iff $r + s$ is even iff r and s are both even or both odd iff $\left(\frac{a}{p}\right)$ and $\left(\frac{b}{p}\right)$ are both $+1$ or both -1 iff $\left(\frac{a}{p}\right)\left(\frac{b}{p}\right) = 1$. □

PROOF OF RULE (4). This is the theorem of Chapter III-3. We proved there that -1 is a square mod p iff $p \equiv 1 \pmod 4$ iff $(p-1)/2$ is even. □

To prove Rules (5) and (6) we need the following generalization of Rule (4).

Euler's Lemma. *If p is an odd prime and $(a, p) = 1$, then $\left(\frac{a}{p}\right) \equiv a^{(p-1)/2} \pmod p$.*

PROOF. Let b be a primitive element mod p, and let $b^r \equiv a \pmod p$. Then $a^{(p-1)/2} \equiv b^{r(p-1)/2} = c$, where, since $c^2 \equiv b^{(p-1)r} \equiv 1^r = 1 \pmod p$, $c \equiv 1$ or $c \equiv -1 \pmod p$. Now $a^{(p-1)/2} \equiv 1 \pmod p$ iff $b^{r(p-1)/2} \equiv 1 \pmod p$ iff $p - 1$ divides $r(p-1)/2$ iff r is even iff a is a square mod p iff $\left(\frac{a}{p}\right) = 1$. In case r is odd, both $\left(\frac{a}{p}\right)$ and $a^{(p-1)/2}$ must be congruent to -1 mod p. That completes the proof of Euler's lemma. □

Note. Euler's lemma also follows from Berlekamp's algorithm, Chapter II-12, E13.

E9. In both (3) and Euler's lemma we used the fact that if e is a primitive element mod p and $e^r \equiv a \pmod p$, then a is a quadratic residue mod p iff r is even. Why is that so?

PROOF OF RULE (5). Using Euler's lemma we need to show that

$$2^{(p-1)/2} \equiv \left\{ \begin{matrix} 1 \\ -1 \end{matrix} \right\} \quad \text{iff} \quad p \equiv \left\{ \begin{matrix} 1 \text{ or } 7 \\ 3 \text{ or } 5 \end{matrix} \right\} \pmod 8.$$

We do it by working with the identity

$$2^{(p-1)/2} \cdot 1 \cdot 2 \cdot 3 \cdots \frac{p-1}{2} = 2 \cdot 4 \cdot 6 \cdots p - 1. \qquad (*)$$

We break up the right-hand side.
Case 1. $p \equiv 1$ or $5 \pmod 8$. Then $(p-1)/4$ is an integer. So the right-hand side of (*) equals

$$2 \cdot 4 \cdots 2\left(\frac{p-1}{4}\right) \cdot 2\left(\frac{p+3}{4}\right) \cdots 2\left(\frac{p-1}{2}\right)$$

$$\equiv 2 \cdot 4 \cdots \frac{p-1}{2} \cdot (-1) \cdot (-3) \cdots \left(-\frac{p-3}{2}\right) \pmod p$$

$$= 1 \cdot 2 \cdots \frac{p-1}{2} \cdot (-1)^{(p-1)/4}.$$

Equating the left and right sides and canceling gives

$$2^{(p-1)/2} \equiv (-1)^{(p-1)/4} = \begin{cases} 1 & \text{if} \quad p \equiv 1 \pmod 8 \\ -1 & \text{if} \quad p \equiv 5 \pmod 8. \end{cases}$$

Case 2. $p \equiv 3$ or 7 (mod 8). Then $(p-3)/4$ is an integer. Then the same kinds of manipulations to the identity (*) above yield

$$2^{(p-1)/2} \cdot 1 \cdot 2 \cdot 3 \cdots \frac{p-1}{2} \equiv 1 \cdot 2 \cdot 3 \cdots \frac{p-1}{2}(-1)^{(p+1)/4} \pmod p,$$

from which one gets

$$2^{(p-1)/2} \equiv (-1)^{(p+1)/4} = \begin{cases} -1 & \text{if} \quad p \equiv 3 \pmod 8 \\ 1 & \text{if} \quad p \equiv 7 \pmod 8. \end{cases}$$

That completes the proof of (5). □

C. Proof of the Law of Quadratic Reciprocity

We are left with (6), the law of quadratic reciprocity. There are reputedly over 150 different proofs of this result. Gauss, in his *Disquisitiones Arithmeticae* (Gauss, 1801), gave the first three. The proof we give is an adaptation by E. R. Berlekamp of a proof of R. G. Swan (see Berlekamp (1968) and Swan (1962)). It uses Stickelberger's theorem. We have chosen this proof because it integrates both themes of this book, the theory of polynomials and number theory. A nice proof which does not require the prerequisites of the proof we give is Gauss's third proof as polished by Eisenstein. It may be found in the elementary number theory texts of Dickson (1929) or McCoy (1965).

PROOF OF RULE (6). (6) involves a relationship between squares in $\mathbb{Z}_p$ and squares in $\mathbb{Z}_q$. To get such a relationship we establish a connection between r, the number of irreducible factors of $x^p - 1$ in $\mathbb{Z}_q[x]$, and e, the order of q mod p. This will imply a connection between p being a square mod q and q being a square mod p, with the aid of Stickelberger's theorem.

We first prove

(i) *If $f(x)$ is any irreducible factor of $(x^p - 1)/(x - 1)$ in $\mathbb{Z}_q[x]$, then the degree of $f(x)$ is e, the order of q mod p.*

PROOF OF (i). Let K be a splitting field of $x^p - 1$ over $\mathbb{Z}_q$. Let α be a root in K of $f(x)$. By Proposition 4, Chapter 15,

$$f(x) = (x - \alpha)(x - \alpha^q)(x - \alpha^{q^2}) \cdots (x - \alpha^{q^{n-1}}),$$

where n is the smallest natural number with $\alpha^{q^n} = \alpha$. We shall show that $n = e$.

First note that since α is a root $\neq 1$ of $x^p - 1$, the order of α in K is > 1 and divides p (a prime), so is exactly p. Thus if $\alpha^r = 1$ for any r, then p divides r.

Let n be the smallest natural number with $\alpha^{q^n} = \alpha$. Then n is the smallest natural number so that $\alpha^{q^n - 1} = 1$, so n is the smallest natural number with p dividing $q^n - 1$. Thus n is the smallest number with $q^n \equiv 1 \pmod{p}$. But that means $n = e$. $\square$

(ii) *The number r of irreducible factors of $x^p - 1$ is $r = 1 + (p - 1)/e$.*

PROOF OF (ii). The degree of $(x^p - 1)/(x - 1)$ is degree $p - 1$, and each irreducible factor of $(x^p - 1)/(x - 1)$ has degree e, by (i). So there are $(p - 1)/e$ irreducible factors of $(x^p - 1)/(x - 1)$ and thus, $r = 1 + (p - 1)/e$. $\square$

We apply Stickelberger's theorem to $x^p - 1$ in $\mathbb{Z}_q[x]$. It asserts that *if*

r is the number of irreducible factors of $x^p - 1$,
d is the degree of $x^p - 1$,
D is the discriminant of $x^p - 1$,

then $r \equiv d \pmod 2$ iff D is a square $(\bmod\ q)$.

Now

$d = p$,
$r = 1 + (p - 1)/e$ where e is the order of $q \bmod p$, by (ii), and
$D = (-1)^{(p-1)/2} p^p$ by Example 4, Chapter 15.

Thus Stickelberger's theorem reads:

$$1 + \frac{p - 1}{e} \equiv p \pmod 2 \;\; iff \; (-1)^{(p-1)/2} p^p \;\; is\ a\ square \bmod q.$$

Since p is odd, this yields:

$$2\ \text{divides}\ \frac{p - 1}{e} \;\; iff \left(\frac{(-1)^{(p-1)/2} p^p}{q} \right) = 1.$$

We analyse the left side:

$$2\ \text{divides}\ \frac{p - 1}{e} \;\; \text{iff}\ e\ \text{divides}\ \frac{p - 1}{2}.$$

Since e is the order of $q \bmod p$,

$$e\ \text{divides}\ \frac{p - 1}{2} \;\; \text{iff}\ q^{(p-1)/2} \equiv 1 \pmod{p}.$$

By Euler's lemma, above,

$$\left(\frac{q}{p} \right) \equiv q^{(p-1)/2} \pmod{p},$$

so

$$q^{(p-1)/2} \equiv 1 \pmod{p} \quad \text{iff} \quad \left(\frac{q}{p}\right) = 1.$$

Thus from Stickelberger's theorem we get

$$\left(\frac{q}{p}\right) = 1 \quad \text{iff} \quad \left(\frac{(-1)^{(p-1)/2}p^p}{q}\right) = 1,$$

or

$$\left(\frac{q}{p}\right) = \left(\frac{(-1)^{(p-1)/2}p^p}{q}\right).$$

Using (1), (2), and (4), this yields

$$\left(\frac{q}{p}\right) = \left(\frac{-1}{q}\right)^{(p-1)/2}\left(\frac{p}{q}\right)\left(\frac{p^2}{q}\right)^{(p-1)/2}$$

$$= (-1)^{((q-1)/2)((p-1)/2)}\left(\frac{p}{q}\right),$$

as was to be proved. □

†E10. Prove that if p is an odd prime, $\sum_{j=1}^{p-1}\left(\frac{j}{p}\right) = 0$.

E11. Euler's lemma is a negative test for primeness: if q is a number and q does not divide $a^{(q-1)/2} - \left(\frac{a}{q}\right)$ for some integer a, then q is not prime. Here one treats $\left(\frac{a}{q}\right)$ as though q were prime. Use this test with $a = 2$ to show that $q = 341$ is not prime, even though $2^{340} \equiv 1 \pmod{341}$.

E12. Show that 7 is a primitive root of any Fermat prime (III-3, E1) of the form $p = 2^{2^k} + 1$ ($k \geqslant 1$), as follows.
   (a) Show that b is a primitive element mod p iff $\left(\frac{b}{p}\right) = -1$.
   (b) Show that $\left(\frac{7}{p}\right) = -1$.

E13. Show that there are infinitely many primes congruent to 7 (mod 12), as follows. Suppose to the contrary that $p_1, \ldots, p_n$ are all such primes. Let

$$t = (2p_1 p_2 \cdots p_n)^2 + 3.$$

   (a) Show that t is divisible by at least one prime $p \equiv 3 \pmod 4$ with $p \neq p_1, \ldots, p_n$ (see I-4, E1).
   (b) Show that if p divides t, then $\left(\frac{-3}{p}\right) = 1$.
   (c) Show that if $\left(\frac{-3}{p}\right) = 1$, then $p \equiv 1 \pmod 3$.
   (d) Use (a) and (c) to conclude that there is a prime p dividing t with $p \equiv 7 \pmod{12}$.

E14. Show that for a, b, c integers, and p an odd prime, $ax^2 + bx + c = 0$ factors mod p iff

$$\left(\frac{b^2 - 4ac}{p}\right) = 1.$$

Duplicate Bridge Tournaments 17

In this chapter we describe a connection between quadratic residues and design of duplicate bridge tournaments.

The connection involves a way of constructing certain square matrices called Hadamard matrices. We begin by looking at them.

A. Hadamard Matrices

A *Hadamard matrix* is an $n \times n$ matrix $\mathbf{S}$ with all entries $+1$ or -1, and with $\mathbf{S} \cdot \mathbf{S}' = n\mathbf{I}$. Here $\mathbf{S}'$ denotes the transpose of $\mathbf{S}$, and $\mathbf{I}$ is the identity matrix.

Here are some examples.

EXAMPLE: $n = 1$. $\qquad\qquad\qquad$ (1).

EXAMPLE: $n = 2$.

$$\begin{pmatrix} 1 & -1 \\ 1 & 1 \end{pmatrix}.$$

EXAMPLE: $n = 4$. Define a 4×4 matrix $\mathbf{S}$ by the following rules. The rows and columns of $\mathbf{S}$ will be labeled $\infty, 1, 2, 3$; $S(i, j)$ denotes the entry in the ith row, jth column. Then

$$S(i, j) = \begin{cases} 1 & \text{if } j = \infty & \text{(the leftmost column)} \\ -1 & \text{if } j \neq \infty, i = \infty & \text{(the rest of the top row)} \\ L(i - j) & \text{otherwise.} \end{cases}$$

Here $L(i-j)$ equals 1 (respectively -1) if $i-j$ is (respectively is not) a square mod 3. When $i-j \neq 0$, $L(i-j)$ is the usual Legendre symbol

$$\left(\frac{i-j}{3}\right).$$

When $i-j=0$, then since $0=0^2$, $L(0)=1$.

Thus the matrix $\mathbf{S}$ is

$$\begin{bmatrix} 1 & -1 & -1 & -1 \\ 1 & L(1-1) & L(1-2) & L(1-3) \\ 1 & L(2-1) & L(2-2) & L(2-3) \\ 1 & L(3-1) & L(3-2) & L(3-3) \end{bmatrix} = \begin{bmatrix} 1 & -1 & -1 & -1 \\ 1 & 1 & -1 & 1 \\ 1 & 1 & 1 & -1 \\ 1 & -1 & 1 & 1 \end{bmatrix}$$

The rows of $\mathbf{S}$ are pairwise orthogonal, that is, $\mathbf{S} \cdot \mathbf{S}' = 4\mathbf{I}$, as is quickly checked. So are the columns of $\mathbf{S}$, since $\mathbf{S} \cdot \mathbf{S}' = \mathbf{S}' \cdot \mathbf{S}$. So $\mathbf{S}$ is a Hadamard matrix.

The idea of using the extended Legendre symbol $L(i-j)$ to construct Hadamard matrices can be extended from the 4×4 case to the $n \times n$ case whenever $4|n$ and $n-1$ is prime. We first illustrate with $n=8$.

EXAMPLE: $n=8$. Number the rows and columns of an 8×8 matrix by ∞, 1, 2, 3, 4, 5, 6, 7. Form the matrix $\mathbf{S} = (S(i,j))$ by defining $S(i,j)$ as follows:

$$\begin{aligned} S(i, \infty) &= 1 && \text{for all } i; \\ S(\infty, j) &= -1 && \text{for all } j \neq \infty; \\ S(i,j) &= L(i-j) && \text{for } i,j \neq \infty, \end{aligned}$$

where $L(i-j) = 1$ (respectively -1) if $i-j$ is (respectively is not) a square mod 7. Again, if $i \neq j$, then

$$L(i-j) = \left(\frac{i-j}{7}\right),$$

the usual Legendre symbol, and $L(0)=1$. Then $\mathbf{S} = (S(i,j))$ is the matrix

$i \backslash j =$	∞	1	2	3	4	5	6	7
∞	1	-1	-1	-1	-1	-1	-1	-1
1	1	1	-1	-1	1	-1	1	1
2	1	1	1	-1	-1	1	-1	1
3	1	1	1	1	-1	-1	1	-1
4	1	-1	1	1	1	-1	-1	1
5	1	1	-1	1	1	1	-1	-1
6	1	-1	1	-1	1	1	1	-1
7	1	-1	-1	1	-1	1	1	1

This is a Hadamard matrix: $\mathbf{S} \cdot \mathbf{S}' = 8\mathbf{I}$ (as is easily verified).

Here is the general example of this type. We assume $4|n$ and $n-1$ is a prime p. Since $p \equiv 3 \pmod 4$, -1 is not a square mod p (Chapter III-2), so for $i \neq i'$ $L(i-i') = -L(i'-i)$. This will be useful in showing our general example is Hadamard. Note also that for $a \neq 0$, $L(a) = \left(\frac{a}{p}\right)$, so we have $L(a)L(b) = L(ab)$ whenever $a, b \neq 0$.

Define an $n \times n$ matrix $\mathbf{S} = (S(i,j))$, where $i = \infty, 1, 2, \ldots, n-1$ are the rows and $j = \infty, 1, 2, \ldots, n-1$ are the columns, by

$$\begin{aligned}
S(i, \infty) &= 1 & &\text{for all } i,\\
S(\infty, j) &= -1 & &\text{for all } j \neq \infty,\\
S(i, j) &= L(i-j) & &\text{for } i, j \neq \infty,
\end{aligned}$$

where again $L(0) = 0$, and

$$L(i-j) = \left(\frac{i-j}{p}\right) \quad \text{for } i \neq j.$$

Theorem. $\mathbf{S}$ *is a Hadamard matrix.*

PROOF. We verify that the rows of $\mathbf{S}$ are pairwise orthogonal.

First suppose $i, i' \neq \infty$. Then the scalar product of the ith and i'th row is

$$S(i, \infty)S(i', \infty) + \sum_{j=1}^{p} S(i,j)S(i',j)$$

$$= 1 + L(i-i)L(i'-i) + L(i-i')L(i'-i') + \sum_{\substack{j=1 \\ j \neq i, i'}}^{p} L((i-j)L(i'-j)).$$

Since $L(0) = 1$ and $L(-1) = -1$ (because $p \equiv 3 \pmod 4$) the second and third terms cancel. Since $L(i-j)L(i'-j) = L((i-j)(i'-j))$ for $j \neq i, i'$, the above scalar product equals

$$1 + \sum_{\substack{j=1 \\ j \neq i, i'}}^{p} L((i-j)(i'-i+i-j)).$$

Let $a = i' - i$, and let $c = i - j$. Then c takes on the values $1, \ldots, p$, omitting p and $-a$. So this sum becomes

$$1 + \sum_{\substack{c=1 \\ c \neq p, -a}}^{p} L(c(a+c)).$$

Since $L(c^2) = 1$, this equals

$$1 + \sum_{\substack{c=1 \\ c \neq -a}}^{p-1} L\left(1 + \frac{a}{c}\right).$$

Set $e = 1 + (a/c)$; as c runs through the numbers 1 through $p-1$, missing $-a$, e runs through the numbers 2 through $p-1$; thus the sum becomes

$$1 + \sum_{e=2}^{p-1} L(e).$$

Since $L(1) = 1$ this becomes

$$\sum_{e=1}^{p-1} L(e).$$

By Exercise E10 of the last chapter, this sum is 0, as we wished to show.

If $i = \infty$, the scalar product of the ∞ and i'th rows is

$$S(\infty, \infty)S(i', \infty) + \sum_{j=1}^{p} S(\infty, j)S(i', j).$$

For $j \neq \infty$, $S(\infty, j) = -1$, so the sum is

$$1 - \sum_{j=1}^{p} S(i', j)$$

$$= 1 - \sum_{j=1}^{p} L(i' - j).$$

Setting $i' - j = c$ this becomes

$$1 - L(0) - \sum_{c=1}^{p-1} L(c) = 0$$

for the same reasons as above. Thus the rows of **S** are orthogonal, and **S** is a Hadamard matrix, completing the proof.   □

E1. Show that there is no 3×3 Hadamard matrix.

*E2. Are there any 6×6 Hadamard matrices?

B. Duplicate Bridge Tournaments

A complete round-robin duplicate bridge tournament works as follows. There are n partnerships, where n is a multiple of 2. A *board* is a deal of the playing cards into four hands of 13 cards each, located in positions N, E, S, W. A partnership plays the N–S hands *head-to-head* against a partnership playing the E–W hands. In a complete tournament each partnership plays once head-to-head against each of the $n - 1$ other partnerships, each on a different board, so there are $n - 1$ boards. Each board is played $n/2$ times.

Scoring works as follows. Each board is played by all partnerships, half playing it N–S, half playing it E–W. After all partnerships have played the board, the partnerships playing E–W are compared and linearly ordered based on how well they did; similarly the partnerships playing N–S. Suppose, for example, $n = 8$, the partnerships are labeled $1, \ldots, 7$ and ∞, and the results on a given board were as follows:

N–S	vs.	E–W	Result
1		2	420 points for N–S

that is, partnership 1 played the N–S hands head-to-head against partnership 2 and the result of the play was that the N–S partnership (= partnership 1) scored 420 points—and partnership 2 gave up 420 points. (We need not go into how individual hands of bridge are played and scored.)

N–S	vs.	E–W	Result
3		4	100 points for E–W
5		6	420 points for N–S
7		∞	400 points for N–S

Then among N–S players, partnerships 1 and 5 did best, then 7, then 3, who did worst. Among their opponents playing E–W, the ordering is reversed: 4 did best, then ∞, finally 2 and 6. The scoring would be to give the worst 0, next 1, next 2, best 3. Ties split the scores. On the illustrated board the scoring would come out as follows:

Partnership	1	2	3	4	5	6	7	∞
Score	2.5	.5	0	3	2.5	.5	1	2

Thus there are three kinds of relationships on a given board. Partnership 1, for example, plays head-to-head against 2, *competes against* 3, 5, 7 but *plays with* 4, 6, ∞. The best result for 1 on this board is for 1 to do well against 2 and have 4, 6, ∞ do well against 3, 5, 7.

In designing a bridge tournament one wants any two partnerships A and B to pay head-to-head on one board, and on the others, compete against each other exactly as many times as they play with each other.

One way to do this is to design the tournament using the Hadamard matrices we constructed.

C. Bridge for 8

We illustrate the general construction by describing an 8 partnership tournament.

Let **S** be the 8×8 Hadamard matrix of Example 8, Section A.

The matrix **S** has the following significance. The row numbers i are the partnerships. Delete the ∞ column; then the remaining column numbers are the boards. The entry $S(i, j)$, called the side function, defines whether partnership i plays board j as N–S (if $S(i, j) = +1$) or E–W (if $S(i, j) = -1$).

If we can arrange the tournament to have partnership i play board j according to the matrix S, then since the rows of S are orthogonal (before deleting the ∞ column), each pair of partnerships i_1 and i_2 will play head-to-head once, play with each other three times, and compete against each other three times. For since $S(i, \infty)S(i_2, \infty) + \sum_{j=1}^{7} S(i_1, j)S(i_2, j) = 0$, being the scalar product of rows i_1 and i_2 of the matrix S, and since $S(i_1, \infty)S(i_2, \infty) = 1$, we have

$$\sum_{j=1}^{7} S(i_1, j)S(i_2, j) = -1.$$

Since $S(i_1, j)S(i_2, j) = 1$ or -1 for each j, there must be three boards j for which $S(i_1, j)S(i_2, j) = 1$, i.e., $S(i_1, j) = S(i_2, j)$, i.e., i_1 and i_2 are playing the same direction (e.g., E–W) and competing against each other, and four boards for which $S(i_1, j) = -S(i_2, j)$, i.e., i_1 and i_2 are playing in opposite directions, i.e., playing with each other or playing head-to-head. Since they play head-to-head only once in the tournament, they must play with each other on three boards.

We have to distribute the partnerships and the boards. To do that is to specify on which *round* partnership i is to play board j. A round is a period of time when certain boards are being played. An efficiently designed tournament would have exactly as many rounds as boards, in this case seven, and would have each partnership playing on a given round.

Here is a way to arrange the partnerships and the boards. Define a function $R(i, j)$, the round function, for $i = \infty, 1, 2, \ldots, 7$, and $j = 1, \ldots, 7$, by:

$$R(\infty, j) = 3j \pmod 7;$$

$$R(i, j) = 3j + (i - j)L(i - j) \pmod 7 \quad \text{if } i \neq \infty.$$

Here $L(i - j)$ is, as before, 1 if $i = j$ and $\left(\frac{i-j}{7}\right)$ if $i \neq j$. Then $R(i, j) = r$ is interpreted to mean that partnership i plays board j in round r.

We must verify of the function R the following.

(1) Given partnership i and board j there is exactly one other partnership $i' \neq i$ playing board j on the same round as i: symbolically,

$$R(i', j) = R(i, j) \quad \text{for exactly one } i' \neq i;$$

for such an i', $S(i', j)S(i, j) = -1$, i.e., one partnership is playing N–S, the other E–W.

(2) Each partnership is playing one and only one board in a given round: $R(i, -)$ is a 1–1 function of j for each fixed i.

E3. Prove (1).

Using the round function we can describe the design. Here it is, in the form of a table.

Round	1	2	3	4	5	6	7
Partnership							
∞	5 −	3 −	1 −	6 −	4 −	2 −	7 −
1	7 +	4 +	1 +	3 −	5 −	6 +	2 −
2	6 −	7 +	3 −	1 +	5 +	2 +	4 −
3	6 +	3 +	5 −	7 −	1 +	4 −	2 +
4	2 +	5 −	3 +	7 +	4 +	6 −	1 −
5	5 +	7 −	2 −	③ +	6 −	4 +	1 +
6	7 −	5 +	2 +	6 +	1 −	3 −	4 +
7	2 −	4 −	5 +	1 −	6 +	3 +	7 +

The circled entry 3 + means that partnership 5 in round 4 plays board 3 as N–S.

The table is obtained by finding for each partnership i and each round r the unique board j such that $R(i, j) = r$, then computing $S(i, j)$.

In setting up the table it is convenient to have a function $B(i, r)$, a board function, so that $B(i, r) = j$ means that in round r, $r = 1, \ldots, 7$, partnership i plays board j. For fixed i, $B(i, r)$ is then the inverse function to the function $R(i, j)$: $B(i, r) = j$ iff $B(i, j) = r$.

Here is the function B. It is defined mod 7:

$$B(\infty, r) = \frac{r}{3} = 5r;$$

$$B(i, r) = \begin{cases} (r - i)/(3 - 1) = 4(r - i) & \text{if } L(3i - r) = 1 \\ (r + i)/(3 + 1) = 2(r + i) & \text{if } L(3i - r) = -1. \end{cases}$$

It is easy to check that for $i = \infty$, $B(\infty, r)$ and $R(\infty, j)$ are inverse functions. When $i \neq \infty$ it is a bit harder.

For $i \neq \infty$, we show $R(i, B(i, r)) = r$ by computing

$$R(i, B(i, r)) = 3B(i, r) + (i - B(i, r))L(i - B(i, r)).$$

First suppose $L(3i - r) = 1$. Then

$$i - B(i, r) = i - (4(r - i)) = 5i - 4r \equiv 4(3i - r) \pmod 7.$$

Since 4 is a square mod 7, $L(i - B(i, r)) = 1$. Thus

$$R(i, B(i, r)) = R(i, 4(r - i)) = 3(4(r - i) + (i - B(i, r))$$

$$= 12(r - i) + 4(3i - r)$$

$$= 8r \equiv r \pmod 7$$

Now suppose $L(3i - r) = -1$. Then $i - B(i, r) = i - (2(r + i)) = -2r$ $-i \equiv 2(3i - r) \pmod 7$. So $L(i - B(i, r)) = L(2(3i - r)) = L(2)L(3i - r)$ $= -1$ (since $L(2) = L(4^2) = 1$). Thus

$$R(i, B(i, r)) = R(i, 2(r + i)) = 3(2(r + i)) - (-2r - i)$$
$$= 8r + 7i \equiv r \pmod 7.$$

E4. Verify for $i \neq \infty$ that $B(i, R(i, j)) = j$.

By showing that $B(i, -)$ is the inverse function of $R(i, -)$, we have in particular verified property (2) above for $R(i, -)$.

To obtain the circled entry in the table above, we compute $B(5, 4)$. Since $L(11) = L(4) = 1$, $B(5, 4) = 4(4 - 5) = -4 = 3$.

D. Bridge for $p + 1$

We consider $p + 1$ players, where p is a prime $\equiv 3 \bmod 4$. The technique used in obtaining the 8 partnership tournament works for designing tournaments with $n = 4m$ players where $n - 1 = p > 3$ is prime: for example, $n = 12, 20, 24, 32, 44, 48 \ldots$. Here is how it is done.

We use the general example of a Hadamard matrix S described in Section A. Thus its entries are defined by the "side" function

$$S(\infty, j) = -1 \qquad \text{for } j = 1, 2, \ldots, p,$$
$$S(i, j) = L(i - j) \quad \text{for } i \neq \infty, j = 1, 2, \ldots, p,$$

where if $i - j \neq 0$,

$$L(i - j) = \left(\frac{i - j}{p}\right),$$

the Legendre symbol.

The interpretation of S is that $S(i, j) = s$ means: partnership i plays board j as $\left\{ \begin{matrix} \text{NS} \\ \text{EW} \end{matrix} \right\}$ if $s = \left\{ \begin{matrix} 1 \\ -1 \end{matrix} \right\}$. As in the 8 partnership case, if we arrange partnerships and boards according to S we will have a well designed tournament.

To complete the design we define the round function $R(i, j)$ as follows. We pick some $b \neq p$ so that $b + 1$ and $b - 1$ are nonzero quadratic residues mod p. Then we set

$$R(\infty, j) = bj$$
$$R(i, j) = bj + (i - j)L(i - j) \quad \text{if } i \neq \infty \pmod p.$$

Note that with $n = 8$ and $p = 7$, $b = 3$ satisfies the conditions we desire. Note also that in the verification $R(i, B(i, r)) = r$ we needed that 2 and 4 were squares mod 7.

E5. Show that for any $p \geqslant 7$, b can be chosen to be either 2, 3, or 5.

We must verify:

(1) $R(i, j) = R(i', j)$ *for exactly one* $i' \neq i$, *and for that* i', $S(i, j)S(i', j) = -1$;
(2) $R(i, -)$, *as a function of one variable with* i *fixed, is* 1–1.

In the verifications all arithmetic is mod p.

PROOF OF (1). If $i = \infty$, $R(\infty, j) = R(i, j)$ iff $bj = bj + (i - j)L(i - j)$ for some i. Now the latter equation is unsolvable if $i \neq j$, for then $(i - j)L(i - j) \neq 0$. Thus $R(\infty, j) = R(i, j)$ exactly when $i = \infty$ or j; in the latter case, $S(\infty, j)S(j, j) = -1$.

If $i, i' \neq \infty$, and $R(i, j) = R(i', j)$, then $j \neq i, i'$ by the last paragraph (otherwise both would equal $R(\infty, j)$). So $R(i, j) = R(i', j)$ iff $bj + (i - j)L(i - j) = bj + (i' - j)L(i' - j)$ iff $(i - j)L(i - j) = (i' - j)L(i' - j)$. If $i' \neq i$, this is true only if $L(i - j) = -L(i' - j)$ and $i - j = -(i' - j)$, or $i' = 2j - i$. So there is at most one $i' \neq i$ satisfying $R(i, j) = R(i', j)$.

Conversely, if $i' = 2j - i$, then $L(i' - j) = L(2j - i - j) = L(-(i - j)) = L(-1)L(i - j) = -L(i - j)$ since $p \equiv 3 \pmod 4$. So $i' = 2j - i$ is the unique solution of $R(i, j) = R(i', j)$ with $i' \neq i$ and neither $= \infty$. Evidently $S(i, j)S(i', j) = -1$. $\square$

PROOF OF (2). To show $R(i, j)$, for fixed i, is 1–1 as a function of j, we define the board function $B(i, r)$, which will, for fixed i, be the inverse of $R(i, j)$:

$$B(\infty, r) = \frac{r}{b} \qquad (b^{-1} \text{ makes sense since } (b, p) = 1);$$

$$B(i, r) = \begin{cases} (r - i)/(b - 1) & \text{if } L(bi - r) = 1 \\ (r + i)/(b + 1) & \text{if } L(bi - r) = -1. \end{cases}$$

E6. Show that $B(i, R(i, j)) = j$ and $R(i, B(i, r)) = r$ for all i, j, r.

Once you show that, (2) is proved.

Using **S** and **R** or **B** one can set up a bridge tournament for any n partnerships where $n = 4m$ and $n - 1$ is prime.

The description of bridge tournament design we have described comes from Berlekamp and Hwang (1972). They also describe designs for n partnerships where $n \neq 4m$. The Hadamard matrices used in the design were discovered by R.E.A.C. Paley in 1933. Hadamard matrices have application also in statistical design of experiments, in telephone circuitry, and in biology. The problem of finding for which n there exist $n \times n$ Hadamard matrices is an active research area in combinatorial mathematics. See W. D. Wallis, et al. (1972), for a long article on Hadamard matrices and many references.

E7. Write down the table for a 12 partnership tournament.

E8. Let

$$H(i, j) = \begin{cases} r, & \text{the round in which partnerships } i \text{ and } j \text{ play head-to-head} \\ 0, & \text{if } i = j. \end{cases}$$

Is there a "nice" general formula for $H(i, j)$?

Algebraic Number Fields

18

Any nonconstant polynomial with rational coefficients has roots in the complex numbers. Those complex numbers which are roots of polynomials with rational coefficients are called *algebraic numbers*. Examples are $\sqrt{2}$, i, $\sqrt[4]{5}$, $2 + \sqrt{3}$.

The discovery that there were algebraic numbers which are not rational was made in the fifth century B. C., when the irrationality of $\sqrt{2}$, the length of the diagonal of a square of side 1, was discovered by an ill-fated Greek mathematician of the school of Pythagoras. It was such a startling discovery to the Pythagoreans that the discoverer was reputedly drowned for the crime of heresy. But once the Greeks got over the shock of realizing that reality was not rational, they discovered a large quantity of irrational algebraic numbers which can be represented as lengths of line segments constructed by straightedge and compass.

There are many complex numbers which are not algebraic, that is, not roots of polynomials with rational coefficients. Those numbers which are not algebraic are called *transcendental*. Proving that a given number is not algebraic is invariably an impressive achievement. The most famous results are that e is transcendental (Hermite, 1873), that π is transcendental (Lindemann, 1882), and that if a and b are algebraic with $a \neq 0$, 1 and b not rational, then a^b is transcendental (Gelfond, 1934; Schneider, 1935). For an exposition of these results, see Niven (1956).

We can get many fields made up of algebraic numbers, in the following way. Let α be an algebraic number, which is a root of a polynomial $p(x)$ in $Q[x]$ of degree d. We consider the evaluation homomorphism

$$\phi_\alpha : \mathbb{Q}[x] \to \mathbb{C}$$

given by $\phi_\alpha(p(x)) = p(\alpha)$. The image of ϕ_α is the set of all polynomials in α with rational coefficients. Call that image $\mathbb{Q}[\alpha]$.

Theorem. $\mathbb{Q}[\alpha]$ *is a field.*

Before proving this let us look at some examples.

EXAMPLE 1. Let $\alpha = i$. Of course i is a root of $x^2 + 1$ so i is an algebraic number. We consider $\mathbb{Q}[i] \subseteq \mathbb{C}$, the set of all polynomials in i with rational coefficients. Claim: $\mathbb{Q}[i]$ consists of elements of the form $a + bi$, a, b rational. For let $p(x)$ be any polynomial in $\mathbb{Q}[x]$. By the division algorithm there is some quotient $q(x)$ in $\mathbb{Q}[x]$ and some remainder $r(x) = a + bx$ in $\mathbb{Q}[x]$ such that in $\mathbb{Q}[x]$,

$$p(x) = (x^2 + 1)q(x) + (a + bx). \tag{1}$$

Now evaluate (1) at $x = i$. Since $i^2 + 1 = 0$, we get an equality of complex numbers,

$$p(i) = a + bi.$$

It is easy to check that $\mathbb{Q}[i]$ is a ring, either by directly checking the axioms, or by using the general principle:

If R, S are rings and $\phi : R \rightarrow S$ is a ring homomorphism, then $\phi(R)$, the image of ϕ, is a ring.

(See E8.) In fact, $\mathbb{Q}[i]$ is a field, because if $a + bi$ is a nonzero element of $\mathbb{Q}[i]$, it has an inverse in $\mathbb{C}$, namely $a/(a^2 + b^2) - ib/(a^2 + b^2)$, and this is again an element of $\mathbb{Q}[i]$.

EXAMPLE 2. Let $\alpha = \sqrt{2}$, and consider $\mathbb{Q}[\sqrt{2}]$. Since $\sqrt{2}$ is a root of $x^2 - 2$, $\sqrt{2}$ is an algebraic number. One shows as with $\mathbb{Q}[i]$ that if $p(x)$ is any polynomial in $\mathbb{Q}[x]$, there are $q(x)$, a, b such that

$$p(x) = q(x)(x^2 - 2) + a + bx;$$

it follows that in $\mathbb{C}$,

$$p(\sqrt{2}) = a + b\sqrt{2}$$

for some a, b in $\mathbb{Q}$. Thus $\mathbb{Q}(\sqrt{2}) = a + b\sqrt{2}$ in $\mathbb{C}$, a, b rational numbers. As before, $\mathbb{Q}(\sqrt{2})$ is a ring.

(Multiplication works using $(\sqrt{2})^2 = 2$ as follows:

$$(a + b\sqrt{2})(c + d\sqrt{2}) = ac + (ad + bc)\sqrt{2} + bd(\sqrt{2})^2$$
$$= (ac + 2bd) + (ad + bc)\sqrt{2}.$$

To check that $\mathbb{Q}[\sqrt{2}]$ is a field we again observe that nonzero elements of $\mathbb{Q}[\sqrt{2}]$ have inverses in $\mathbb{Q}[\sqrt{2}]$:

$$1/(a + b\sqrt{2}) = \frac{a - b\sqrt{2}}{(a + b\sqrt{2})(a - b\sqrt{2})} = \frac{a}{a^2 - 2b^2} - \left(\frac{b}{a^2 - 2b^2}\right)\sqrt{2}.$$

(Note that $a^2 - 2b^2$ is never zero unless $a = b = 0$, for otherwise $\sqrt{2}$ would be a rational number.)

EXAMPLE 3. Let $\omega = (-1\sqrt{-3})/2$, a complex number satisfying $\omega^3 = 1$, and let $\alpha = \omega(\sqrt[3]{2})$. Then $\mathbb{Q}[\alpha]$ is all polynomials in $\alpha = \omega(\sqrt[3]{2})$, that is, all rational linear combinations of $1, \alpha, \alpha^2, \alpha^3, \ldots, \alpha^n, \ldots$, a subring of the complex numbers.

Evidently α is a root of the irreducible polynomial $x^3 - 2$ in $\mathbb{Q}[x]$, so $\alpha^3 = 2$. Thus we can describe any power of α. If m is any positive integer, write $m = 3n + k$, with $k = 0$, 1 or 2. Then $\alpha^m = \alpha^{3n+k} = 2^n\alpha^k$. So every power of α is a $\mathbb{Q}$-multiple of 1, α, or α^2. It is then very easy to see that every polynomial in α with rational coefficients can be written as a $\mathbb{Q}$-linear combination of 1, α and α^2.

E1. Show that if $a_1 + b_1\alpha + c_1\alpha^2 = a_2 + b_2\alpha + c_2\alpha^2$ in $\mathbb{Q}[\alpha]$, then $a_1 = a_2, b_1 = b_2, c_1 = c_2$.

We can show that $\mathbb{Q}[\alpha]$ is a field, that is, every nonzero element of $\mathbb{Q}[\alpha]$ has an inverse in $\mathbb{Q}[\alpha]$, by using Bezout's lemma.

Let $f(\alpha) = a + b\alpha + c\alpha^2 \neq 0$ be in $\mathbb{Q}[\alpha]$. Since $p(x) = x^3 - 2$ is irreducible in $\mathbb{Q}[x]$, the greatest common divisor of $p(x)$ and $f(x) = a + bx + cx^2$ is 1. So there are $r(x)$ and $s(x)$ in $\mathbb{Q}[x]$ with $p(x)r(x) + f(x)s(x) = 1$. Setting $x = \alpha$, we get that $f(\alpha)s(\alpha) = 1$ (since $p(\alpha) = 0$), so $f(\alpha)$ has an inverse in $\mathbb{Q}[\alpha]$.

Before stating the following theorem, recall (Chapter III-8B) that if α is algebraic over $\mathbb{Q}$, then there is a unique monic irreducible polynomial $p(x)$ in $\mathbb{Q}[x]$ with α as a root. We called $p(x)$ the minimal polynomial of α over $\mathbb{Q}$.

E2. Prove that there is such a polynomial $p(x)$ without looking at Chapter III-8.

Theorem. *Let α in $\mathbb{C}$ be an algebraic number. Then $\mathbb{Q}[\alpha]$ is a field. In fact, let $p(x)$ be the minimal polynomial of α over $\mathbb{Q}$. Then $\mathbb{Q}[\alpha]$ is isomorphic to $\mathbb{Q}[x]/p(x)$.*

PROOF. It is easy to check that the set of polynomials in α, $\mathbb{Q}[\alpha]$, is a commutative ring with identity. Let $f(\alpha)$ be a nonzero element of $\mathbb{Q}[\alpha]$, where $f(x)$ is a polynomial in $\mathbb{Q}[x]$. The $p(x)$ does not divide $f(x)$, so $1 = f(x)r(x) + p(x)s(x)$ for some polynomials $r(x)$ and $s(x)$ in $\mathbb{Q}[x]$ (since $p(x)$ is irreducible). The $1 = f(\alpha)r(\alpha)$. So $f(\alpha)$ has an inverse. Since every nonzero element of $\mathbb{Q}[\alpha]$ has an inverse, $\mathbb{Q}[\alpha]$ is a field.

To prove that $\mathbb{Q}[\alpha]$ is isomorphic to $\mathbb{Q}[x]/(p(x))$, we let $\phi : \mathbb{Q}[x]/(p(x)) \to \mathbb{Q}[\alpha]$ by $\phi([f(x)]) = f(\alpha)$, where $[f(x)]$ denotes the congruence class of the polynomial $f(x)$. Since $\phi(a) = a$ for a in $\mathbb{Q}$, and

$\phi([x]) = \alpha$, a root of $p(x)$, ϕ is well defined (see Chapter 10, Proposition 4). It is easy to see that ϕ is a homomorphism. Since $\mathbb{Q}[x]/(p(x))$ is a field, ϕ is 1–1 (Chapter 10, Exercise E14). Since any polynomial $f(\alpha)$ in $\mathbb{Q}[\alpha]$ is the image under ϕ of the corresponding congruence class $[f(x)]$, ϕ is onto. So ϕ is an isomorphism. That completes the proof.   □

Note. We could have omitted the first paragraph of this proof. For since $\mathbb{Q}[\alpha]$ is isomorphic to a field, it also must be a field.

A consequence of the isomorphism is that every element of $\mathbb{Q}[\alpha]$ can be written as a polynomial in α of degree $< n$, the degree of $p(x)$.

Note. The is nothing special about $\mathbb{Q}$ in this last example. If F is any subfield of $\mathbb{C}$, and α is a complex number which satisfies an irreducible polynomial in $F[x]$ of degree n, then $F[\alpha]$ is a field.

EXAMPLE. When $F = \mathbb{R}$, $\alpha = i$, then $\mathbb{R}[i] = \mathbb{C}$ itself.

Consider how many fields we have discovered in this course. At the start you probably only knew three of them, $\mathbb{Q}$, $\mathbb{R}$ and $\mathbb{C}$. How many do you know now?

In introducing the theorem of this chapter, we said that we were going to describe fields of algebraic numbers. In fact, if α is an algebraic number, then all the elements of $\mathbb{Q}[\alpha]$ are algebraic numbers. This is the content of Proposition 2, Chapter III-8.

Using some linear algebra one can prove also the following.

*E3. (a) If γ is algebraic over $\mathbb{Q}[\alpha]$ (that is, γ is a root of a polynomial with coefficients in $\mathbb{Q}[\alpha]$) where α is algebraic over $\mathbb{Q}$, then γ is algebraic over $\mathbb{Q}$.

(b) If α, β are algebraic numbers, then $\alpha + \beta$, and $\alpha \cdot \beta$ are algebraic numbers.

*E4. The set of all algebraic numbers is a field.

The following exercises are easier.

E5. Show that the following are algebraic numbers:
(a) $3 + 2\sqrt{2}$;
(b) $3 + 2(\omega\sqrt[3]{2})^2$;
(c) any element of $\mathbb{Q}[i]$.

E6. (a) Show that if α is an algebraic number, then $-\alpha$ is an algebraic number.
(b) Show that if α is an algebraic number, then α^{-1} is an algebraic number.

E7. Find the inverse of $1 + 2\sqrt[3]{3} + 4(\sqrt[3]{3})^2$ in the form $a + b(\sqrt[3]{3}) + c(\sqrt[3]{3})^2$.

E8. Prove that if R and S are rings and $\phi : R \to S$ is a ring homomorphism, then $\phi(R)$, the image of R in S, is a ring.

E9. Describe the smallest field containing $\mathbb{Q}$ and the complex number $(2 + \sqrt[3]{5})^{1/2}$.

E10. Let ζ be an imaginary 7th root of 1. Describe all elements of $\mathbb{Q}[\zeta]$.

*E11. (a) By induction on n prove that there is a polynomial $f(x)$ of degree $\leqslant n$ with integer coefficients such that $f(\cos \theta) = \cos n\theta$.
   (b) By setting $\theta = \pi/n$, prove that $\cos(\pi/n)$ is an algebraic number.
   (c) Show that $\sin(\pi/n)$ is an algebraic number.

19 Isomorphisms, II

We showed in the last chapter that if α is a complex number which is a root of the irreducible polynomial $p(x)$ in $\mathbb{Q}[x]$, then $\mathbb{Q}[x]/(p(x)) \cong \mathbb{Q}[\alpha]$ by the function which sends $[f(x)]$ to $f(\alpha)$.

Since $\mathbb{Q}[x]/(p(x)) \cong \mathbb{Q}[\alpha]$, why not say $\mathbb{Q}[x]/(p(x)) = \mathbb{Q}[\alpha]$? The difference is that $\mathbb{Q}[x]/(p(x))$ is a formally constructed system, while $\mathbb{Q}[\alpha]$ is a subset of $\mathbb{C}$, so is more "real" (and, in particular, may be thought of as a set of points in the plane). It turns out also that there may be more than one subfield of $\mathbb{C}$ which is isomorphic to $\mathbb{Q}[x]/(p(x))$. For example, $x^3 - 2$ has three roots in $\mathbb{C}$, $\sqrt[3]{2}$, $\omega\sqrt[3]{2}$, and $\omega^2\sqrt[3]{2}$, where $\omega = e^{(2i\pi/3)}$ satisfies $\omega^3 = 1$. We can define three ring homomorphisms from $\mathbb{Q}[x]/(x^3 - 2)$ into $\mathbb{C}$, given by sending $[f(x)]_{x^3-2}$ to $f(\sqrt{2})$, $f(\omega\sqrt[3]{2})$, or $f(\omega^2\sqrt[3]{2})$. So, in particular,

$$\mathbb{Q}[x]/(x^3 - 2) \cong \mathbb{Q}[\sqrt[3]{2}],$$
$$\mathbb{Q}[x]/(x^3 - 2) \cong \mathbb{Q}[\omega\sqrt[3]{2}],$$
$$\mathbb{Q}[x]/(x^3 - 2) \cong \mathbb{Q}[\omega^2\sqrt[3]{2}].$$

Now $\mathbb{Q}[\sqrt[3]{2}]$ and $\mathbb{Q}[\omega\sqrt[3]{2}]$ are really different subsets of $\mathbb{C}$: the first is a subset of $\mathbb{R}$, while the second satisfies $\mathbb{Q}[\omega\sqrt[3]{2}] \cap \mathbb{R} = \mathbb{Q}$.

The topic of isomorphisms of fields leads into the beautiful subject of Galois theory of fields, for which we refer you to a more advanced book on algebra (such as Herstein (1964)), noting only that Galois theory is the tool one uses to prove the theorem of Abel cited in Chapter II-3.

E1. Show that the only 1–1 homomorphism from $\mathbb{Q}$ to $\mathbb{C}$ is the identity homomorphism i, $i(a/b) = a/b$ in $\mathbb{C}$.

E2. Find a field of the form $\mathbb{Q}[x]/p(x)$ which has five different ring homomorphisms into $\mathbb{C}$, exactly three of which have their image inside the real numbers.

E3. If $\mathbb{Q}[x]/p(x) = F$ is a field, and s_2 is the number of distinct ring homomorphisms of F into the complex numbers whose image is not contained in the real numbers, show that s_2 is an even number

E4. Let $p(x) = x^4 + x^3 + x^2 + x + 1$ and consider $\mathbb{Q}[x]/p(x) = K$. Let $\zeta = e^{2\pi i/5}$, and let $F = \mathbb{Q}[\zeta]$.
  (a) Show that every ring homomorphism of K into $\mathbb{C}$ actually has its image in $F = \mathbb{Q}[\zeta]$, and so yields an isomorphism of K with F.
  (b) Show that there are exactly four different isomorphisms of K and F.
  (c) Let $\text{Aut}(F)$ be the set of isomorphisms of F with itself: f is in $\text{Aut}(F)$ iff $f : F \to F$ is an isomorphism. Show that $\text{Aut}(F)$ is closed under composition of functions and is an abelian group with four elements.

E5. For F any algebraic number field, let $\text{Aut}(F)$ be as in E4(c). Show that $\text{Aut}(F)$ is a group.

*E6. Let $p(x)$ be in $\mathbb{Q}[x]$, $\deg p(x) = n$, let α in $\mathbb{C}$ be a root of $p(x)$, and let $F = \mathbb{Q}[\alpha]$. Suppose that every ring homomorphism from F to $\mathbb{C}$ has its image in F. Show that
  (a) F is a splitting field for $p(x)$,
  (b) $\text{Aut}(F)$ has exactly n elements.

E7. Find an example of $p(x)$ in $\mathbb{Q}[x]$ of degree 3 to which E6 applies.

20 Sums of Two Squares

Inside any algebraic number field K is the set of numbers of K which satisfy a monic polynomial with coefficients in $\mathbb{Z}$ (not just in $\mathbb{Q}$). The set of these numbers is called the ring of integers of K, and turns out to be a ring containing $\mathbb{Z}$, but not a field.

Here is a particularly interesting example. Inside $\mathbb{Q}[i]$ is the set $\mathbb{G}$ of integers of $\mathbb{Q}[i]$, called by the special name of the Gaussian integers. We want to know exactly which numbers are in $\mathbb{G}$.

Proposition. $\mathbb{G} = \{a + bi \mid a, b \text{ in } \mathbb{Z}\}$, *that is,* $\alpha = a + bi$ *satisfies a monic polynomial with coefficients in $\mathbb{Z}$ if and only if a and b are in $\mathbb{Z}$.*

PROOF. First, if $\alpha = a + bi$ is in $\mathbb{Q}[i]$, then α is a root of the polynomial

$$p_0(x) = (x - (a + bi))(x - (a - bi)) = x^2 - 2ax + (a^2 + b^2),$$

which is in $\mathbb{Q}[x]$ since a, b are in $\mathbb{Q}$. If $b \neq 0$, $p_0(x)$ is irreducible, because it has no roots in $\mathbb{Q}$.

Suppose $\alpha = a + bi$ where a and b are in $\mathbb{Z}$. Then $p_0(x)$ is a monic polynomial with coefficients in $\mathbb{Z}$, so α is in $\mathbb{G}$.

Conversely, suppose α is in $\mathbb{G}$. We want to show that a and b are in $\mathbb{Z}$.

Since α is in $\mathbb{G}$, α satisfies some monic polynomial $p(x)$ with coefficients in $\mathbb{Z}$. Since $p_0(x)$ is irreducible and has α as a root, $p_0(x)$ must divide $p(x)$.

Let $p(x) = p_0(x)q(x)$. Since $p(x)$ is a polynomial with integer coefficients, Gauss's lemma tells us that there are rational numbers r, s such that $rp_0(x)$ and $sq(x)$ are in $\mathbb{Z}[x]$ and $p(x) = (rp_0(x))(sq(x))$. Since $p(x)$ is monic, the leading coefficient of $rp_0(x)$ must be $+1$ or -1. But $p_0(x)$ is monic. So r must be $+1$ or -1, which means that since $rp_0(x)$ has coefficients in $\mathbb{Z}$, $p_0(x)$ has coefficients in $\mathbb{Z}$.

Since $p_0(x)$ has coefficients in $\mathbb{Z}$, $2a$ and $a^2 + b^2$ must be in $\mathbb{Z}$. Hence $4(a^2 + b^2) - (2a)^2 = 4b^2$ is in $\mathbb{Z}$. Since $2b$ is in $\mathbb{Q}$ and $(2b)^2$ is in $\mathbb{Z}$, $2b$ must be in $\mathbb{Z}$. Write $a = r/2$, $b = s/2$ with r, s in $\mathbb{Z}$. Now

$$a^2 + b^2 = \left(\frac{r}{2}\right)^2 + \left(\frac{s}{2}\right)^2 = \left(\frac{r^2 + s^2}{4}\right)$$

is in $\mathbb{Z}$. So 4 divides $r^2 + s^2$. But this can only happen in case r and s are both even. (Why?) Thus b and a are in $\mathbb{Z}$.

Thus $\mathbb{G}$ is the set of all complex numbers of the form $a + bi$, a, b in $\mathbb{Z}$, as we wished to prove. $\qquad\qquad\qquad\qquad\qquad\qquad\qquad\qquad\qquad\square$

It is easy to check that $\mathbb{G}$ is a commutative ring. It turns out that $\mathbb{G}$ is almost as "nice" as $\mathbb{Z}$, in the sense that just like $\mathbb{Z}$ or $F[x]$, $\mathbb{G}$ has a division algorithm, a Euclidean algorithm, unique factorization, etc. We shall develop these facts in the process of proving a very elegant result of Gauss, which illustrates in a small way how knowledge of algebraic numbers and finite fields contributes to knowledge of ordinary numbers.

Theorem. *An odd prime number p (in $\mathbb{Z}$) is the sum of two squares if and only if p is congruent to* 1 mod 4.

Try some examples.
We shall divide the proof up into a number of steps. Some of them will be exercises.

PROOF. Part of the proof is easy.
(i) Suppose p is an odd prime and $p = a^2 + b^2$. Then p cannot be congruent to 3 (mod 4) because $a^2 + b^2$ cannot be congruent to 3 (mod 4) for any integers a, b.

E1. Why?

The converse, showing that if p is congruent to 1 (mod 4), then p is the sum of two squares, is much more involved, and uses the Gaussian integers $\mathbb{G}$ in an essential way.

(ii) Any complex number has a *norm* associated with it which is a real number: $N(a + bi) = (a + bi)(a - bi) = a^2 + b^2$. $N(a + bi)$ is the square of the length of the complex number $a + bi$ thought of as a vector in real 2-space.

(a) The norm of the product of two complex numbers is equal to the product of the norms of the two numbers.
(b) If $a + bi$ is in $\mathbb{G}$, then $N(a + bi)$ is a positive integer, unless $a = b = 0$.
(c) Say that $a + bi$ is a *unit* of $\mathbb{G}$ if $a + bi$ is in $\mathbb{G}$ and there is another element of $\mathbb{G}$, $c + di$, such that their product $(a + bi)(c + di) = 1$. Then a Gaussian integer is a unit if and only if it has norm $= 1$, and so the only units of $\mathbb{G}$ are 1, -1, i and $-i$.

E2. Prove these assertions.

(iii) For G we have a

Division Theorem. *If $\alpha \neq 0$, β are in G, there exist θ and ρ in G with $N(\rho) \leqslant N(\alpha)/2$ such that $\beta = \alpha\theta + \rho$.*

Note. The quotient and remainder are not unique, however, since, for example,

$$7 + 6i = 4(2 + i) + (-1 + 2i) = 4(2 + 2i) + (-1 - 2i).$$

PROOF OF THE DIVISION THEOREM. To find θ, ρ, we first observe that $\beta = \alpha\theta + \rho$ if and only if $\bar{\alpha}\beta = \bar{\alpha}\alpha\theta + \bar{\alpha}\rho$ where $\bar{\alpha} = a - bi$ is the complex conjugate of $\alpha = a + bi$; also $N(\rho) \leqslant N(\alpha)/2$ iff $N(\bar{\alpha}\rho) \leqslant N(\bar{\alpha}\alpha)/2$.

Thus since $\bar{\alpha}\alpha = c$, an integer > 0, it suffices to prove the division theorem with $\alpha = c$, a positive integer. So we shall prove that given β in G, $c > 0$ in $\mathbb{Z}$, there are θ, ρ with $\beta = c\theta + \rho$ and $N(\rho) \leqslant N(c)/2 = c^2/2$.

Let $\beta = a + bi$. We want to choose $\theta = e + fi$ so that $N(\beta - c\theta) \leqslant c^2/2$, that is,

$$(a - ce)^2 + (b - cf)^2 \leqslant c^2/2.$$

Now the numbers $a - ce$, for e in $\mathbb{Z}$, are c units apart on the real line. So one of them is within $c/2$ of 0. Thus there is some integer e_0 such that $a - ce_0 \leqslant c/2$. Also, there is some integer f_0 so that $b - cf_0 \leqslant c/2$. But then

$$N((a + bi) - c(e_0 + f_0 i)) = (a - ce_0)^2 + (b - cf_0)^2$$
$$\leqslant c^2/4 + c^2/4 = c^2/2.$$

Setting $\rho = (a - ce_0) + (b - cf_0)i$, $\theta = e_0 + f_0 i$ proves the division theorem. $\square$

(iv) Given the division theorem we get a Euclidean algorithm, a greatest common divisor of two elements of G, a Bezout's lemma, etc.

E3. Why can't the Euclidean algorithm go on forever in G?

E4. Show that even though the remainder in the division algorithm need not be unique, the greatest common divisor of two elements of G must be unique, in the sense that if δ_1, δ_2 are both greatest common divisors of two numbers α and β of G, then $\delta_1 = \varepsilon\delta_2$, where ε is an unit of G.

We shall call a *prime* of G an element with norm > 1 which cannot be factored into a product of two other elements of G with norms > 1.

E5. Show that if σ is a prime of G and σ divides $\alpha\beta$, then σ divides α or σ divides β.

Let us return now to the problem of when a prime is the sum of two squares.

(v) An element n of $\mathbb{Z}$ is a sum of two squares in $\mathbb{Z}$ if and only if n is the norm of an element of G.

For $n = a^2 + b^2$ means that $n = N(a + bi)$. This gives a criterion for deciding when a prime is a sum of two squares in $\mathbb{Z}$—when it is a norm of an element of G.

(vi) Recall the fact about the field $\mathbb{Z}_p$ of congruence classes mod p, that $\mathbb{Z}_p$ has a primitive element s, so that every nonzero element of $\mathbb{Z}_p$ is a power of s. If p is congruent to 1 (mod 4), and if s is a primitive element of $\mathbb{Z}_p$, then $s^{(p-1/4)}$ is an element of $\mathbb{Z}_p$ whose square $= [-1]_p$ in $\mathbb{Z}_p$.

Thus there is some b in $\mathbb{Z}$ such that $b^2 + 1 \equiv 0 \pmod p$ (this was Theorem 1 of Chapter III-4; see also III-4, E7, which gives an explicit b, namely $b = ((p-1)/2)!$).

(vii) Now we can prove the converse part of the theorem.

Let p be a prime of $\mathbb{Z}$ with $p \equiv 1 \pmod 4$. We want to show that p is the sum of two squares in $\mathbb{Z}$.

We know that there is some b in $\mathbb{Z}$ which satisfies $b^2 + 1 \equiv 0 \pmod p$. Thus $b^2 + 1 = pd$ for some integer d. Thus in G, $(b + i)(b - i) = pd$. If p were a prime in G, then p would have to divide $b + i$ or $b - i$, by Exercise E5. But this is not true: for if $p(a + ci) = (b + i)$, with a, c in $\mathbb{Z}$, then $pc = 1$, which is impossible. The case $p(a + ci) = b - i$ is just as impossible.

So p cannot be a prime in G. Thus p must factor in $G : p = \alpha\beta$, with $N(\alpha) > 1$, $N(\beta) > 1$. Taking norms, we get $p^2 = N(p) = N(\alpha)N(\beta)$. Since $N(\alpha)$ and $N(\beta)$ are integers > 1 which divide p^2, therefore $N(\alpha) = N(\beta) = p$. By part (v), p is therefore the sum of two squares. That proves the theorem. □

Notice that E5 played a crucial role in the proof, and E5 depended on knowing that in G there is a division theorem and a Euclidean algorithm. So all parts of the proof were needed.

E6. State and prove a theorem about uniqueness of factorization of elements of G into products of primes of G.

Note. For a very different proof of Exercise E6 than the one I hope you give, see Ruchte and Ryden (1973).

†**E7.** Consider the set of complex numbers of the form

$$\mathbb{E} = \{a + b\sqrt{-5} \mid a, b \text{ in } \mathbb{Z}\}.$$

(a) Show that $\mathbb{E}$ is the set of integers of $\mathbb{Q}[\sqrt{-5}]$.
(b) What are the units of $\mathbb{E}$?
(c) In $\mathbb{E}$ we can factor 9 in two ways as follows:

$$9 = 3 \cdot 3 = (2 + \sqrt{-5})(2 - \sqrt{-5}).$$

Prove that in $\mathbb{E}$, the numbers 3, $2 + \sqrt{-5}$, and $2 - \sqrt{-5}$ are all irreducible. (*Hint*: Try using the norm of a complex number, and the fact that $N(\alpha\beta) = N(\alpha)N(\beta)$.)
(d) Conclude that unique factorization does not hold in $\mathbb{E}$.

21 On Unique Factorization

In this book, we have been studying fields, such as $\mathbb{Q}$, $\mathbb{R}$, $\mathbb{Z}_p$, and simple extensions of these, and commutative rings such as $\mathbb{Z}$, $F[x]$, or, in the last chapter, $\mathbb{G}$.

For each of $\mathbb{Z}$, $F[x]$ (F a field), and $\mathbb{G}$ we proved a theorem about uniqueness of factorization into products of primes. The statement of each such theorem was as follows. Let R be the commutative ring we were studying ($R = \mathbb{Z}$, $F[x]$, or $\mathbb{G}$). A *unit* of R is an element u for which there is a v in R with $uv = 1$. An *irreducible element* of R is a nonzero nonunit p which does not factor into $p = ab$ with a, b nonunits in R. Two elements p_1, p_2 of R are *associates* if $p_1 = up_2$ for some unit u of R. Then uniqueness of factorization says: *any nonzero nonunit of R factors uniquely, up to associates, into a product of irreducible elements of R.* This means that for any nonunit a ($\neq 0$) of R there are irreducible elements $p_1, \ldots, p_r$ of R, not necessarily all distinct, such that

$$a = p_1 p_2 \cdots p_r;$$

if also

$$a = q_1 q_2 \cdots q_s$$

with $q_1, \ldots, q_s$ irreducible, then $s = r$ and there is a 1–1 correspondence between the set $\{p_1, \ldots, p_r\}$ and the set $\{q_1, \ldots, q_s\}$ such that if p_i corresponds to q_j, then p_i and q_j are associates.

Note. This theorem holds even if R is a field for the trivial reason that in a field there are no nonzero nonunits.

In Chapter 20, Exercise E7, we gave an example of a commutative ring $\mathbb{E}$ for which uniqueness of factorization does not hold. That example is more typical of commutative rings than examples of rings with uniqueness of factorization. This fact was first discovered by mathematicians working with algebraic numbers.

Given any field of algebraic numbers $\mathbb{Q}[\alpha] \subseteq \mathbb{C}$, we can investigate the ring of integers O_α of $\mathbb{Q}[\alpha]$, and ask whether O_α has unique factorization. This question has been answered for many algebraic number fields.

For example, consider $\mathbb{Q}[\sqrt{d}\,]$ where d is a negative square-free integer. We already looked at the case $d = -1$ in the last chapter, and $d = -5$ in Exercise E7 of the last chapter. When $d = -1$ there was uniqueness of factorization; when $d = -5$ there was not. It turns out that the ring of integers of $\mathbb{Q}[\sqrt{d}\,]$ has uniqueness of factorization for only nine negative values of d—$d = -1, -2, -3, -7, -11, -19, -43, -67$, and -163— and for no other $d < 0$. (That those were the only ones was proved only in 1966 by H. M. Stark.)

An example that is highly interesting both intrinsically and historically is the ring $\mathbb{Z}[\zeta]$, polynomials in ζ with coefficients in $\mathbb{Z}$, where $\zeta = e^{2\pi i/p}$ is a pth root of 1, and p is an odd prime. Then $\mathbb{Z}[\zeta]$ turns out to be the ring of integers of the algebraic number field $\mathbb{Q}[\zeta]$, and is of particular interest because of Fermat's last theorem (Chapter III-2).

Suppose we wanted to show that there are no x, y, z, all nonzero with $x^p + y^p = z^p$ in $\mathbb{Z}$. If there are such integers, we can assume that the greatest common divisor of x, y and z is 1, and that $p \nmid z$. (For if $p | z$, then $p \nmid y$ and $x^p + (-z)^p = (-y)^p$, so we could replace y, z by $-z$, $-y$.)

Now in $\mathbb{Z}[\zeta]$,

$$z^p = x^p + y^p$$

$$= (x + y)(x + \zeta y)(x + \zeta^2 y) \cdots (x + \zeta^{p-1} y). \qquad (*)$$

If $\mathbb{Z}[\zeta]$ has unique factorization, it can be shown that the factors on the right side are pairwise relatively prime, from which it follows, again by unique factorization, that because of (*), for each i, $i = 0, 1, \ldots, p - 1$, $x + \zeta^i y$ is the associate of a pth power of an element of $\mathbb{Z}[\zeta]$.

These arguments are the first steps in a fairly subtle proof by Kummer, in 1840, of

Theorem. *If $\mathbb{Z}[\zeta]$ has uniqueness of factorization, where $\zeta = e^{2\pi i/p}$, p an odd prime, then there are no integers x, y, z, all nonzero, with $x^p + y^p = z^p$.*

When Kummer found this proof, he came quickly to the discovery that $\mathbb{Z}[\zeta]$ does not always have unique factorization. In fact, $\mathbb{Z}[\zeta]$ is known to have unique factorization only for $\zeta = e^{2\pi i/p}$, $p \leqslant 19$. Kummer's attempts to find a substitute for unique factorization in his proof led to the notion of an *ideal*—an ideal being for Kummer something invented to act as a substitute in a ring R for a greatest common divisor of two numbers when the two numbers did not have a greatest common divisor in R.

The concept of ideal has been refined to become the following

Definition. An *ideal* of a commutative ring R is a subset I of R such that

(i) for any a, b in I, $a + b$ is in I, and
(ii) for any a in I, r in R, ra is in I.

An ideal I is *principal* if there is some d in I so that every element of I is a multiple of d.

The reason the concept of ideal has not arisen before in this book is because of

Proposition. *In $\mathbb{Z}$, $F[x]$ (F a field) and G, every ideal is principal.*

PROOF FOR $\mathbb{Z}$. Given an ideal I, if I contains a nonzero number, then I contains a positive number (for if b is in I, so is $-b$). Let d be the smallest positive number in I. Then every number of I is a multiple of d. For if e is in I, $e \neq 0$, let $(e, d) = r$; then $r \leqslant d$. Now by Bezout's lemma, there are x, y in $\mathbb{Z}$ such that $xe + yd = r$. But since d, e are in I, so is $xe + yd = r$. Since d was the least positive number in I, and $r \leqslant d$, then $r = d$ and so d divides e.   □

The proofs for $F[x]$ and G are essentially the same.
In $\mathbb{Z}[\sqrt{-5}]$, the set of all elements of the form $(2 + \sqrt{-5})a + 3b$, a, b in $\mathbb{Z}[\sqrt{-5}]$, is an ideal which is not principal.
The subject of ideals in rings of integers of algebraic number fields and other commutative rings is well-developed; we leave it for you to study elsewhere, except for a few exercises. For a full discussion of Fermat's last theorem, see Borevich and Shafarevich (1966, pp. 156–164, 378–381, or Ribenboim (1979).

E1. If $\phi : R \to S$ is a ring homomorphism from a commutative ring R to a commutative ring S, prove that the set of all r in R such that $\phi(r) = 0$ in S is an ideal of R.

E2. What are the ideals of a field?

E3. Prove that every ideal is principal in
(a) $F[x]$, F a field,
(b) G, the Gaussian integers.

E4. Are there ideals of $\mathbb{Z}_{15}[x]$ which are not principal? (Cf. II-2, Exercise E28.)

E5. Prove in R, $R = \mathbb{Z}$ or $F[x]$, that if $a_1, \ldots, a_n$ are any nonzero elements of R, then the smallest ideal of R containing $a_1, \ldots, a_n$ is the principal ideal consisting of multiples of the greatest common divisor of $a_1, \ldots, a_n$.

Exercises Used in Subsequent Chapters

III-2, E13   (III-5)
III-3, E1   (III-16, E12)
III-4, E7   (III-20)
III-10, E7   (III-14)
III-10, E13   (III-14)
III-10, E14   (III-11, III-13A, III-14, III-18)
III-11, E1   (III-14)
III-11, E2   (III-14)
III-16, E10   (III-17A)
III-20, E7   (III-21)

Comments on the Starred Problems

I-2, E8. Let $P(n)$ be an assertion. If you know of it that (a) and (b) hold (the hypotheses of induction (1)), show that (a') and (b') (the hypotheses of induction (2)) hold. Then by induction (2), $P(n)$ is true for all n. Thus, assuming induction (2), induction (1) follows.

I-2, E11. (a) Let $P(n)$ be a statement for which (a) and (b) (the hypotheses of induction (1)) hold, and let S be the set of $n \geqslant n_0$ for which $P(n)$ is false. Show: if S is empty $P(n)$ holds for all $n \geqslant n_0$; if S is nonempty, and n_1 is the least element of S, then $P(n_1 - 1)$ is true but $P(n_1)$ is false, contradicting (b) (or (a) if $n_1 = n_0$).

(b) To show well-ordering implies induction (2) is the same as in part (a). To show the converse, let S be a set of numbers $\geqslant n_0$ with no least element, let $P(n)$ be "n is not in S," and proceed as in the proof of well-ordering in the text.

I-2, E16. (c) Do induction on n: expand $(n + 1)^{r+1}$ by the binomial theorem, subtract $n + 1$, and, using induction, substitute for $n^{r+1} - n$. For (d), do induction on r. See also Levy (1970) for a different approach to this problem. Our approach is from Chrystal (1898–1900).

I-2, E17. One way to do part (b) is by induction on m. The case $m = 2$ is the binomial theorem. For $m > 2$,

$$(x_1 + \cdots + x_m)^n = \sum (x_1 + \cdots + x_{m-1})^{n - a_m} x_m^{a_m} \binom{n}{a_m};$$

Now

$$(x_1 + \cdots + x_{m-1})^{n - a_m} = \sum \binom{n - a_m}{a_1 \cdots a_{m-1}} x_1^{a_1} \cdots x_{m-1}^{a_{m-1}}$$

so the coefficient of $x_1^{a_1} \cdots x_{m-1}^{a_{m-1}} x_m^{a_m}$ is

$$\binom{n}{a_1 \cdots a_m} = \binom{n}{a_m} \binom{n - a_m}{a_1 \cdots a_{m-1}}.$$

Now use induction and the definition of $\binom{n}{a_m}$.

I-3, E7. The key is whether or not the ratio of the larger to the smaller number exceeds the golden mean (Chapter 2, E10). See: Cole and Davie (1969), and Spitznagel (1973).

I-3, E39. Try $x = c/([ab, c], c)$. The prime factorization of x is that part of the factorization of c which involves primes not dividing a or b.

I-3, E42. See Chapter 7, Exercise E12.

I-4, E3. See III-4A.

I-4, E4. (a) Any prime dividing $2^{2^m} + 1$ and $2^{2^n} + 1$ $(n > m)$ is odd and must divide $2^{2^n} - 2^{2^m} = 2^{2^m}(2^{2^n - 2^m} - 1)$, and

$$2^{2^n - 2^m} - 1 = 2^{2^m(2^{n-m} - 1)} - 1 = (2^{2^m} - 1)(1 + 2^{2^m} + 2^{2^m 2} + 2^{2^m 3} + \cdots$$
$$+ 2^{2^m(2^{n-m} - 2)}).$$

Show that each of these last factors is relatively prime to $2^{2^m} + 1$.

I-5, E9. First observe the identity:

$$(x_n + x_{n-1} + \cdots + x_1 + x_0)^3$$
$$= x_n^3 + \left[3x_{n-1}(x_n)^2 + 3x_{n-1}^2 x_n + x_{n-1}^3 \right]$$
$$+ \left[3x_{n-2}(x_n + x_{n-1})^2 + 3x_{n-2}^2(x_n + x_{n-1}) + x_{n-2}^3 \right]$$
$$+ \left[3x_{n-3}(x_n + x_{n-1} + x_{n-2})^2 + 3x_{n-3}^2(x_n + x_{n-1} + x_{n-2}) + x_{n-3}^3 \right]$$
$$+ \cdots .$$

Each term in brackets involves one new variable beyond those used before. We let $x_i = r_i a^i$ for fixed a and use this identity to determine r_n, then $r_{n-1}, \ldots,$ as follows.

Let m be a number in base a. To find the largest number r with $r^3 \leqslant m$, write $r = r_n a^n + r_{n-1} a^{n-1} + \cdots + r_1 a + r_0$.

Write $m = s_n a^{3n} + k_n$, $k_n < a^{3n}$.

Choose r_n maximal so that $r_n^3 \leqslant s_n$.

Let $m_{n-1} = m - r_n^3 a^{3n} = s_{n-1} a^{3n-3} + k_{n-1}$.

Let $q_n = r_n$.

Choose r_{n-1} maximal so that

$$b_{n-1} = 3r_{n-1}q_n^2 a^2 + 3r_{n-1}^2 q_n a + r_{n-1}^3 < s_{n-1}.$$

Let $m_{n-2} = m_{n-1} - b_{n-1}a^{3n-3} = s_{n-2}a^{3n-6} + k_{n-2}$.

Let $q_{n-1} = (r_n a + r_{n-1})$.

Choose r_{n-2} maximal so that

$$b_{n-2} = 3r_{n-2}q_{n-1}^2 a^2 + 3r_{n-2}^2 q_{n-1}a + r_{n-2}^3 < s_{n-2}.$$

Let $m_{n-3} = m_{n-2} - b_{n-2}a^{3n-6} = s_{n-3}a^{3n-9} + k_{n-3}$.

Let $q_{n-2} = (r_n a^2 + r_{n-1}a + r_{n-2})$, etc.

If $s_n < a^3$ one can prove that $r_n < a$, $r_{n-1} < a$, etc.

For most bases a this is hopeless to do in practice, but not in base 2, since all r_i are either 0 or 1, and it is easy to decide: at each step try $r_{n-i} = 1$; if the resulting b_{n-i} is s_{n-1}, the choice $r_{n-i} = 1$ is correct, otherwise $r_{n-i} = 0$.

For a very fast nondigital algorithm for computing nth roots, see Fine (1977). For a classical procedure for computing nth roots by writing the expansion of $1/n$ in base 2 and taking square roots, see Chrystal (1898–1900, Chapter XI).

I-6, E24. Replacing a, b by a/d, b/d where $d = (a, b)$, if necessary, we can assume that $(a, b) = 1$. If two successive terms are equal, manipulate the equality to show $a + b$ divides $n + 1$. Conversely, assume $n = (a + b)s + 1$ and substitute this into the desired equality of the kth and $(k + 1)$st terms to find an appropriate k. See Lehmer (1914).

I-7, E12. Modulo $(a + b)$, subtracting b is the same as adding a. So modulo $(a + b)$, $s_{k+1} = (k + 1)a$. Use the fact that $(a, b) = 1$ to conclude that $s_m = s_n$ iff $m \equiv n$ (mod $a + b$). If $(a, b) = d$, one gets as residues mod $a + b$ all multiples of d.

I-11, E14. For part (b), $\phi(p^n)$ is the number of numbers $< p^n$ less the number of multiples of p which are $< p^n$. One way to do part (c) is to write down all the numbers $\leqslant ab$ in a rectangular array, e.g.

$$
\begin{array}{ccccccc}
1 & 2 & 3 & 4 & \cdots & a \\
a + 1 & a + 2 & a + 3 & a + 4 & & 2a \\
2a + 1 & & & & & \\
\vdots & & & & & \vdots \\
(b - 1)a + 1 & & & & & ba
\end{array}
$$

Try to show that each column topped by a number prime to a consists entirely of numbers prime to a, and that each such column consists of a complete set of representatives of $\mathbb{Z}_b$ (Chapter 7, preceding E10). See also I-14, E4.

I-11, E16. By Fermat's theorem, for $p \neq 2, 5$, $10^{p-1} \equiv 1$ (mod p), so p divides $10^p - 1$. Now $10^p - 1 = 9u$ where u is in the set. If $p \neq 3$, p divides u, But 3 is no problem since 3 divides 111.

I-12, E9. The hard way to do this problem is to find a number t so that the order of 10 mod t is at least 61. One strategy for finding such a t is to let t be a prime so that $t - 1$ has few divisors. The easy way to do this problem is to write down an appropriate repeating decimal expansion and then convert it to a fraction. See also III-2, Proposition 3.

I-13, E2. If $p = $ probability of error in a given digit, then the probability P of at most two errors in an 8-digit word is

$$
P = (1 - p)^8 + 8(1 - p)^7 p + \frac{8 \cdot 7}{2}(1 - p)^6 p^2
$$

$$
= (1 - p)^6 (1 + 6p + 21p^2).
$$

If $p = .1$, $P \sim .96$. If $p = .01$, $P \sim .999988$. If $p = .02$, $P \sim .99918$.

I-13, E3. Take the matrix $\mathbf{H}$ of code II, write it beside itself to make a 4×16 matrix, then put the row

$$(0 \quad 0 \quad 0 \quad 0 \quad 0 \quad 0 \quad 0 \quad 0 \quad 1 \quad 1 \quad 1 \quad 1 \quad 1 \quad 1 \quad 1 \quad 1)$$

under it.

I-14, E14. Given $m = p_1^{e_1} \cdots p_r^{e_r}$, let $k \geqslant \max\{e_1, \ldots, e_r\}$. Let

$$
a = p_1^{j_1} \cdots p_r^{j_r} t
$$

where $(t, m) = 1$. Let

$$
q = [\phi(p_1^{e_1}), \ldots, \phi(p_r^{e_r})].
$$

Show $a^{k+q} \equiv a^k \pmod{p_i^{e_i}}$ for each i: if $j_i = 0$, a is prime to $p_i^{e_i}$ and Euler's theorem applies; if $j_i > 0$, then a^k is divisible by $p_i^{e_i}$ so $a^{k+q} \equiv a^k \equiv 0 \pmod{p_i^{e_i}}$. So $a^{k+q} \equiv a^k \pmod{p_i^{e_i}}$ for all i, hence also mod m.

I-14, E15. If m is square-free, $m = p_1 p_2 \cdots p_r$, $\phi(m) = (p_1 - 1) \cdots (p_r - 1)$ is a multiple of $[p_1 - 1, p_2 - 1, \ldots, p_r - 1]$.

I-15, E4. Since $561 = 3 \cdot 11 \cdot 17$, it suffices to observe, as in the proposition of Chapter 14, that $561 = 1 + $ a multiple of $q = [2, 10, 16]$. (See Chapter III-3: 561 is a pseudoprime.)

II-1, E4b. If $a_0 \neq 0$, define the coefficients b_n inductively, by assuming $b_0, \ldots, b_{n-1}$ are defined, writing down the equation for the coefficient of x^n in the power series $1 + 0x + \cdots$, namely,

$$0 = a_n b_0 + a_{n-1} b_1 + \cdots + a_1 b_{n-1} + a_0 b_n,$$

and solving for b_n.

II-2, E15. Multiply both polynomials by $x - 1$ to get $x^{n+1} - 1$ and $x^{m+1} - 1$. Set $m + 1 = a$, $n + 1 = b$, and suppose $a < b$. Then

$$x^b - 1 = (x^a - 1)(x^{b-a} + \cdots + x^{b-qa}) + (x^r - 1)$$

if $b = qa + r$. This is the induction step in an argument by induction on the length of the Euclidean algorithm for a and b. You get $(x^b - 1, x^a - 1) = x^d - 1$, where $d = (a, b)$. To get the result, divide all polynomials by $x - 1$. This exercise can also be done by factoring these polynomials in $\mathbb{C}[x]$.

II-5, E5. $\dfrac{17}{180} = \dfrac{1}{3} + \dfrac{1}{9} - \dfrac{1}{2} - \dfrac{1}{4} + \dfrac{2}{5}$.

II-5, E9. Let $f(x) = c_n x^n + \cdots c_1 x + c_0$, $g(x) = b_n x^n + \cdots + b_0$ (where $n = \max(\deg f, \deg g)$ and some of the c_i's or b_j's may equal zero if $\deg f \neq \deg g$). Then

$$\frac{f(1/x)}{g(1/x)} = \frac{c_n + \cdots + c_n x^{n-1} + c_0 x^n}{b_n + \cdots + b_1 x^{n-1} + b_0 x^n} = \frac{f_0(x)}{g_0(x)}.$$

Here $g_0(x) = g(1/x)x^n$, $g_0(1/x) = g(x)/x^n$. To get the expansion as an infinite $x - $ al, write

$$f_0(x) = g_0(x)a_0 + r_0(x) \quad \text{with } \deg r_0 < n$$
$$xr_0(x) = g_0(x)a_1 + r_1(x)$$
$$xr_1(x) = g_0(x)a_2 + r_2(x)$$
$$\vdots$$
$$xr_{k-1}(x) = g_0(x)a_k + r_k(x).$$

Then

$$\frac{f_0(x)}{g_0(x)} = a_0 + \frac{a_1}{x} + \cdots + \frac{a_k}{x^k} + \frac{1}{x^k}\frac{r_k(x)}{g_0(x)}$$

for each $k > 0$. So

$$\frac{f(x)}{g(x)} = a_0 + a_1 x + \cdots + a_k x^k + x^k \frac{r_k(1/x)}{g_0(1/x)}$$

and $r_k(1/x)/g_0(1/x) = x^n r_k(1/x)/g(x)$. Since $r_k(x)$ has degree $<n$, $x^n r_k(1/x) = s_k(x)$, a polynomial in x. So $x^n r_k(1/x)/g(x) = s_k(x)/g(x)$. Since $g(0) \neq 0$, all the derivatives up to the kth of $x^k s_k(x)/g(x)$ are 0 when $x = 0$. Now

$$f(x)/g(x) = a_0 + a_1 x + \cdots + a_k x^k + x^k s_k(x)/g(x).$$

The rth coefficient of the Taylor series for $f(x)/g(x)$ is $1/r!$ times the rth derivative of $f(x)/g(x)$ evaluated at $x = 0$. For $r < k$ this derivative is easily seen to be $r! a_r$. Since k can be chosen as large as you like, for any r a_r is the coefficient of x^r in the Taylor series for $f(x)/g(x)$, and the series of Exercise E9 is the Taylor series for $f(x)/g(x)$.

II-6, E6. Suppose $p^n + q^n = r^n$, p, q, r in $\mathbb{R}[x]$, all of degree ≥ 1. We can assume that p, q, r have no common factor. Then, taking derivatives and dividing by $n - 1$,

$$p^{n-1} p' + q^{n-1} q' = r^{n-1} r'.$$

So

$$p \cdot p^{n-1} + q \cdot q^{n-1} + r(-r^{n-1}) = 0,$$
$$p' \cdot p^{n-1} + q' \cdot q^{n-1} + r'(-r^{n-1}) = 0$$

Then

$$(pq' - qp')p^{n-1} + (rq' - qr')(-r^{n-1}) = 0$$

and

$$(pr' - rp')p^{n-1} + (qr' - rq')q^{n-1} = 0.$$

Since $(p, r) = 1$ and $(p, q) = 1$, it follows that

$$r^{n-1} \quad \text{divides } pq' - qp',$$
$$p^{n-1} \quad \text{divides } qr' - rq',$$
$$q^{n-1} \quad \text{divides } pr' - rp'.$$

Let $\deg p = a$, $\deg q = b$, $\deg r = c$. Then

$$\deg(pq' - qp') \leq \max(\deg p + \deg q', \deg p' + \deg q)$$
$$\leq a + b - 1,$$

so $(n - 1)c \leq a + b - 1$. Similarly,

$$(n - 1)b \leq a + c - 1,$$
$$(n - 1)a \leq b + c - 1.$$

Summing, get $(n - 1)(a + b + c) \leq 2(a + b + c) - 3$, so $n - 1 < 2$, $n \leq 2$.

II-7, E3. Suppose $f_n = $ g.c.d. of f and f'. Then f_n divides $f_0, f_1, \ldots, f_n$. Let $g_i = f_i/f_n$. Then $g_n = 1$. Since $f_{j-1}(x) = f_j(x)q_j(x) - f_{j+1}(x)$, $g_{j-1}(x) = g_j(x)q_j(x) - g_{j+1}(x)$. Now since $\text{sign}(g_i(x)) \cdot \text{sign}(f_n(x)) = \text{sign}(f_i(x))$, $w(x)$, computed using the f_j's, is the same as $w(x)$ computed using the g_j's as long as x is not a root of f_n. Since $g_{j-1} = q_j g_j - g_{j+1}$, the argument in the text shows that $w(x)$ computed using the g_j's counts distinct roots of g_0. But the roots of g_0 are the distinct roots of f_0, so $w(x)$ counts the distinct roots of f_0 and can be computed using the f_j's.

II-11, E7. There are 4^{d+1} polynomials. If $f(n_i) = r_i$, $i = 0, \ldots, d$, and $\mathbf{s} = (s_0, \ldots, s_d)$ is a vector of divisors of $(r_0, \ldots, r_d)$, then so is $-\mathbf{s} = (-s_0, \ldots, -s_d)$. If $a_s(n_i) = s_i$, then $-a_s(n_i) = -s_i = a_{-s}(n_i)$. Since $a_s(x)$ is uniquely determined by $\mathbf{s}$, $-a_s(x) = a_{-s}(x)$ and the interpolators pair off as associates.

II-12, E11. The remainder on dividing x^p by $f(x) = x^2 - q$ is $q^{(p-1)/2}x$. So q is a square in $\mathbb{Z}_p$ iff $f(x)$ factors in $\mathbb{Z}_p[x]$ iff there is $b \neq 0$ with $f(x)$ dividing $(a + bx)^p - (a + bx)$ iff $b \neq 0$ and $b(q^{(p-1)/2} - 1) = 0$ in $\mathbb{Z}_p$ iff $q^{(p-1)/2} = 1$ in $\mathbb{Z}_p$.

III-3, E3. (a) has the form $2p + 1$; (b) is $2^{15} - 1$; (c) is $2^{2^4} + 1$; (d) is a prime factor of a composite Fermat number.

III-3, E4. It is a fact that $\phi(n)/n$ can be as close to zero as you wish. This follows from Theorem 328 of Hardy and Wright (1960), the proof of which is difficult.

III-5, E6. Let $n = 2^{e_0}p_1^{e_1} \cdots p_r^{e_r}$. Let $a_0, \ldots, a_r$ be elements of maximal order in $\mathbb{Z}_{2^{e_0}}, \ldots, \mathbb{Z}_{p_r^{e_r}}$ respectively. Then a_0 has order 2^s, where $s = e_0 - 1$ if $e_0 = 1$ or 2, and $s = e_0 - 2$ for $e_0 \geqslant 3$; and a_i has order $\phi(p_i^{e_i}) = p_i^{e_i - 1}(p_i - 1)$ if $i = 1, \ldots, r$; and $h(n) = \text{l.c.m.}[2^s, \phi(p_1^{e_1}), \ldots, \phi(p_r^{e_r})]$. We consider two cases.

  Case 1. Two of the terms inside the brackets in the expression for $h(n)$ are divisible by 2. Let t be the maximum power of 2 dividing $h(n)$. Let i be such that 2^t divides $\phi(p_i^{e_i})$; $i = 0$ if $2^t = 2^s$. Let $j \neq i$ be some index such that 2 divides $\phi(p_j^{e_j})$ (if $i \neq 0, j$ can $= 0$ if $s > 0$). Let x be a number satisfying the congruences

$$x \equiv a_0 \pmod{2^{e_0}},$$
$$x \equiv a_1 \pmod{p_1^{e_1}},$$
$$\vdots$$
$$x \equiv a_j^2 \pmod{p_j^{e_j}},$$
$$\vdots$$
$$x \equiv a_r \pmod{p_r^{e_r}}.$$

Then it is not hard to show that the order of $x \bmod n$ is exactly $h(n)$, and that $x^{h(n)/2} \not\equiv -1 \pmod{n}$. To see the latter, note that if $x^{h(n)/2} \equiv -1 \pmod{n}$, then $(a_j^2)^{h(n)/2} \equiv -1 \pmod{p_j^{e_j}}$, which is impossible. So use x as a spacing number.

  Case 2. Only one term is divisible by 2. Then either $r = 0$, $s \geqslant 3$ and the spacing number 3 works by Exercise E8 of Chapter III-2, or else there is a primitive element mod n. In that case the $h(n)/2$ power of any primitive element must be $\equiv -1 \pmod{n}$.

III-6, E3. Show that there are no real roots, by calculus (show the minimum value of the polynomial is > 0); then show there is no factorization into factors of degree 2 by "brute force" (II-8), using the inequalities on a and b.

III-8, E2. In any field F of 16 elements, $x^{16} = x$ for all x in F by Fermat's theorem. So any field of 16 elements will do.

III-13, E11. I do not know a nice way to do this problem, except if n is a prime power.

III-13, E13. Certainly $N_n^p/p^n < 1/n$. The worst case is when $n = 2 \cdot 3 \cdot 5 \cdot 7 \cdot 11 \cdot 13 \cdots$; then

$$\frac{N_n^p}{p^n} = \frac{1}{n}\left(1 - \frac{1}{p^2} - \frac{1}{p^3} - \frac{1}{p^5} + \frac{1}{p^6} - \frac{1}{p^7} + \frac{1}{p^{10}} - \cdots\right),$$

which is an expansion in base p. The smallest it can be is when $p = 2$ and then it is larger than when we truncate the sum at 7, in which case we get

$$\frac{1}{n}\left(1 - \frac{1}{4} - \frac{1}{8} - \frac{1}{32} + \frac{1}{64} - \frac{1}{128}\right) \approx \frac{1}{n}(.60).$$

Thus

$$.60\left(\frac{1}{n}\right) \leqslant \frac{N_p^n}{p^n} < \frac{1}{n},$$

and the limit behaves like $1/n$.

III-15, E12. For part (a) replace ϕ_p, the function on $\mathbb{Z}_q$ given by $\phi_q(x) = x^q$, by complex conjugation in the proof, and copy the proof just as it appears in the text. The irreducible case is easy, since any irreducible polynomial with real coefficients has degree 1 or 2. For part (b), proceed as follows. From part (a), $D(f) > 0$ iff $D(f)$ is a square iff $d \equiv r$ (mod 2), where r is the number of irreducible factors of f. Let r_1 = number of irreducible factors of degree 1, r_2 the number of degree 2. Then $r_1 + 2r_2 = d$ and r_1 = number of real roots. From $d \equiv r$ (mod 2) follows $r_1 + 2r_2 \equiv r_1 + r_2$ (mod 2); hence r_2 is even, $r_2 = 2k$, and so $r_1 = d - 4k$.

III-18, E3 and E4. If α is algebraic over $\mathbb{Q}$, then $\mathbb{Q}[\alpha]$ is an r-dimensional vector space over $\mathbb{Q}$ (where r = degree of the minimal polynomial of α over $\mathbb{Q}$). Then if γ is algebraic over $\mathbb{Q}$, γ is algebraic over $\mathbb{Q}[\alpha]$. If γ is algebraic over $\mathbb{Q}[\alpha]$, then $\mathbb{Q}[\alpha][\gamma]$ is a vector space of dimension s over $\mathbb{Q}[\alpha]$, where s is the degree of the minimal polynomial of γ over $\mathbb{Q}[\alpha]$. But then the set of elements $\alpha^i \gamma^j$ with $i = 0, 1, \ldots, r-1, j = 0, 1, \ldots, s-1$, is a $\mathbb{Q}$-basis of $\mathbb{Q}[\alpha][\gamma]$. Now if K is a field containing $\mathbb{Q}$ and is a d-dimensional vector space over $\mathbb{Q}$, then every element β of K must satisfy a polynomial with coefficients in $\mathbb{Q}$ of degree $\leqslant d$, since $1, \beta, \beta^2, \ldots, \beta^d$ are $d+1$ elements of K and so cannot be linearly independent over $\mathbb{Q}$; any equation of linear dependence describes a polynomial with coefficients in $\mathbb{Q}$ with β as a root. These remarks are the lemmas needed to prove E3 and E4.

III-18, E11. For part (a) it may be convenient to prove by induction on n that both $\cos(n\theta)$ and $\sin(n\theta)\sin(\theta)$ are polynomials in $\cos\theta$, using the formulas for the sine and cosine of the sum of two angles. Part (b) is easy, and for part (c), let $\theta = (\pi/2) - (\pi/n)$ in part (a).

III-19, E6. If $p(x) = (x - \alpha_1)(x - \alpha_2) \cdots (x - \alpha_n)$ in $F[x]$, $F = \mathbb{Q}[\alpha_1]$, then any homomorphism σ from F to $\mathbb{C}$ sends r in $\mathbb{Q}$ to itself (Exercise E1) and is determined by what it does to α_1. But $\sigma(\alpha_1)$ must be another root in $\mathbb{C}$ of $p(x)$, hence must be α_i for some i. So σ maps F to F, and the number of homomorphisms σ, which equals the number of elements of $\text{Aut}(F)$, must equal the number of roots of $p(x)$ in F, and this is n, the degree of $p(x)$.

References

Note: Monthly = American Mathematical Monthly

Abhyankar, S. (1976). Historical ramblings in algebraic geometry and related algebra, *Monthly* **83**, 409–448.

Apostol, T. M. (1976). *Introduction to Analytic Number Theory*. Springer-Verlag, Berlin and New York.

Bender, E. A., and Goldman, J. R. (1975). On the application of Möbius inversion in combinatorial analysis, *Monthly* **82**, 789–802.

Berlekamp, E. R. (1967). Factoring polynomials over finite fields, *Bell System Technical Journal* **46**, 1853–1859.

Berlekamp, E. R. (1968). *Algebraic Coding Theory*. McGraw–Hill, New York.

Berlekamp, E. R., and Hwang, F. K. (1972). Constructions for balanced Howell rotations for bridge tournaments, *J. Comb. Theory* **A12**, 159–166.

Blake, I. F., and Mullen, R. C. (1975). *The Mathematical Theory of Coding*. Academic Press, New York.

Borevich, Z. I., and Shafarevich, I. R. (1966). *Number Theory*. Academic Press, New York.

Boyle, W. S. (1977). Light-wave communication, *Scientific American* **237**, 40–48.

Cardano, G. (1545). *Artis Magnae, sive de Regulis Algebraicis*. Witmer, T. R., trans. and ed. (1968). M.I.T. Press, Cambridge, Mass.

Chrystal, G. (1898–1900). *Algebra, an Elementary Textbook*, 7th ed. A. and C. Black, London. Repr. (1964) Chelsea, New York.

Cole, A. J., and Davie, A. J. T. (1969). A game based on the Euclidean algorithm and a strategy for it, *Math. Gazette* **53**, 354–357.

Collins, G. E. (1973). Computer algebra of polynomials and rational functions, *Monthly* **80**, 725–744.

Dejon, B., and Henrici, P. (1969). *Constructive Aspects of the Fundamental Theorem of Algebra.* John Wiley Ltd., London.

Dickson, L. E. (1929). *Introduction to the Theory of Numbers.* Univ. of Chicago Press, Chicago.

Diffee, W., and Hellman, M. (1976). New directions in cryptography, *IEEE Trans. on Information Theory* **IT22**, 644–654.

Euclid (300 B. C.). The Elements. Heath, T. L., trans. (1925–56). Dover, New York.

Fefferman, C. (1967). An easy proof of the fundamental theorem of algebra, *Monthly* **74**, 854–55.

Fine, N. J. (1977). Infinite products for *k*th roots, *Monthly* **84**, 629–30.

Fisher, R. A. (1935). *The Design of Experiments.* Oliver and Boyd, Edinburgh.

Gardner, M. (1977). Mathematical games, *Scientific American* **237**, 120–124.

Gauss, C. F. (1801). *Disquisitiones Arithmeticae.* Clarke, A. A., trans. (1966) Yale Univ. Press, New Haven.

Gerst, I., and Brillhart, J. (1971). On the prime divisors of polynomials, *Monthly* **78**, 250–266.

Goldstein, L. J. (1971). Density questions in algebraic number theory, *Monthly* **78**, 342–351.

Goldstein, L. J. (1973). A history of the prime number theorem, *Monthly* **80**, 599–614.

Hardy, G. H., and Wright, E. M. (1960). *An Introduction to the Theory of Numbers*, 4th ed. Oxford Univ. Press, New York.

Henrici, P. (1974). *Applied and Computational Complex Analysis*, Vol. I. John Wiley, New York.

Herstein, I. N. (1964). *Topics in Algebra.* Blaisdell, New York.

Hill, L. S. (1931). Concerning certain linear transformation apparatus of cryptography, *Monthly* **38**, 135–154.

Hooley, C. (1967). On Artin's conjecture, *J. Reine Angew. Math.* **225**, 209–220.

Ireland, K., and Rosen, M. (1972). *Elements of Number Theory.* Bogden and Quigley, Tarrytown-on-Hudson, N. Y.

Knuth, D. E. (1969). *The Art of Computer Programming, Vol. II: Seminumerical Algorithms.* Addison–Wesley, Reading, Mass.

Kuhn, H. W. (1974). A new proof of the fundamental theorem of algebra, *Mathematical Programming Studies* **1**, 148–158.

Lawther, H. P., Jr. (1935). An application of number theory to the splicing of telephone cables, *Monthly* **42**, 81–91.

Lehmer, D. N. (1914). A theorem in number theory connected with the binomial formula, *Monthly* **21**, 15.

Levinson, N. (1969). A motivated account of an elementary proof of the prime number theorem, *Monthly* **76**, 225–245.

Levy, L. S. (1970). Summation of the series $1^n + 2^n + \ldots + x^n$ using elementary calculus, *Monthly* **77**, 840–47; **78**, 987.

Mann, H. B. (1949). *Analysis and Design of Experiments.* Dover, New York.

McCoy, N. H. (1965). *The Theory of Numbers.* MacMillan, New York.

Morrison, M. A., and Brillhart, J. (1975). A method of factoring and the factoring of F_7, *Math. of Computation* **29**, 183–205.

Niven, I. (1956). *Irrational Numbers.* Carus Math. Monog. No. 11, The Mathematical Assoc. of America, Washington, D.C.

Niven, I., and Zuckerman, H. S. (1972). *An Introduction to the Theory of Numbers*, 3rd ed. John Wiley, New York.

Ore, O. (1948). *Number Theory and its History.* McGraw–Hill, New York.

Peterson, W. W., and Weldon, E. J. (1972). *Error Correcting Codes*, 2nd ed. M.I.T. Press, Cambridge, Mass.

Ribenboim, P. (1979). *Fermat's Last Theorem.* Springer-Verlag, Berlin and New York.

Ruchte, M. F., and Ryden, R. W. (1973). A proof of uniqueness of factorization in the Gaussian integers, *Monthly* **80**, 58–59.

Shanks, D. (1962). *Solved and Unsolved Problems in Number Theory.* Spartan Books, New York.

Shapley, D., and Kolata, G. B. (1977). Cryptology: scientists puzzle over threat to open research, publication, *Science* **197**, 1345–49.

Simmons, G. (1970). On the number of irreducible polynomials of degree d over $GF(p)$, *Monthly* **77**, 743–45.

Spitznagel, E. L., Jr. (1973). Properties of a game based on Euclid's algorithm, *Math. Magazine* **46**, 87–92.

Stark, H. M. (1971). *An Introduction to Number Theory.* Markham, Chicago.

Swan, R. G. (1962). Factorization of polynomials over finite fields, *Pacific J. Math.* **12**, 1099–1106.

Terkelsen, F. (1976). The fundamental theorem of algebra, *Monthly* **83**, 647.

Wallis, W. D., Street, A. P., and Wallis, J. S. (1972). *Combinatorics: Room Squares, Sum-Free Sets, Hadamard Matrices.* Springer Lecture Notes in Mathematics No. 292, Springer-Verlag, Berlin and New York.

Zagier, D. (1977). The first 50 million prime numbers. *Mathematical Intelligencer* **0**, 7–19.

Zelinsky, D. (1973). *A First Course in Linear Algebra*, 2nd ed. Academic Press, New York.

Index

abelian group 92
Abel's theorem 138
abstract Fermat theorem 92
algebraic 238
algebraic number 309
argument of complex number 4
arithmetic progression 47
Artin's conjecture 210
associates 133, 183, 320
associativity 63
automorphism 261, 315

base 37, 146
base *a* expansion 44, 101, 148, 212
basis 81
BCH code 242
Berlekamp's factoring algorithm 187, 198, 295
Bertrand's postulate 36
Bezout's identity 22, 133
Bezout's lemma 22, 133
binomial theorem 14, 176, 195, 214
board 302
bound on roots of polynomial 141, 161, 194, 196, 197
brute force technique 170

cancellation in congruences 49, 51, 66, 173
casting out nines 50, 60, 105

characteristic *p* 176, 258
characteristic zero 175, 258
Chinese remainder theorem 112, 216, 291
  for polynomials 180
closed 63
common divisor 20
commutativity 63
complete set of representatives 61
complex conjugation 143, 261, 262
complex numbers 4, 136, 237, 260, 312
complex numbers mod *p* 223
congruence classes 56, 57, 234
congruent modulo *n* 47
congruent modulo *p(x)* 173
constant polynomial 129
cube root 41
cubic polynomial 137, 143, 290
cyclic group 210

decimal expansion 44
degree 126, 129
derivative 161, 182, 283
determinant 75
differentiation 157
dimension of null space 82, 189
Dirichlet's theorem 33
discriminant 283
distributive law 63
divides 20, 130
division algorithm = division theorem